Pushing the Envelope

Pushing the Envelope

The American Aircraft Industry

Donald M. Pattillo

Ann Arbor

The University of Michigan Press

338.4762913
P32p

*To my parents,
who showed me a wider world*

Copyright © by the University of Michigan 1998
All rights reserved
Published in the United States of America by
The University of Michigan Press
Manufactured in the United States of America
♾ Printed on acid-free paper

2001 2000 1999 1998   4 3 2 1

No part of this publication may be reproduced, stored in a retrieval system,
or transmitted in any form or by any means, electronic, mechanical, or otherwise,
without the written permission of the publisher.

A CIP catalog record for this book is available from the British Library.

Library of Congress Cataloging-in-Publication Data

Pattillo, Donald M., 1940–
    Pushing the envelope : the American aircraft industry / Donald
M. Pattillo.
      p. cm.
    Includes bibliographical references and index.
    ISBN 0-472-10869-7 (acid-free paper)
    1. Aircraft industry—United States—History. 2. Aerospace
industries—United States—History. I. Title.
HD9710.U52P34 1998   97-45390
                                                                    CIP

Acknowledgments

This book represents the culmination of a lifelong fascination with aviation and the aircraft industry. I continued that interest through my studies up to and including a doctorate in business. As I pursued an academic career that never hinged on aviation or the industry, I still managed to remain a scholar of the industry even as my professional efforts centered elsewhere. Finally I reached the point at which I knew I must write a history of the aircraft industry, irrespective of any professional recognition or advancement that might result. I was further motivated by my knowledge that a comprehensive history of the aircraft manufacturing industry had never been accomplished.

In conducting my research, there were numerous individuals who provided valuable assistance in many areas and are deserving of my gratitude. In industry, they include Steve Dondero and William Wachino of Grumman Corporation, Doug Oliver of Lockheed, Tom Lubbesmeyer of the Boeing Historical Archives, and Craig Thompson of Rockwell. Anne Millbrooke of the United Technologies Archive was especially helpful. In museums, Bill Camp and Joshua Stoff of the Cradle of Aviation Museum, Nassau County, New York, were generous with their time and effort. Archie diFante of the U.S. Air Force Historical Research Center at Maxwell Air Force Base provided valuable assistance. Lieutenant Colonels Larry Emmelhainz and Mike Farr of the Air Force Institute of Technology at Wright-Patterson Air Force Base, historian Al Misenko of the Aeronautical Systems Division at Wright-Patterson, Dave Minard of the U.S. Air Force Museum, Pat Reilly of the New Jersey Aviation Museum, and Ned Preston, historian of the Federal Aviation Administration, provided valuable help and encouragement in several areas. Will Mahoney of the National Archives was helpful with several sources there, and Paul Silberman of the Garber Facility, National Air and Space Museum, provided assistance with the Fairchild and Keys papers. In other areas, Anthony Velocci, senior editor of *Aviation Week,* was most encouraging. Virginia Lopez and David Napier of the Aerospace Industries Association, and most especially Billie Perry of the AIA Library, were extremely helpful. David Lewis and Bill Trimble, professors of business history at Auburn University, were encouraging and provided useful ideas. Harry Gann, historian at Douglas Aircraft Company of McDonnell Douglas Corporation, provided answers to several questions. I also benefited from conversations with Dr. Roger Bilstein of the University of Houston–Clear Lake City.

There were several librarians, names unknown, at the Hunterdon County Library in New Jersey, who located and ordered numerous books and other references on my behalf. The same applies to assistance I received at the Library of Congress with the Martin, Gross, and Arnold papers, and at the Alexander Library of Rutgers, the State University of New Jersey, with numerous government documents.

I owe special thanks to Colin Day of the University of Michigan Press for his initial interest and unflinching support, and to his very capable and pleasant associates, Heather Lengyel and Melissa Holcombe, for their help and attention. I am also grateful for the astute comments of three anonymous readers of my initial manuscript. Their efforts enabled me to produce a much improved work.

It is widely understood that any book is only as good as its sources. I have endeavored for factual correctness in my work, which has been no easy task in a field that seems to contain so much contradictory information. Yet the responsibility for any errors of fact, interpretation, or implication is mine and mine alone.

Finally, I am indebted to my wife, Sharon. As a professional librarian she provided practical assistance with numerous reference questions and data sources. But possibly most important, she alone could begin to understand what this project meant to me at this stage of my life.

Contents

List of Tables ix

Introduction 1

1 • Origins and Pioneers 5

2 • The Industry and the First World War 28

3 • Building an Industry: The 1920s 39

4 • The Industry's Golden Age 64

5 • Progress amid Struggle: The Mid-1930s 87

6 • Industry Survey, 1935–40 105

7 • The Wartime Buildup Begins: 1940–42 117

8 • The Industry in Wartime 127

9 • The Postwar Industry Adjustment 149

10 • Industry Survey, 1945–54 165

11 • The Cold War Industry and the Transition to Aerospace 192

12 • Aerospace and Defense Converge 221

13 • The Advanced-Technology Era 261

14 • An Industry under Stress 285

15 • The 1980s and Beyond 317

16 • Uncertainty after the Cold War 344

Epilogue: What Manner of Man? 367

Appendix: Chronology of the Aircraft Industry   373

Notes   383

Select Bibliography   421

General Index   437

Index of Aircraft, Engines, Missiles, and Space Vehicles   453

Illustrations   *following page 86*

List of Tables

1. U.S. Air Appropriations, 1908–19	29
2. U.S. Aircraft Production, 1912–19	32
3. U.S. Air Appropriations, 1920–27	45
4. U.S. Aircraft Production, 1920–27	48
5. U.S. Aircraft Exports, 1913–27	61
6. Prewar Production and Exports of Aircraft	94
7. U.S. Aircraft Exports, 1928–39	100
8. U.S. Air Appropriations, 1928–40	119
9. U.S. Aircraft Production, 1928–41	125
10. Aircraft Production and AAF Procurement during the Second World War	130
11. U.S. Aircraft Production, 1942–52	151
12. Industry Employment, 1939–54	154
13. U.S. Aircraft Exports, 1940–48	156
14. Largest Aircraft Manufacturers, 1954	178
15. Evolution of the U.S. Air Force	196
16. Industry Employment, 1955–61	203
17. Largest Aerospace Manufacturers, 1967	229
18. U.S. Aircraft Production, 1953–68	240
19. Aerospace Exports, 1950–67	251
20. Largest Aerospace Manufacturers, 1979	303
21. Leading Exporters among Aerospace Companies, 1988	314
22. Transport Aircraft Shipments	319
23. U.S. Aircraft Production, 1969–87	325
24. Largest Aerospace Manufacturers, 1989	331

Introduction

The history of flight and of the broad field of aviation has been well researched. In particular, many significant aircraft and many pioneer aviators and designers have been given their due. There are also histories of individual aircraft companies, airlines, military arms and units, aerial feats, and the technical development of aviation. There is no aspect of the field that has not been competently addressed by historians.

What has been lacking, however, is a comprehensive analytical survey of the aircraft-manufacturing industry, from its origins to the present. No single work has presented a unified, cohesive history of aircraft manufacture. That is the purpose of this book. Previous published works covered only certain periods or certain aspects of the industry. Furthermore, much of what is available concerning specific companies is in the form of laudatory histories commissioned by the subject firms, or otherwise in the nature of muckraking exposes, especially with regard to defense contracting. Not having received a comprehensive, balanced treatment, the aircraft industry consequently has suffered in public understanding.

This book covers the aircraft business from its modest beginnings early in the century through its evolution into the few aerospace giants of today. The book further provides a conceptual understanding of the American aircraft industry's social and economic progress, scientific and technological advancement, and contributions to national security and American global preeminence, while striving for balance and objectivity. The industry has been essential to victory in war, to maintaining peace in the Cold War era, to rapid advances in transportation and communications, and to national economic performance, especially in skilled employment and exports. The contribution of the industry to the economy often is overlooked. The book concludes with a survey of how the industry is consolidating for the future.

A point of emphasis is that the aircraft, now aerospace, industry does not conform to any broad industrial model. The automobile industry, with which it has sometimes been compared, was sustained by a large middle class and long-term economic growth, since its output was affordable by large numbers of consumers. The aircraft/aerospace industry, however, is not and cannot become a consumer products industry. It is both a military and a commercial industry, but its major customer is the government. That the industry exists largely as creature of governmental needs is both its greatest strength and its greatest vulnerability. The industry has always been cyclical in that it is affected by the overall economy yet is affected more than others by changes in the military-political environment and in the legal-regulatory environment.

Only the commercial-airliner sector of the industry, which developed in the early 1930s, may approach a definition of a consistent, long-term growth market. But even this market has been subject to volatile shifts in demand and has always presented major financial risks to manufacturers. Of the several competitors in the 1950s, after the McDonnell Douglas acquisition only Boeing remains, and Boeing alone has approached consistent profitability in the field.

The history of the aircraft industry, in common with most technological industries, largely coincides with the twentieth century and has been linked closely with American political and economic leadership. Two broad recurring themes for the history may be discerned. First, it is a history of men, especially of those flight pioneers or entrepreneurs who founded companies, assembled management teams, and built the industry. Second, its history is one of almost continuous struggle for survival of individual manufacturers. The record of most firms in the industry is of reorganizations, mergers, spin-offs, and diversification attempts, in many instances to acquire needed capital and to lessen vulnerability to the cyclical nature of the business. For some the record is outright financial failure.

The small prewar aircraft industry still developed the infrastructure that enabled the production and technological achievements that carried the nation to victory in the Second World War. Today the industry is a major industrial employer and the leading industrial exporter, but the number of active aircraft manufacturers is at an all-time low and may fall further. Even the largest aerospace firms of today remain vulnerable to economic and political shifts. In an earlier era, the crash or technical failure of a new design could easily spell doom for a small, struggling firm. In recent times the loss of a single prime contract could doom a multibillion dollar company in the aircraft field. At the end of the Cold War, with cancellations or assignment of lower priorities to military programs and with deficit reduction efforts, the industry once again is struggling for survival.

Despite its rather checkered history, aerospace may well be the country's most successful distinct industry in the technical or product sense, adapting to changing technology and economic conditions and meeting competitive challenges better than any other. Yet the industry has not been so successful politically or in gaining public esteem. Its history also has been one of self-inflicted wounds, especially in the often desperate

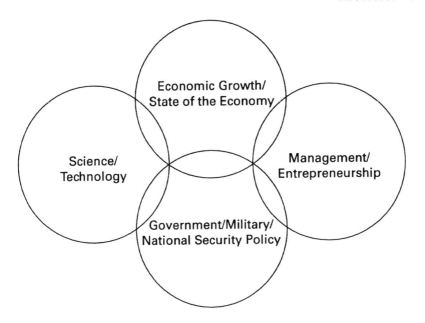

Industry development variables. Exogenous variables are: (1) public/social view of the military, the arms race, the industry; (2) export markets, international protectionism; (3) international technology diffusion.

attempts of firms to win development contracts and export orders, to conceal developmental difficulties or cost overruns threatening cancellation, and to gain or maintain political favor. Defense contracting, the arms export trade, and commercial-airliner marketing seldom have been known for their scrupulous ethics and have cast the industry as an object of wide public suspicion. This troubling aspect of the industry is also detailed and examined.

This book focuses on manufacturers of complete aircraft and does not emphasize firms involved in production of components and parts, or modification or conversion of existing aircraft. Thus the primary coverage is of what is defined as the airframe industry. Engines or propulsion systems, missiles, and other weapons are covered only to the extent that their activities relate directly to airframe companies and aircraft production. Also excluded are many firms that contain the terms *aircraft* or *aviation* in their names but do not manufacture, and may have never manufactured, complete aircraft. Bendix and Rohr are prominent examples. Further, from the 1950s many long-established aircraft manufacturers diversified into related high-technology fields, including defense electronics and the space program, with some exiting aircraft entirely. That transformation into aerospace also is traced, but specific firms are treated in detail only for the period in which they were active aircraft manufacturers. Finally, only those firms judged to have been significant producers of aircraft, or to have made a significant impact on aeronautical development, are included. Those involved with private or general aviation, statistically a minor sector of the broad industry, are the subject of another book.

The book is neither an aeronautical nor a production history but is primarily a business history. The structure is necessarily chronological, tracing individual firms and industry developments over certain periods. But technical achievements and organizational developments are considered within the broad context of aeronautical progress and the military and political developments of the day that affected aircraft demand and production. Included are such well-reported episodes as the Wright patent suits, the First World War production controversy, the founding of Douglas in 1920, the rise of Lockheed from bankruptcy in 1932, and the development of first-generation jet airliners, all part of the lore of American aviation. However, a historical survey of the industry invites reconsideration, and as the philosopher Santayana stated, "History is always written wrong, that is why it is always necessary to rewrite it."

Despite its instability and vulnerability to policy shifts, the industry remains critical to the contemporary economy. Yet the small industry was recognized as vital for both commerce and national security as early as the 1920s. Then with its success in the Second World War and in postwar missile and space development, it provided instruction in the management of other large technological undertakings. Government-funded military technology has been transferred directly to commercial aviation and commercial aircraft, in which the United States has long been dominant. The history of the industry has been an epic of much of the American dream: the individualistic pioneer spirit, the two-guys-in-a-garage entrepreneurial spirit, and the optimism in overcoming obstacles and in accomplishing what many held to be impossible. For these reasons alone the aircraft industry deserves greater appreciation and respect from the public than it has enjoyed to date. It is the author's hope that this book will be a major step in that direction.

Chapter 1 ✈ Origins and Pioneers

The development of manned, powered, heavier-than-air, controlled flight, and consequently of the aircraft industry, almost coincides with the twentieth century. As an emerging field, aviation activities tended to be combined into an all-encompassing industry rather than divided into an aircraft construction industry and other distinct aviation functions. Pioneer aviators generally designed and built both airframes and engines, conducted flying schools and operated airfields, and undertook such revenue-producing activities as exhibition flying, passenger carrying, and instruction. It was not until the 1930s that these aspects of aviation became largely separate as business entities.

The major credit for the development of airplanes, and consequently for a new industry, must go to the Wright brothers. Although not formally trained as scientists or engineers, they nevertheless were intuitive researchers and technicians of a high order. Wilbur Wright solved the problem of flight control, which had bedeviled more learned experimenters, by developing the system of wing warping that enabled the pilot not only to fly but *control* his flight.[1] The control system probably was the Wright brothers' single greatest contribution to aviation. Others would do more to advance the theory of flight. Their simple system of linking dual propellers to a single engine by a chain drive came from their knowledge of bicycles.[2] Yet when the Wright brothers first successfully tested their flyer on December 17, 1903, at Kitty Hawk, they could not have foreseen that they were establishing the basis for a great industry. Indeed, they foresaw no military application for their invention other than for observation, and no commercial future other than passenger carrying. Only a dozen years later, however, aeronautical applications would far surpass what they had envisioned.

The Wrights and the Army

The Army Signal Corps, after more than two years of official disinterest and only after some political pressure, created an Aeronautical Division on August 1, 1907, then issued a specification for a flying machine on December 23, 1907.[3] To that point flying machines were strictly experimental, and the Signal Corps remained primarily interested in balloons. As early as 1905, however, the Wright brothers realized that with their improved model, capable of sustained flight, they had progressed far enough beyond initial experimentation to have a salable product.[4] On October 16 of that year they suspended flying to concentrate on patenting and marketing their flyer. They did not resume flying until May 6, 1908, by which time others had challenged their lead. It was also evident to others that the Wright's obsession with secrecy and protection of their invention harmed sales prospects.[5] Further, their first official offer to the army, at a price of one hundred thousand dollars, including the training of a pilot, simply was unrealistic. The eventual twenty-five-thousand-dollar army contract, on February 10, 1908, represented the first instance of an airplane being designed and built for delivery to a customer at a profit.[6] This in turn defines business. The contract also established the basis on which the industry would operate, in that it was subject to competitive bidding, with the initial customer, to become the major customer, being the government.

Not having expected numerous bids for the contract, the army accepted three finalists, conditional upon their ability to deliver by a given date. The others were Augustus M. Herring, who earlier had worked with Octave Chanute and was later associated with Curtiss, and J. F. Scott of Chicago. The Scott bid was not credible, however, and it was held that Herring had been inactive in the field for years.[7] Herring adamantly maintained for a time that he was building a machine that would meet the specification, but despite being given an extension by the army, he finally forfeited his contract due to nondelivery.

Delivery of the Wright flyer to the army may also be considered the founding date of the American aircraft industry. Both the Wright brothers and Glenn Curtiss were interested from the outset in the business of building and selling airplanes, despite the lack of a clear market. But even with the army order, few could envision broader applications. The delivery of the Wright Model A, at the basic contract price of twenty-five thousand dollars with provision for a twenty-five-hundred-dollar bonus for each mile per hour of speed in excess of the specification, was made on schedule on August 20, 1908. This provision, incidentally, presaged the incentive contracts of a later era of military procurement.

Orville Wright demonstrated the flyer's capabilities to the Signal Corps and trained three pilots, Lieutenants Benjamin Foulois, Frank Lahm, and Thomas E. Selfridge. After several successful demonstration flights, tragedy struck on September 17, 1908, as a defective propeller blade cracked, setting off a vibration that forced the flyer out of control. It crashed headlong into the ground, severely injuring Orville and fatally injuring Lieutenant Selfridge, who thus became the first air fatality.[8]

Orville Wright was to carry the effects of his injuries the rest of his life. The Signal Corps, to its credit, maintained faith in the design and extended the contract for a replacement.

On June 28, 1909, the Wrights delivered the replacement flyer to the army, which officially accepted it on August 2. With success of the second aircraft, which earned a five-thousand-dollar bonus for a 2 mph bettering of the specified speed (40 mph), the Wrights attracted interest from Wall Street. With the backing of several investors, they incorporated the Wright Company on November 22, 1909. Capital eventually reached one million dollars, and the board of directors included such names as Belmont, Collier, Vanderbilt, Ryan, and Alger. Headquarters was in New York, with Wilbur Wright as president, and a new factory was built in Dayton. The firm purchased all patent rights, for which the brothers received one hundred thousand dollars plus a 10 percent royalty on all sales.[9]

The Wright Company briefly enjoyed the position of largest aircraft manufacturer in the world. The company was also involved with exhibitions and flight schools and was the first to produce a standardized design, the Model B, which of course lent itself to more efficient series production. Concerns even were expressed that the Wright Company, with its worldwide patents, might lead to a global monopoly of aviation.[10] The 1909 edition of *Jane's* stated that the Wright Company had built ten airplanes and had twelve on order by the end of 1909.[11] By 1911 the factory had the capacity to construct two airplanes per month.

An Industry Emerges

Aircraft manufacturing began to grow from 1908, and a major developmental influence was that aviation inherently was an international field. American aeronautics reflected heavy international influences. The Wrights were inspired by the German Otto Lilienthal in the 1890s. France first became aware of the Wright experiments with heavier-than-air machines through the travels of Octave Chanute in 1902. Europe subsequently showed more enthusiasm for the new field than did the United States. The Federation Aeronautique Internationale (FAI) was established in Paris on October 12, 1905, to manage and facilitate international cooperation and aeronautical development, and to establish and certify aviation records. The United States was a founding member.

The Wrights, after earlier rejections, were finally granted a U.S. patent for a "flying machine" on May 23, 1906, after having been awarded patents from Great Britain, France, and Belgium in 1904.[12] After their initial failure to interest the army in an airplane in 1905, the Wright brothers secretly negotiated with a skeptical France in 1906 for purchase of their patents, followed by trips to Great Britain and Germany in 1906 and 1907 to interest and assist others in their designs. Eventually it became necessary for the brothers to divide their efforts in order to complete both the army contract and French negotiations. Orville remained in the United States to deal with the army, while Wilbur traveled to France.[13] French rights were finally agreed to in March 1908. Short Brothers also purchased a license during 1908, and a sale of rights to two groups in Germany followed in 1909. In each instance contracts were based on technical descriptions and without actual demonstrations, indicative of the brothers' secrecy.

Wilbur Wright spent several productive months in France and Italy, from August 1908 to March 1909, demonstrating the flyer, competing in aviation contests, and stimulating interest in the potential of aviation. Not incidentally, Wilbur became a celebrity and earned a considerable sum in the process.[14] Most noteworthy was his new distance record of seventy-seven miles, set on December 31, 1908. Orville, recuperated from his crash, and their sister Katherine joined Wilbur in France in January 1909, and Orville later demonstrated the flyer in Germany. On their return to the United States in May, necessitated by the pending delivery of the replacement machine to the army, the brothers were welcomed as heroes.[15]

The first international aeronautical salon was held in Paris in 1908, followed by the great international air exhibition at Reims in August 1909. Glenn Curtiss was the star. Aeronautical developments were passed from country to country soon after their discovery, and the initial American lead was soon lost. Many French aviators, moreover, regarded their country as the true birthplace of flight, and by 1911 France was generally regarded as the leading aeronautical nation, both in design and production. That lead was due as much to the indifference of the U.S. government, however, as to the interest and determination of the French.

Internationally licensed production, initiated by the Wrights, became widely practiced. A major portion of U.S. production in the First World War consisted of license-produced British and French designs. Early industry plans also centered on exports, and American manufacturers exported to Europe from the 1913–14 period. Glenn Curtiss exported a major portion of his production, especially flying boats, before U.S. entry into the war.

Similarly, foreign investment in the American industry began early in its history. The Mitsui Company, Ltd., financed the establishment of the Standard Aero Corporation in Plainfield, New Jersey, in 1916.[16] Anthony Fokker of the Netherlands established a factory in New Jersey in 1924.

Also indicative of the international nature of the industry was that many significant contributors to the industry, including the Thomas brothers, Giuseppe M. Bellanca, Igor Sikorsky, and Alexander de Seversky, were immigrants. The designers Armand Theiblot of Fairchild, Alexander Kartveli of Seversky/Republic, Michael Stroukoff of Chase/Stroukoff Aircraft, and Edgar Schmued, chief designer of the P-51 Mustang, also were immigrants. The noted aeronautical engineer Alexander Klemin was British, and Clement Keys, a major industry financier, was Canadian.

From the 1930s the work of the Hungarian émigré scientist Dr. Theodor von Kármán had a profound impact on aeronautics and later on astronautics. After the Second World War the German rocket scientists and technicians brought to the United States greatly advanced missile and space programs. As the aircraft industry continued to develop, its international dimensions expanded concurrently.

The aircraft/aerospace industry, now some ninety years old, has long been a major factor in the economy. Yet long-term survival was never assured for any company. The remaining giants responsible for the majority of the industry's output, despite their modern corporate structures and leadership in advanced technology, originated with, or were formed from, companies established before the Second World War. No major aircraft firm has been established since and survived. Many pioneering firms exited production since the war, and it is perhaps ironic that relatively few aircraft were produced carrying the Wright nameplate.

The question of which was the first American aircraft company, while of dubious relevance, still is debated. Some credit the distinction to Glenn Curtiss. Others believe that Professor Edson Gallaudet built the first aircraft factory, while still others confer that distinction on the Wright brothers or the Wittemann brothers. The first aircraft company in the world, however, was that of the Voisin brothers of France, whose factory began production in 1907.[17] Gabriel Voisin developed the first practical airplane in France and produced and sold a number of experimental airplanes. He never acknowledged the Wrights as the first to fly. Also noteworthy is that the Short brothers, Wright licensees, were the first series manufacturers in the world, in that they produced six aircraft of the same design from 1909.

Certainly there were numerous American experimenters and constructors in the first decade of powered flight, but business, in the form of sales of aircraft, grew slowly. In 1914, for example, sixteen firms were listed as aircraft manufacturers, but collectively they built only forty-nine aircraft, valued at some $789,000, of which thirty-four were exported.[18] It was not until 1914 that any manufacturer received an order for more than six airplanes at a time. Claiming to be an aircraft manufacturer was relatively easy, but it was quite a different task to sustain production in the field.

Aircraft manufacturing was somewhat difficult to classify as a conventional industry due to the nature of its pioneers. In almost all cases they were motivated by an intense interest in airplanes and flying, and by a desire to participate in the exciting new field, rather than by any opportunity for profit or for building a business empire. Thus they tended to have more in common with a Thomas Edison than with an Andrew Carnegie or a John D. Rockefeller. The early experimenters and builders tended to be young and from modest circumstances. Few were formally educated. Even though many eventually prospered financially, the lack of a primary profit motivation in the founders alone distinguishes the young aircraft industry from many others.

Curtiss and the Wright brothers, while entrepreneurial, were not professional managers. The Wrights possessed no inclination to actively manage their company, and Curtiss's business judgment was at times nearly disastrous. Most certainly they were risk takers, in both the physical and financial senses, but did not demonstrate the broad strategic or conceptual grasp necessary for building and sustaining a large enterprise. In many cases professional managers succeeded where the company founders had failed.

The early aircraft market was primarily one of wealthy sportsmen and adventurers attracted to the excitement and novelty of flight. From 1909 flying was regarded as a spectator sport, as the public would pay for aerial contests and exhibitions. There was little vision of the airplane as a weapon of war or as a means of transportation. A related problem was that banks generally were reluctant to finance a new aircraft venture. Many early entrants were forced to seek sponsorship by men of wealth in order to initiate or sustain their companies. A few, including Bill Boeing and Sherman Fairchild, had sufficient means to finance their aviation ventures, while others, such as Donald Douglas, Claude Ryan, and John Northrop, attracted wealthy backers. Many early managerial conflicts were linked to differences between those with primarily financial concerns and those concerned with aeronautical development and progress. A comparable conflict persists in the contemporary industry, as beleaguered manufacturers fight to remain in aircraft in the face of declining contract opportunities and daunting financial risks, rather than face a future, even if more stable, as subcontractors or component manufacturers.

Curtiss and His Associates

It does appear that the first aircraft-manufacturing company in the United States, albeit short-lived, was that of Glenn Curtiss. A native of Hammondsport, New York, Curtiss was a high school dropout who showed an early mechanical inclination and an obsession with speed. This led him first to motorcycles and then, after learning of the Wright brothers' experiments, to airplanes. With success in motors he formed the Glenn H. Curtiss Company in 1906 at the age of twenty-eight, followed by the Curtiss Motor Vehicle Company in 1907. Neither was specifically involved with aviation at the time. It is not clear when Curtiss first became aware of the Wright experiments, but he did contact the Wrights on May 16, 1906, offering to sell them an engine developed by the Aerial Experiment Association.[19]

Curtiss's initial involvement in aircraft was in fact through the Aerial Experiment Association (AEA), formed on October 1, 1907. Other members were Alexander Graham Bell, J. A. D. McCurdy, Thomas Baldwin, and Lieutenant Selfridge, "on loan" from the army. Financial backing was provided by Bell's wife. The AEA constructed and first flew the *Red Wing* model on March 12, 1908. An advertisement in *Scientific American* offering the *White Wing* for sale at a price of five thousand dollars apparently was fictitious, although it succeeded in upsetting the Wrights, already concerned about possible patent infringement.[20] The *White Wing,* incidentally, was the first airplane to fly using ailerons. Curtiss first flew his *June Bug* design on June 20, 1908, and his test on July 4 was the first officially witnessed flight in the United States, winning the *Scientific American* trophy.[21]

Despite the accomplishments of the AEA, Bell remained essentially research oriented, while Curtiss wished to enter aircraft production. The AEA planned to form a stock company to lend a more permanent structure to its endeavors, but Curtiss had been approached in the meantime by Augustus M. Herring with a business proposition.[22] Already worried about possible legal action by the Wrights, Curtiss was attracted by Herring's claims of financial connections in New York and of important patents predating those of the Wrights. Even as the AEA was debating its future, Curtiss and Herring formed the Herring-Curtiss Company on March 19, 1909, to produce and sell its designs for seventy-five hundred dollars each. The move raised questions about Curtiss's ethics, and the AEA was dissolved after, although Bell and Curtiss remained on good terms.

Courtland Field Bishop of the New York Aero Club and Thomas

Baldwin of the AEA were made directors of the Herring-Curtiss Company. Capitalization was set at $360,000, but the corporate structure was so loose that Herring, putting up only a nominal amount of cash, emerged with a majority of the stock.[23] Curtiss folded his Glenn H. Curtiss Company into the new company, adding motors and motorcycles to aircraft production. At the time his manufacturing arm was prospering, with 100 employees and a 1908 profit of $120,000.[24] Despite outside investors, Glenn Curtiss was in fact the major financial contributor to the Herring-Curtiss Company, the millionaire Bishop having invested only twenty-one thousand dollars. Thus was born the first aircraft-manufacturing company in the United States, though not of the scale of the Wright Company founded several months later. It still may be argued that the Wright Company was the first serious, professional manufacturer. *Jane's* reported the Herring-Curtiss Company as delivering six airplanes in 1909.[25]

It soon became evident, however, that Herring-Curtiss was a misbegotten enterprise. Herring, despite his extensive experience, was highly suspect as an aeronautical designer or innovator. At best he exaggerated his knowledge and accomplishments; at worst he was an outright fraud. His claimed patents and financial connections proved to be nonexistent, and thus he was of no value to Curtiss. Curtiss, who initially had regarded Herring as a genius, became disenchanted with him even before setting off for the Reims exhibition in July 1909. By that time the company was building a new factory in Hammondsport with no means of payment.[26] Herring assured all that the necessary funds could be raised by selling his stock. The board finally demanded that Herring produce his patents or at least applications, which of course he was unable to do. Curtiss even offered to buy out Herring's interest for ten thousand dollars to settle the matter.[27]

Even while undergoing his difficulties with Curtiss, Herring was secretly negotiating with W. Starling Burgess, another pioneer manufacturer, on yet another venture.[28] A Herring-Burgess design was tested in 1910, but Herring soon developed a dispute with Burgess's partner Greely S. Curtis, and Burgess eventually canceled the agreement.[29] In the meantime, adverse publicity and the initial injunction brought by the Wrights against the Herring-Curtiss Company effectively killed business, and creditors demanded payment. Left with no business other than exhibitions, the Herring-Curtiss Company entered involuntary bankruptcy on April 10, 1910.[30] Curtiss was not entirely unhappy at this turn of events, viewing the bankruptcy as a means of ridding himself of Herring as well as of escaping further action by the Wrights. The bankruptcy decree was issued on December 2, 1910, and Curtiss thereupon formed the Curtiss

Aeroplane Company, with capital of twenty thousand dollars and succeeded in buying back his original factory at auction. The initial Wright injunction had been lifted on June 14, 1910, clearing the way for Curtiss to do business again. On December 12, 1910, Curtiss also formed a new Curtiss Motor Company, with capital of six hundred thousand dollars, of which he held 50 percent, for the purposes of producing motorcycles and motors as well as aircraft engines.[31] Herring, while discredited at the time, remains a baffling figure in early aviation, still regarded by some as a significant pioneer. He was to remain a legal thorn in Curtiss's side the rest of his life. The Herring-Curtiss Company, incidentally, was not legally dissolved until 1926, after Herring's death.

Glenn Curtiss, free of Herring and the initial Wright injunction, regarded his new Curtiss Aeroplane Company as a defense vehicle against the Wrights, since the suits were against the Herring-Curtiss Company as well as himself. He was awarded a patent for an "improved flying machine" in 1911. With successful tests of his seaplane, he delivered the first model, the A-1 Triad, to the navy on July 1, 1911. Earlier, on March 17, he delivered his first Model D airplane to the army, which became the army's second flyer. Winning further military orders, he was able in 1914 to develop the large *America* flying boat with the goal of making the first aerial crossing of the Atlantic. The outbreak of war prevented the attempt, but the design attracted the attention of the British, paving the way for profitable exports for Curtiss.

On a trip to England in 1914 Curtiss hired B. Douglas Thomas as his first design engineer, and Thomas was to design the famed JN-4 Jenny trainer.[32] The Thomas-designed Model J, predecessor of the Jenny, first flew on May 10, 1914. The influx of war orders required a stronger labor supply, leading to a move to Buffalo late in 1914.[33] The motor operation remained in Hammondsport. The new Churchill Plant in Buffalo covered 120,000 square feet, and capacity increased further in 1915 with the establishment of the Curtiss Aeroplane and Motor Company, Ltd., in Toronto.[34] The company secured a $15 million British order at the end of 1915, and Glenn Curtiss was becoming wealthy. By that time he had attained the position of largest airplane manufacturer in the United States.[35] In addition to his plants in Hammondsport, Buffalo, and Toronto, Curtiss was to add the Burgess plant in Marblehead, Massachusetts.

W. Starling Burgess, a successful boat and yacht builder, with his associate Greely S. Curtis incorporated the Burgess Company and Curtis, Inc., in June 1910, capitalized at eighty thousand dollars. The firm produced Wright and British Dunne designs under license, as well as its own designs. Even with the license, the company became a target of patent infringement suits over modifications to Wright designs.[36] The Wright license permitted production only of exact copies, and royalties were paid on that basis.[37] Greely Curtis withdrew his interest in 1914, and the company was renamed the Burgess Company, although Curtis remained as treasurer.[38] The company merged into the Curtiss Aeroplane and Motor Company in 1916. Curtiss acquired all Burgess stock, and Burgess joined the Curtiss board.[39]

The First Aviation Boom, 1909–12

As late as 1908 only the Wright brothers and members of the AEA had actually flown airplanes in the United States, but activity soon expanded. With the buildup of production by the Wright and Curtiss companies, and with the popularity of exhibition flying, public awareness increased markedly. New experimenters and constructors entered the field, although few then foresaw the later practical applications of the airplane or any mass-market potential.

Progress continued apace, however. Despite the flimsiness of structures and the low power and unreliability of engines, new records were set frequently, and new competitions and prizes for aviation feats were established. On July 25, 1909, Louis Bleriot successfully flew across the English Channel, despite almost crashing with an overheating engine. Glenn Curtiss won the Gordon Bennett speed trophy in France on August 29, 1909, with a prize of ten thousand dollars. Curtiss briefly held both land and air speed records. On October 4, 1909, Wilbur Wright made his famous flight up the Hudson River and back. On May 29, 1910, Curtiss flew the 151 miles between Albany and New York City, the first flight between cities, winning another prize of ten thousand dollars. More ominously, in 1909 the Italian officer Giulio Douhet first wrote of the use of airplanes in war. In the United States the first primitive experiments with aerial gunnery and bomb dropping took place in 1911 and 1912.

Pioneer aviators, most prominently the Wright brothers, Curtiss, and Glenn Martin, devoted major activity to flying schools. Hundreds of aspirants undertook flight training in 1910 and 1911. Eddie Stinson, who learned to fly with the Wright brothers in 1911, enjoyed the distinction in the 1920s of having taught more pilots to fly than any other man.[40] Eugene Ely made the first airplane takeoff from a ship in 1911, the same year in which the first seaplanes were successfully tested. Lincoln Beachey raised the world altitude record to 11,642 feet on August 20, 1911. Calbraith Rodgers completed the first transcontinental flight on November 5, 1911, in his Wright *Vin Fiz*, taking a total of forty-nine days. In 1912 the British first used a monocoque structure, in which the fuselage bore a portion of the aircraft weight rather than simply serving to extend the tail far from the wings. Also in 1912, radio was used in an airplane for the first time.

Progress was enhanced by associations that promoted aviation and producers of aircraft. The Aero Club of America was formed on November 30, 1905. The U.S. Aeronautical Society, founded in New York on June 10, 1908, was an early supporter of air exhibitions and brought together many pioneers to exchange ideas. By 1909 numerous aero clubs had been formed, principally in the Northeast and Midwest. An Aeronautical Manufacturers Association was formed in Connecticut on April 21, 1911.

Interest in aviation and its commercial possibilities increased sharply after the 1909–10 period. There was a high degree of self-selection by new entrants into aircraft during this period. Barriers to entry were virtually nonexistent, as there were no legal restrictions, and aircraft construction was a labor-intensive activity requiring little in the way of expensive tooling or facilities.[41] Further, technical knowledge was rapidly disseminated and easily acquired. E. L. Jones, reporting in *Jane's,* estimated that by 1913 some two thousand people had built airplanes, although only a handful could be described as established manufacturers.[42] With the passage of time it is evident that Jones's figure was wildly inflated, although aircraft construction was a widespread phenomenon. Most aircraft of the period were the equivalent of what became known later as homebuilts. Some two hundred airplanes had in fact been produced in the United States by 1914, but few had been manufactured in series, and most aspiring firms turned idle facilities to production of other materials during periods of inactive aircraft demand. Among those early builders were some who would make their mark later, such as Clyde Cessna, who built his first airplane in 1911, and the brothers Allan and Malcolm Loughead, who flew their first airplane in 1913.

The lifespan of most early aircraft ventures was, of course, brief. A major reason was the narrow market, involving only sport or exhibition aircraft and small numbers for the military services. The boom from 1909 to 1912 was in research and development, innovation, and extension of flight capabilities, not in manufacturing. In fact, revenues from exhibition flying and aviation meets for both the Wright and Curtiss companies greatly exceeded revenues from aircraft sales. Yet no firm, whatever the determination of its founder or the merits of its designs, could survive without continuing aircraft orders. It was clear, however, that aviation for the early participants was neither a science nor an industry, but a passion.

Another reason for halting development of the industry was a lack of government policy for either the military or commercial spheres.

Although the United States was the birthplace of the airplane and was the first nation to possess a military air arm, these advantages were soon lost. The Signal Corps did not receive a second airplane until 1911, when $125,000 was appropriated for purchase of a further five aircraft.[43] One unfortunate result of this lack of government support was that when the United States entered the First World War, other countries were far ahead in design and production. This lag was reflected clearly in government aviation appropriations, in which major European powers were spending millions, while the U.S. budget was minimal. In 1913, for example, the French government appropriated $7,400,000 for aviation, while the United States spent only $125,000, about the same as Bulgaria.[44]

In 1913 the U.S. Army Flying Service owned twenty-one planes, five Burgess, six Curtiss, and ten Wright. The army had twenty-two aviators and the navy four.[45] Through 1912 some forty-two pilots, military and civilian, had been killed, including Selfridge, Moisant, Ely, and Quimby. Also in 1912 there were some four hundred known pilots, including the Wrights, Curtiss, Burgess, Beachey, Gallaudet, Martin, the Thomas brothers, Allan Loughead, Willard, and Vought.[46] Aviation lost pioneers even as the field began. Octave Chanute, who had assisted the Wright brothers but became somewhat estranged from them over questions of credit for the accomplishments, died on November 24, 1910, at age seventy-eight; then Wilbur Wright died on May 30, 1912, at age forty-five.

Despite the industry's difficulties in gaining a market foothold, aviation had progressed impressively in the three years since the Wright and Herring-Curtiss firms were formed. After 1912 the numbers of aircraft, pilots, flying fields, and schools in operation across the country reached a level in which aviation had become more than a novelty, and new builders were emerging to challenge the Wright and Curtiss companies. But legal obstacles also were to appear.

The Wright-Curtiss Feud

Preceding formation of both the Herring-Curtiss and Wright companies, the notorious Wright patent suits cast a shadow over the industry for years. Characteristic of the Wrights was their distinction between profit making and purely experimental or scientific use of their patents. Uses restricted to aeronautical progress were welcome, but if others tried to employ Wright developments for profit, they demanded to be compensated with royalties. Glenn Curtiss first ran afoul of the Wrights when Lieutenant Selfridge, then of the AEA, requested technical assistance from the Wrights in a letter of January 15, 1908, which they cheerfully granted.[47] Disturbed over Curtiss's apparent commercial plans, the Wrights warned Curtiss by letter in July 1908 of potential patent infringement, stating that they would be glad to discuss a license for commercial use of their patents. Curtiss responded that his work was experimental and denied that he was stealing Wright patents in designing and building any aircraft for sale. Later the Wrights learned that an aircraft built with the assistance they had provided had in fact been sold. This sale, on June 26, 1909, to the Aeronautical Society of New York for seventy-five hundred dollars, was the first in the United States.[48]

Curtiss believed that the Herring-developed separate ailerons for lateral control exempted him from any patent liability.[49] The Wrights remained suspicious, however, and in August 1909 filed patent infringement suits against Curtiss and others, both in the United States and Europe.[50] Frequently it has been mistakenly believed that Curtiss was their sole target, but suits were filed against many individuals whom the Wrights felt were making profitable use of their patents. Curtiss was, however, the most prominent target, and the object later of the Wrights' greatest bitterness. Infringement suits were also filed through the Wright licensees, resulting in a dozen domestic suits and some twenty filed abroad. The French in particular argued that there was no infringement since their use of Wright developments was based on prior disclosure, before the Wrights were granted their original patent.

Injunctions filed against the Herring-Curtiss Company and Glenn Curtiss early in 1910 effectively prohibited the company from building and selling airplanes. While these injunctions were later lifted on appeal, the Wrights pursued further suits, and the matter became a major public controversy. Many felt that the original Wright patent was so broad that it would not be upheld. There was also a view that the airplane was a public invention and should not be the province of any individual. The suits were finally resolved, however, in a federal court of appeals ruling on January 13, 1914, which upheld the Wrights. The court ruled that the Wright patent was all-encompassing, including Herring-developed ailerons, thus disallowing the central point of Curtiss's defense. Orville felt vindicated, but litigation continued as Curtiss resisted, even talking of taking his case to the Supreme Court. By that time, however, aeronautics had progressed to the point that the original patent was effectively obsolete.

Even with the legal victory, all was not going smoothly within the Wright Company. Grover Loening, joining the Wright Company as an engineer in 1913, found Orville, who had succeeded Wilbur as president, to be detached and indecisive, perhaps due to his brother's death. Alphaeus Barnes, installed by the board as corporate secretary/treasurer, openly regarded the Wrights and their associates as Ohio hicks and provoked numerous conflicts.[51] The board, moreover, was interested in an effective monopoly on aircraft production under patent protection, while Orville remained interested primarily in royalties.[52]

Chafing under the direction of a board with priorities different from his own, Orville Wright from July 1914 moved to buy the interests of all investors except for that of his friend and supporter Robert J. Collier.[53] Although forced to borrow heavily to do so, Orville now enjoyed a free hand in his company, which he exercised by firing Alphaeus Barnes, filing a new patent suit against Curtiss, and filing for new patents based on improvements to his basic designs. Then on October 15, 1915, he in turn sold to a New York investment syndicate led by William B. Thompson and Frank Manville of the Johns-Manville Corporation of New Jersey. The new owners proved to be more compatible with his views, but Orville was now inactive in the firm he had founded only six years earlier. Having received some $1.5 million for his interest, however, he possessed the capital to do as he wished.[54]

Ironically, with Wilbur's untimely death in 1912 and Orville's sale in 1915, the Wright brothers contributed little to flight beyond its first decade. Orville, while innovative, still was reluctant to adopt ideas of others, persisting with his original concepts even after design improvements and formidable competition appeared at home and overseas. As one illustration, casualty rates of pilots training on planes built along the Wright-developed pusher propeller model were appalling. Glenn Martin introduced his tractor-engined trainer, which proved to be much safer, and won orders from the Signal Corps. Orville, however, resisted this development even after its superiority had been demonstrated.[55] Further, his Model C design was found after investigation to possess basic design flaws.[56] Orville remained active in aviation for the remainder of his life, but chiefly in an advisory capacity rather than as an active designer. His final contribution to aeronautical progress was the split flap, patented in 1921 and employed effectively in Second World War aircraft.[57]

The Wrights' obsession with the patent infringement suits probably diminished them. The Wright family felt that the strain of the lawsuits had contributed to Wilbur's illness and early death, increasing their bitterness against Curtiss. But litigation cost the Wrights much public popularity, and even victory in court did not result in any permanent benefit. All parties incurred heavy expenses that might have been better applied to aeronautical development. The beleaguered Curtiss actually gained public sympathy, and his companies still managed to prosper during litigation. The entire episode provided an instructive contrast between the secrecy of the Wrights and the more open Curtiss, but all that endured was bitterness between Wright and Curtiss.

The brothers apparently had expected all aeronautical development to be frozen during the litigation. But experimentation proceeded, although many designers and constructors were highly aware of potential patent liability. Many pursued designs that differed radically from those of the Wrights, in part to distance themselves from any patent infringement question. Whether aviation would have progressed during 1909–14 more than it did had the infringement threat not existed remains speculative.

Curtiss, moreover, held several valuable patents on his own. He perfected the aileron, which succeeded the Wright wing-warping system, and pioneered the seaplane. His "hydroaeroplane," tested on January 26, 1911, at San Diego, led to the initial navy order. He clearly was the father of naval aviation. With the court ruling, however, Curtiss was required to remit a 20 percent royalty on all airplanes sold. Other manufacturers named in the ruling were treated leniently by Orville Wright, but not Curtiss. Most paid royalties to the Wright Company, but Curtiss resisted successfully until the litigation ended. Curtiss even enlisted the Smithsonian Institution, which held the Langley designs, in an effort to show that Langley's developments predated those of the Wrights. A Langley replica was flown, but it was determined that it included modifications based on Wright developments, and the test posed no threat to the Wright patents.[58]

While the patent litigation remained active, the United States entered the war against Germany in 1917. The Army and Navy Joint Technical Advisory Board, formed to expedite war production, imposed a policy of pooling of aircraft patents of all manufacturers, effectively ending the Wright patent infringement suits and any manufacturing restrictions.[59]

Other Pioneers and Their Firms

Edson Gallaudet

Professor Edson Gallaudet, son of the founder of Gallaudet University for the deaf, earned a doctorate at Johns Hopkins University and became a professor of physics at Yale. An early experimenter in aviation despite the derision of his learned colleagues, Gallaudet organized an engineering office in Norwich, Connecticut, in 1908. Although its stated purpose was consulting engineering, the later General Dynamics Corporation, of which Gallaudet was a distant predecessor, credited this firm as being the first aircraft factory in America.[60] Although Gallaudet was an early designer and innovator and built his first airplane in 1909, it is doubtful that he was the first series builder.

He then organized Gallaudet Engineering Company in 1910, for contract work. After financial difficulties, the firm was reorganized as the Gallaudet Aircraft Corporation on January 19, 1917, in East Greenwich, Rhode Island, under the control of investment bankers, including Averill Harriman.[61] Some of Gallaudet's designs, especially the streamlined *Bullet* of 1912, advanced aeronautics, but he did not attain mass production until 1918 with a war contract to build Curtiss floatplanes.[62] He continued low-volume production until 1923, when all rights were sold to Major Reuben Fleet, and Edson Gallaudet exited aviation.

Glenn Martin

One pioneer would make his mark on the industry over decades: Glenn L. Martin. Alone among the pioneers, his career extended well into the jet age. As with many others he began in another field, which he soon abandoned for aviation. A native of Kansas, Martin moved to California as a youth, his parents having made the move for economic reasons as well as for his mother's health. Martin was an instant success as an auto dealer in Santa Ana, but after learning of the work of the Wright brothers, devoted more and more time to building his own flyer. His second effort in 1909 succeeded, and Martin gained a strong reputation as an exhibition flyer, earning a substantial income in the process.

Martin moved late in 1911 to organize an aircraft company. Attracting one hundred thousand dollars from local investors, the Glenn L. Martin Company was officially founded in Santa Ana on August 16, 1912. Martin relocated to Los Angeles by the end of the year, where he designed and produced several successful aircraft. Martin won orders for his Model TT tractor-engined trainer from the Signal Corps, besting Wright and Curtiss competition, and delivered seventeen military trainers in 1914.[63] Production of even that modest order taxed the capabilities of the small company but provided a major boost to its prospects. Martin was assisted first by Charles H. Day and later by Charles F. Willard, notable pioneers themselves. Then in 1912 he hired the eighteen-year-old Lawrence D. Bell, who had come to California as a mechanic for his older brother Grover, an exhibition pilot. In 1915, in a move of even greater significance, Martin brought in the young Donald Douglas, succeeding Willard, who had joined the Signal Corps, as chief engineer.[64] Bell became plant manager also in 1915.

On August 7, 1916, the new Wright Company, although losing money at the time, approached Martin concerning a merger. Martin saw a stronger firm resulting and was also attracted to an association with the prestigious Wright name, although Orville Wright was no longer involved.[65] The Wright-Martin Aircraft Company formed in September 1916 included the Simplex Automobile Company of New Brunswick, New Jersey, also involved with aviation, the newly formed General Aeronautic Company, and the Wright Flying Field. On November 4, 1917, the company, under control of the Simplex interests, raised the original capital from $1 million to $10 million, an impressive sum for the day.[66] Headquarters was moved to New Jersey. The Martin Los Angeles factory was retained for the time, but the original Wright factory in Dayton was closed. Glenn Martin was appointed vice president for aircraft production, and the company announced a goal of developing superior aircraft. Donald Douglas did not follow Martin but left for the War Department in December 1916. Management was under George W. Goethals and Company, a prominent engineering firm, in the person of George H. Houston.[67] The banker and investor Richard F. Hoyt was secretary, and Guy W. Vaughan, later to head Curtiss-Wright, joined the firm in 1917.

Martin soon became disenchanted, however, finding himself with little influence and unhappy with management's increasing orientation toward engines. With the transfer of the Wright patents on December 16, 1916, it became evident that the chief goal of the company was to extract royalties from other manufacturers. No new aircraft designs emerged, and the original Martin factory was closed later in 1917. Glenn Martin withdrew after only ten months and reentered aircraft by forming a new Glenn L. Martin Company in Cleveland, Ohio, on September 10, 1917. Martin attracted the support of Cleveland businessmen for the move, and a new factory was completed in 1918.[68] With the assistance of Donald Douglas and Larry Bell, who rejoined him in Cleveland, Martin designed the MB-1 heavy bomber, which first flew on August 17, 1918. While too late for the war, the bomber became a service success.

The Wright-Martin company profited with license-produced Hispano-Suiza engines but declined after the Armistice. Since it had failed in aircraft, the company was reorganized as the Wright Aeronautical Company in October 1919, with activity oriented toward engines.

The Thomas Brothers

The brothers William T. and Oliver W. Thomas, born and educated in England, settled in the Hammondsport of Glenn Curtiss. W. T. Thomas learned to fly with Curtiss in 1909, then formed Thomas Brothers Company in November 1909. The brothers then incorporated the Thomas

Brothers Airplane Company in nearby Bath, New York, in 1912, and began manufacture. In early 1914 they added their Thomas School of Aviation, and in December 1914 moved both firms to Ithaca, New York. On January 31, 1917, the businesses were merged with the financially strong Morse Chain Company of Ithaca, headed by Frank L. Morse, to become Thomas-Morse. B. Douglas Thomas, originally hired by Curtiss and unrelated to the brothers, became chief design engineer. Successful designs followed, and Thomas-Morse became a major wartime manufacturer. Its MB-3A was the first American fighter to attain production, but the firm declined in the 1920s with the postwar aircraft collapse.

William Boeing

William Edward Boeing, born in Michigan in 1881 in affluent circumstances, was educated in Europe as a youth. He attended Yale but dropped out to become a successful lumberman in Seattle, Washington. He took his first flight in a Curtiss seaplane on July 4, 1914, and from that point became enraptured by aviation. He enrolled in Glenn Martin's flying school in Los Angeles in October 1915 and thereafter purchased a Martin flyer for his personal use.

Boeing, convinced that he could build superior airplanes, determined to enter aircraft manufacturing. He enlisted the assistance of his friend Conrad Westervelt, a naval engineer then assigned to Seattle, who obtained technical information from Professor Jerome C. Hunsaker of MIT. The prototype Boeing and Westervelt (B & W) seaplane, powered by a Hall-Scott engine, made its first flight on June 15, 1916. Westervelt by that time had been reassigned to Washington, over his protests, but encouraged Boeing to develop seaplanes. This led Boeing to incorporate the Pacific Aero Products Company on July 15, 1916, with himself as president and with his cousin Edgar N. Gott, an engineer, as vice president. With U.S. entry into the war, the venture was renamed Boeing Airplane Company on April 18, 1917. Two young engineers later to lead Boeing, Phillip G. Johnson and Clairmont T. Egtvedt, joined the company in May. Only two B & W seaplanes were sold, but the new firm succeeded with wartime production of fifty of its Model C floatplane for the navy, followed by an order for fifty Curtiss-designed HS-2L trainers, which it produced more efficiently than Curtiss.

Curtiss and Dayton Wright

Glenn H. Curtiss, with his production and sales success, captured the attention of Wall Street as had the Wrights earlier. Late in 1915 the investment firm of Imbrie and Company purchased control of the Curtiss companies. They were combined as the Curtiss Aeroplane and Motor Company on January 13, 1916, with headquarters in Buffalo, and with capital of $9 million. Curtiss realized a sum of $5 million on the sale.[69] A month later the interests of the Burgess Company were acquired. Curtiss Aeroplane and Motor then established Curtiss Engineering Company and the Burgess Company as subsidiaries. With the sale of the Wright Company to new investors, a Wright-Curtiss merger was considered briefly, but the proposal did not advance due to antitrust concerns.

The Imbrie acquisition attracted the Canadian-born Clement M. Keys. Keys, only forty and successful with his own investment-banking firm, developed a strong interest in the industry's potential. After advising Curtiss, he joined the firm as an unsalaried vice president in November 1916. Keys was strongly oriented toward organization and financial structure of the industry and transformed Curtiss from an essentially single-owner company to a public corporation.

Glenn Curtiss admittedly was not inclined to be a captain of industry and was totally astray in finance. The new owners recognized the need for professional management and for additional financing to expand production. Keys made progress with both but soon realized that major new investment would be necessary. He thereupon attracted auto magnate John North Willys of the Willys-Overland Company to Curtiss. Willys, whose upstate New York background was similar to Glenn Curtiss's in many respects, invested $2 million in July 1917 and took over as president. Keys left temporarily but returned in December 1917 at Willys's behest. To meet the demands of war business, the company built the $4 million North Elmwood plant in Buffalo, the largest aircraft plant in America.[70] Wartime peak employment was eighteen thousand. Curtiss then was able to devote himself to his first love, experimentation, and became president of Curtiss Engineering at Garden City, Long Island. This subsidiary, the first in the industry devoted exclusively to research and development, developed the NC flying boats. They were to be Glenn Curtiss's final contribution to aviation.

W. Starling Burgess left the Curtiss company when he was commissioned in the U.S. Navy and entered active service. Greely S. Curtis joined General Electric as an engineer and played no further part in the aircraft industry.[71] The Burgess Company was absorbed into Curtiss in 1919. Curtiss was the largest wartime aircraft manufacturer and possibly the most successful, if any could have been regarded as fully successful. Growth came with a heavy debt burden, and Curtiss experienced major difficulties, as did others, with license production of foreign designs.

On April 11, 1917, after the U.S. declaration of war, a number of prominent Ohio investors, including Charles F. Kettering and Edward A. Deeds of Dayton Engineering Laboratories Company (DELCO), formed the Dayton Wright Company. Orville Wright lent his name and served as consulting engineer but had no role in management. The investors were determined that the city of Dayton, although it had lost the original Wright Company, would continue its aviation prominence. The original plant, which had survived the disastrous flood of 1913, was retained, and two new plants also were constructed. The investors were convinced that small firms could not survive in the aircraft business, and sought influence in order to win military contracts, employing the influential Howard E. Coffin of Hudson Motor Company for that purpose.[72] Dayton Wright also was among the first to acquire professional management, in this instance the Harold E. Talbotts, father and son and prominent industrialists. First winning a contract to produce four hundred Standard SJ-1 pursuit planes, Dayton Wright eventually produced 3,506 aircraft, mostly DH-4s, second only to Curtiss. But a further order for four thousand DH-4s, with a prospective $4 million profit, was canceled at war's end.[73]

Grover Loening

Grover Loening, to win prominence as the developer of the short-hulled flying boat, was one of the most successful early designers. Loening had first associated with the small Queen Aeroplane Company of Willis McCornick in 1911 as an engineer. He applied his expertise to the design of aircraft for the company, but his first flying-boat design was wrecked in a storm, and orders were not forthcoming.[74] Loening then attracted an offer from Orville Wright to move to Dayton as chief engineer, which he accepted in July 1913. Although loyal to Wright, Loening soon became frustrated by the bitter conflicts between Wright and his New York associates and departed a year later to become chief engineer of the Aviation Section of the Signal Corps. There he became frustrated with the military bureaucracy as well. He was contacted by former governor Foss of Massachusetts, owner of the Sturtevant Blower Works, to help organize and manage a new Sturtevant Aircraft Company.[75] He arrived in July 1915 and began developing original designs, although Sturtevant was to become primarily a subcontractor.[76]

With U.S. entry into the war, Loening attracted backing from the investor Henry M. Crane and others to launch his own company, with a primary focus on naval aircraft.[77] While never among the largest manufacturers, Loening won acclaim for his designs, and his firm made an impact on aeronautical progress. His first design, the small *Kitten,* was intended as a naval scout airplane, but became the basis for the M-8, one of the first American fighters.

Loening was a prominent defender of the young aircraft industry, especially against inroads by the automobile industry. While dealing with the Naval Aircraft Factory during the war Loening became acquainted with Lieutenant Leroy Grumman, a Cornell engineering graduate, whom he later hired and who went on to found his own firm. Loening also possessed another distinction: He received the first master's degree in aeronautics, from Columbia University in 1910, and was thus the first degreed aeronautical engineer.

Chance Vought

The brilliant young designer Chance (originally Chauncey) Milton Vought, a native New Yorker, founded the Lewis and Vought Company on June 18, 1917, and attracted immediate attention with his advanced designs. Vought, a rare combination of pilot, designer, and businessman, learned to fly in 1910, afterward worked for the Illinois Aero Club, and designed his first airplane in 1914.[78] He later worked for the new Simplex Aircraft subsidiary, then briefly as chief engineer for the Wright Company after its affiliation with Simplex.[79] He launched his own company with the assistance of the wealthy sportsman and aviation enthusiast Birdseye B. Lewis, to take advantage of business opportunities presented by the war. Birdseye Lewis was killed in 1918 in an air crash in France, while serving on General Pershing's staff, but that event did not appear to have a material effect on the company. In fact, it is difficult to determine just what Lewis's role in the company was, since he was not identified as an officer, director, or investor in any company documents.[80] The young company managed to survive the mass cancellations at the end of the war, and Vought became a prolific designer in the 1920s.

Others

Aircraft firms grew steadily in number in the prewar period, but series production was sporadic. That picture changed dramatically with the entry of the United States into the war. New firms were formed that, while not enjoying long lives, were to make significant contributions to the war effort, and in some instances to aeronautical progress. One of the first was the St. Louis Aircraft Company, formed as a subsidiary of the St. Louis Car Company by a business group in that city in the fall of 1917, at the instigation of the government.[81] It contracted for the Curtiss Jenny trainer. There were also the aforementioned Sturtevant Aircraft Company of

Boston, founded in December 1915, and the L-W-F Company of Long Island, also founded in December 1915. Charles F. Willard, a founder of L-W-F, worked for both Curtiss and Martin before becoming involved with the new venture and later became the first of the great barnstormers. L-W-F, while named after its founders Edward Lowe Jr., Charles Willard, and Robert G. Fowler, another pioneer aviator, also was considered to be named after the construction it developed, the laminated wood fuselage. It pioneered the monocoque fuselage in the United States, a major advance over earlier open fuselages.[82] Lowe provided most of the financing, while Fowler attracted Willard. Lowe soon secured total control, renaming the firm L-W-F Engineering, and Fowler and Willard departed in 1916. Wartime production included the Model V trainer and a reconnaissance model for the Signal Corps.

The Aeromarine Plane and Motor Company of Keyport, New Jersey, was founded in March 1914 by New York businessman Inglis M. Uppercu. Uppercu initially financed the small Boland Aeroplane Company and its founder Frank E. Boland in 1908. Boland died in a crash in 1913, and his widow sold control to Uppercu, who renamed the firm Aeromarine and became president. Uppercu developed the Boland designs and later made major contributions to the flying-boat field. Charles F. Willard became chief engineer after his departure from L-W-F, and a military order enabled Aeromarine to begin mass production. Originally locating in Nutley, the company moved to Keyport in February 1917. Aeromarine expanded to 125,000 square feet of floor space and nine hundred employees at the war peak. Wartime output was some three hundred aircraft, and its factory was one of the best in the industry.[83]

John A. Sloan, son-in-law of Thomas Edison and a flying-boat pioneer, moved his firm to Plainfield, New Jersey, late in 1915. After financial difficulties it was taken over by the Japanese Mitsui bank and reincorporated as Standard Aero Corporation in May 1916, and managed by investor Harry B. Mingle. The affiliated Standard Aircraft Corporation of Elizabeth, New Jersey, was formed immediately upon American entry into the war on April 6, 1917, to provide greater capacity. Chief engineer Charles H. Day was instrumental in developing the widely used SJ-1 trainer. The Engel Aircraft Company of Niles, Ohio, was formed in December 1916, following an earlier effort in Depew, New York, by A. J. Engel, who had begun with Curtiss. With new investors entering on March 5, 1917, the company was positioned to pursue war production.[84] Engel was followed closely by the Springfield Aircraft Corporation of that Massachusetts city, on September 27, 1917.[85] Another wartime manufacturer was Ordnance Engineering Company (Orenco) of Long Island. None would survive the postwar cancellations.

The brothers Charles, Paul, and Walter Wittemann of Staten Island, New York, began experimenting in aviation in 1905. Their Wittemann Brothers Company, formed in 1912 or 1913 on the family estate, undertook limited production. The firm became Wittemann Aircraft Company in 1915. Needing space for expansion, the company relocated to Hasbrouck Heights, New Jersey, in 1917 as the Wittemann-Lewis Company, reflecting the interest of Samuel C. Lewis, a family friend. The airstrip later became Teterboro Airport. Lewis soon withdrew, and the successor Wittemann Aircraft Corporation received a contract to modify DH-4s for airmail service but did not become a significant wartime producer.

Chapter 2 ✈ The Industry and the First World War

Aircraft orders and financial support for the industry were meager in the prewar years, especially when compared with efforts of many European countries. Production remained largely stagnant until the United States entered the war, but that event changed permanently the field of aviation.

Military Aviation Development

Military aviation organization become more formally developed in the prewar years. Primarily in response to the rash of air accidents, the Aviation Section of the Signal Corps was formed on July 28, 1914, succeeding the original Aeronautical Division.[1] At the time of declaration of war it was still very small, with only two fields, a fleet of largely second-rate aircraft, and supported by an industry with extremely limited production capacity. The First Aero Squadron had been formed on March 5, 1913, however, and was steadily expanded. It first saw action in 1916 with General Pershing's expedition into Mexico against Pancho Villa.

The U.S. Navy first entered the aeronautical field with an appropriation of twenty-five thousand dollars, to purchase one aircraft, on July 1, 1911. Following the recommendation of a special board that had reported in October 1913, an Office of Naval Aeronautics was established on July 1, 1914.[2] The Naval Appropriations Act of August 29, 1916, provided $3.5 million for aviation and authorized the Naval Flying Corps. At that time the navy had only twenty-one seaplanes, but it soon expanded.

Despite the slow development of government support for aviation and the aircraft industry, one development in particular proved to be of major significance to the infant industry. The National Advisory Committee for Aeronautics (NACA) was formed on March 3, 1915, largely in reaction to complaints of a lack of government support for aviation. It was to "direct and conduct research and experimentation in aeronautics" and received modest appropriations from Congress. The committee consisted of fifteen members, appointed by the president. Orville Wright served for many years. The NACA Langley Memorial Aeronautical Laboratory, begun in 1917, became the most advanced aeronautical test and experimentation facility in the world.[3]

Aircraft appropriations gradually increased. The army received $13 million for aviation in 1915, and Curtiss attained in 1916 what could be considered the first true mass production of aircraft in the United States, building ninety-three JN-4 Jenny trainers and exporting flying boats to Great Britain.[4] Even then, ominously, there were signs of army-navy rivalry in military procurement. At times the services would attempt to play manufacturers against each other. The industry would attempt to do the same with the services. An aircraft rejected by one service might be ordered by the other, to the short-run benefit of the industry.[5]

At the time of America's entry into the war, the army had ordered just 532 aircraft over a period of ten years from fourteen manufacturers, and seven contractors had delivered 227.[6] There were just fifty-five serviceable airplanes in the Signal Corps inventory when war began, most deployed along the Mexican border. None could be described as true combat airplanes.[7] Development of combat aircraft was very slow for all nations, as it proved difficult to synchronize a machine gun to fire through a spinning propeller.

The Naval Flying Corps at the time of U.S. entry into the war also was small, with a strength of only forty-eight officers, 239 enlisted men, and fifty-four planes.[8] Only one naval air station was operational.[9] But war strength would increase manyfold.

War Production: Plans and Performance

The Signal Corps and NACA jointly drew up aircraft production plans in the event of war and submitted their report to the War Department on March 29, 1917, not knowing that war would be declared just eight days later.[10] The U.S. entry on April 6, 1917, presented the young aircraft

TABLE 1. U.S. Air Appropriations, 1908–19 (in thousands of dollars)

Year	Army	Navy	Post Office	NACA
1908–9	30			
1911–12	125	25		
1913[a]	100	10		
1914	175	10		
1915	200	10		5
1916	800	1,000		5
1917	18,082	3,500		88
1918	734,750	61,133	100	112
1919	952,305	220,383	100	205

Source: Aircraft Year Book, various editions.
a. In this year, by comparison, France spent $7.4 million, Russia $5 million, Germany $5 million, and Britain $3 million (Purtee, Army Air Service, 11).

industry with its greatest opportunity as well as its greatest frustration. As of that date just 243 military aircraft were on order, yet many manufacturers had blindly optimistic plans for virtually overnight expansion into a capacity to produce thousands.[11] NACA independently surveyed the capabilities of all manufacturers in May 1917, and the French and British pushed for major contributions of American aero squadrons and manpower. The reality, in addition to a lack of true combat aircraft, was that the combined production capacity of the few firms in the industry was modest. Further, the Signal Corps had no experience in organization and production.

Although capacity expanded rapidly and design advanced, two wartime policy decisions caused the industry much grief: to produce proven British and French designs under license rather than proceed with untested American designs, and to turn major responsibility for mass production over to the larger and more experienced automobile industry. These decisions ultimately led to disappointing results and to a relationship of mutual suspicion between the aircraft industry and the government. This wartime structure also became the only instance in which the production organization was determined not by Congress but by industry. Government regulation over the industry at the time was largely informal, with most influence extending from its status as the major customer.[12]

In addition to the Army and Navy Joint Technical Advisory Board, which handled patent questions in war production, the Aircraft Production Board (APB) was organized on May 16, 1917, and established a production goal of 22,625 airplanes and engines. Advisory only, the APB reported to the Council of National Defense, which had been established by the Defense Act of 1916. It was dominated by the automobile industry and chaired by Hudson vice president Howard E. Coffin. Coffin brought in Edward A. Deeds of Dayton Wright, DELCO, and National Cash Register (NCR), who was considered the real power on the board.[13] In the same month Colonel Raynall C. Bolling led a technical mission to Europe to select aircraft for U.S. production. The Bolling mission selected four designs, but only the British DH-4 was to see service.[14] The APB soon became the subject of bitter contention, especially with its focus on license production of the DH-4 powered by the Liberty engine. Coffin found himself regarded with suspicion both by the military and the industry. Nonetheless, such firms as Curtiss, Standard Aero, and Thomas-Morse managed to play a major part in war production along with the automobile manufacturers, who would supply most of the engines.

Financing and further reorganization followed the declaration of war. On July 24, 1917, the Congress passed an appropriation of $640 million for aircraft, the largest sum ever appropriated for anything to that time, although not with the support of the General Staff.[15] This was followed by an even greater $840 million appropriation by the end of the year. The Aircraft Board was created on September 27, 1917, in succession to the APB, with executive power to direct the purchase and manufacture of all army and navy aircraft.[16] Howard Coffin also headed the Aircraft Board. The board reported to the War and Navy Departments, and the Signal Corps thus reentered the procurement picture.[17]

On September 3, 1917, the First Aero Squadron, preceded by the first naval aviation elements, arrived in France. Marine Corps aviation also began at that time, with the formation of the First Marine Aeronautical Company in October 1917. It was not until April 14, 1918, however, that an American unit, the Ninety-fourth Pursuit Squadron, saw combat. In the meantime, in February 1918, the first de Havilland DH-4s came off assembly lines in the United States. By this time military aviation had outgrown the Signal Corps aviation section, and the Signal Corps bureaucracy was also found to be a hindrance to efficient production. Subsequently, on May 21, 1918, an autonomous Aeronautical Division was created, controlling the Division of Military Aeronautics and the Bureau of Aircraft Production, which replaced the Aircraft Board.[18] The air arm became the U.S. Air Service overseas.

Aircraft production, overly ambitious originally, fell hopelessly behind schedule. Howard E. Coffin, suffering criticism for that shortfall, also experienced numerous conflicts with the military and was forced to resign. Recriminations were widespread and formal investigations begun. While American designs generally were below world standards, some showed great promise, and the delays attributed to license production might have been avoided if American designs had been ordered. Further, production delays of licensed designs meant that they had become obsolete by the time they were available in quantity. Production of the SPAD fighter with the Liberty engine was an abject failure. There was widespread suspicion in Congress and elsewhere that fraud and conspiracy were afoot, and that aircraft profits were excessive.

By April 18, 1918, three separate investigations into the industry's performance were under way. In addition to the Aircraft Board's own investigation, the Senate Military Affairs Committee conducted its investigation of the apparent production failures, and President Wilson appointed a committee under Charles Evans Hughes, the Republican presidential candidate in 1916 and later chief justice, to conduct another.[19] Whatever their appropriateness or effectiveness, they had established a precedent.

It became widely accepted that the original aircraft production goals were unrealistic, and that insufficient appreciation was given to the differing nature of automobile and aircraft production. Further, many accusations were wildly exaggerated. There were claims that more than one billion dollars was spent to result in only a few hundred aircraft. While more than $1,687,000,000 was appropriated for aircraft during the war, only $598,000,000 was actually spent, net of cancellation costs and salvage, and unused funds were returned to the Treasury.[20] Further, accusations focused only on the actual number of aircraft sent to Europe, not on the huge quantities of trainers and other types remaining in the United States. It remained a daunting task to determine precisely the value received for war appropriations and the average cost per aircraft.

John D. Ryan, previously head of Anaconda Copper, was appointed director of the reorganized Division of Military Aeronautics, succeeding Coffin. The Bureau of Aircraft Production, successor to the Aircraft Board, remained subordinate.[21] Public criticism of the industry continued in addition to the official investigations. Not a single aircraft of American design saw war action in Europe, and only limited numbers of license-built DH-4s reached the front. Already considered obsolete, DH-4s were rushed into service to show tangible results of the production program. By the Armistice, the army had accepted 11,754 airplanes built in the United States and Canada, and a further 2,317 from war contracts were delivered afterward.[22] Of some 602 American aircraft available in France, only 213 were actually at the front at the time of the Armistice, of which 196 were U.S. built.[23] Ryan's efforts were showing results, but American totals still lagged those of Great Britain and France.

In retrospect it appears that timing was the major problem. The government simply lacked mobilization experience, and industry capacity was too small, but these shortcomings affected the entire war effort. It was

TABLE 2. U.S. Aircraft Production, 1912–19

	Total	Military	Civil
1912	45	16	29
1913	43	14	29
1914	49	15	34
1915	178	26	152
1916	411	142	269
1917	2,148	2,013	135
1918	14,020	13,991	29
1919	780	682	98

Source: *Aerospace Facts and Figures, 1959,* 6–7.

never realistic to expect that U.S. production would have a major impact before 1919, but production and deliveries were in fact approaching the original goals by the Armistice. After the Armistice John D. Ryan resigned as Director of Military Aeronautics. Major General Charles T. Menoher, although not a pilot, replaced him, aided by Brigadier General William Mitchell as assistant chief of the Air Service.

One reason for the failures of the aircraft production effort lay with the relationship of the automobile industry to the new aircraft industry. Henry Ford, with a confidence bordering on arrogance, maintained that his industry could produce all aircraft and engines needed by the services. That it did not do so may be attributed to the fact that the mass-production techniques used in the auto industry simply were not readily adaptable to aircraft. Aircraft production at the time, with its use of formed wood and wire, was more comparable to the manufacture of pianos than of automobiles.

The often troubling relationship between the auto and aircraft industries would continue. Both Curtiss and Wright-Martin were controlled by auto interests and were joined, of course, by Ford and Packard for war production. Aircraft companies actually produced the majority of airframes during the war period, but the automobile industry did attain some success with engine production. There was a far higher degree of commonality between aircraft and automobile engines of the period than between aircraft structures and automobiles. Despite the judgment of failure of the industry's wartime performance, in one respect it may have been too successful: If it had not built so many DH-4s to war contracts, they would not have flooded the postwar market as cheap surplus, limiting demand for new aircraft.[24]

The industry pioneer Grover Loening was prominent among those feeling that existing aircraft firms were unjustly ignored, and that sound American aircraft designs had been passed over in favor of existing foreign designs.[25] Such promising designs as the Martin MB bomber, the Vought trainer, and the Loening M8 and Thomas-Morse MB-3 fighters were ignored. Loening also discerned a conspiracy by the automobile industry to dominate the aircraft industry even before the First World War.[26] He led a sharp defense of aircraft manufacturers against accusations that they were incapable of producing to military requirements during the war.[27] Edward A. Deeds, who had been commissioned a colonel and placed in charge of aircraft procurement in the Bureau of Aircraft Production, was criticized especially on this issue as well as on the overall production record, and at one point was even considered for a military court martial.[28] He was suspected of favoring his old company, Dayton

Wright, in contract awards, and the Hughes committee recommended charges of conflicts of interest and making false and misleading statements.[29] By present-day standards Deeds definitely appeared to have a conflict of interest, but was eventually exonerated. The Hughes committee cited many contractual improprieties, but the new Manufacturers Aircraft Association was not found to have acted improperly.

Congress, of course, had its own views of the aircraft industry and of its role in the war effort. Widespread suspicions arose, and remained through the 1930s, of a conspiratorial "aircraft trust" to limit price competition and the entry of new competitors. Howard Coffin, while head of the APB and the later Aircraft Board, spent more time fighting "aircraft trust" charges than in getting airplanes built. Investigations of the wartime production failures and of the disposition of the government's large appropriations continued for some time, but the entire episode became regarded as a fiasco. Still, congressional investigations of contract performance, including but not limited to aircraft, became a permanent feature of the industrial landscape.

One conspicuous technical and production success of the war effort was the Liberty engine. The Liberty essentially was developed by Jesse G. Vincent of Packard Motor Car Company in cooperation with the Hall-Scott Motor Company of San Francisco. The Liberty became the standard wartime aircraft engine and was produced by several manufacturers. Its basic success was tempered by the fact that it was forcibly grafted onto certain aircraft, such as the DH-4, for which it was ill suited, but it remained an important aircraft engine into the 1920s.

Another wartime success was the Naval Aircraft Factory. The declaration of war led to a major naval aviation expansion along with that for the army, and large orders were placed with Curtiss, Boeing, and Aeromarine. Along with the expansion came the establishment (facilitated by Assistant Secretary of the Navy Franklin D. Roosevelt, but over the protests of the aircraft industry) of the Naval Aircraft Factory (NAF). Authorizing legislation became effective on July 27, 1917, and the factory, at the Philadelphia Shipyards, was completed on November 28, 1917.[30] Policy at that time was that the navy should not rely totally on private industry to meet its aircraft needs, and it was intended also that the NAF serve as a check on private industry in costs and profits.[31] Government intrusion into private enterprise notwithstanding, the NAF became a successful designer and manufacturer of naval airplanes. Its strong production record included completion of fifty Curtiss H-16s by July 1, 1918, ahead of schedule, and total war production was 183 flying boats.[32] With industry support, the Naval Flying Corps reached a strength of 2,107 aircraft by the Armistice.[33]

By the end of the war *Jane's* listed some thirty-one aircraft manufacturers for the United States.[34] Probably no more than seven, Curtiss, St. Louis, Springfield, Dayton Wright, Fisher Body, Standard, and Thomas-Morse, contributed significantly to war production. Still, at its war peak industry employment was 175,000.[35] But with the cancellation of war contracts, the market contracted to the point that most wartime producers had left the market by 1920, never to return. Of the sixteen firms constituting the membership of the Manufacturers Aircraft Association (MAA) as of war's end, only three firms, Curtiss, Martin, and Boeing, were still active in 1929, ten years later, although Dayton Wright and Thomas-Morse were represented through their acquisition by Consolidated.[36]

Concerns for the future of the manufacturing industry led to the combined military-civilian American Aviation Mission to Europe in 1919. The mission wished to preserve some of the gains made in the war production program. Its ultimate goal was to place the U.S. industry first in the world.[37] This group, consisting of prominent figures in both industry and the military, held discussions with many leaders of Great Britain, France, and Italy about the future of the industry. The resulting Crowell Report was issued on July 19, 1919. Its recommendations, which included a single cabinet-level department for air matters, standards for airfields, weather, and flying training, and consistent production support for the industry, were a significant prelude to the development of a postwar aviation policy.[38]

The war had led, of course, to rapid progress in aircraft and engine design, but this progress was not so dramatic as that with production. The industry had learned, albeit with great difficulty, to produce large numbers of standardized aircraft. Lingering suspicions with regard to pricing and performance, however, were not conducive to the development of a healthy industry and, to an extent, may be held responsible for the struggles of the industry in the 1920s. Eventually such suspicions faded, but the effects lingered. While the industry was to face many difficulties in the new decade, wartime experience would serve it well.

Industry Associations and Military Organization

The First World War aviation buildup also affected industry associations. The rush to mass production in an environment of litigation required considerable effort to resolve the numerous patent and royalty questions. The

Aircraft Manufacturers Association was founded on February 13, 1917, primarily to address the patent issue and other technical and business concerns of the industry. Howard Coffin and Harry B. Mingle first proposed an organization in 1916, feeling that an association built around a patent pool would benefit the industry. The government also supported the association as a counter to the Wright-Martin Company's attempt to collect royalties from other manufacturers. It was succeeded by the Manufacturers Aircraft Association on July 24, 1917, formed specifically to administer the cross-licensing agreements brought about by the industry war effort, as well as overall government-industry dealings.

The MAA established a patent-pooling arrangement by which any manufacturer could produce designs of others and pay a license fee into the pool for later distribution. All members relinquished patent rights and were to pay royalties into the pool of two hundred dollars for each plane produced until Curtiss and Wright-Martin, which held most patents, had received $2 million each. Upon reaching that amount, members were released from any patent liability and could enjoy unlimited use of all aeronautical patents.[39] After numerous objections, however, payments were reduced to one hundred dollars per plane and $1 million to each firm. The patent pool also increased suspicions of an "aircraft trust," however, but with less than $2 million paid, inconsequential compared to total appropriations, such a conspiracy seemed unfounded. Actually, the MAA served to enhance free competition. Further, in order to avoid any suspicion of conflicts of interest, the MAA appointed Samuel S. Bradley, a respected industrialist with no previous ties to the industry, as general manager. He was to serve with distinction for more than a decade.[40] Yet the benefits of the MAA still were limited, since military patents had no legal standing and only commercial patents could be enforced.[41] After the war the MAA undertook public promotion of aviation, although several companies still withheld membership, feeling that it was too automobile dominated. The MAA eventually became a vital source of industry information for the government.[42]

The Aeronautical Chamber of Commerce, formed on January 1, 1922, served as the national trade association. Grover C. Loening served as its first president. Samuel S. Bradley was also appointed general manager of the ACC, and the MAA and ACC became closely linked operationally. The approximate British counterpart was the Society of British Aircraft Constructors (SBAC), founded on March 29, 1916. The stated purposes of the ACC were to work within the industry for improvement of conditions and practices conducive to public appreciation and support, and represent the industry in correcting or stimulating the public viewpoint.[43] In operation the ACC promoted commercial air transport more than the industry, causing some manufacturers to withhold membership. The National Aeronautic Association was formed on October 7, 1922, in succession to the Aero Club of America, and became the American representative to the FAI.

Military aviation was aided by the Air Service Act of 1918. The Aircraft Board was officially dissolved on March 19, 1919, and the successor Bureau of Aircraft Production was placed under the U.S. Air Service. Once again aircraft production was under the control of the military rather than a civilian body.[44] The act also established the Engineering Division, at McCook Field in Dayton, Ohio. While naturally oriented toward military applications, the division's research contributed to aeronautical progress, which helped the industry along many fronts. It was the American equivalent of Britain's Royal Aircraft Establishment (RAE).

The Defense Act of 1920 brought major changes to the national military structure. It created the U.S. Army Air Service effective June 4, 1920, costing the air arm a degree of autonomy. Major General Mason M. Patrick became chief of the Army Air Service in 1921 and was to serve for six years.

Naval aviation was reorganized under the Bureau of Aeronautics, established on July 12, 1921. Rear Admiral W. A. Moffett, previously director of the Naval Flying Corps, headed the BuAer, and was to serve in that post for twelve years. He became regarded as the architect of naval air power. The BuAer controlled the specifications and procurement of naval aircraft and equipment, and both the Atlantic and Pacific Fleet air arms. The navy remained oriented toward seaplanes, but development of aircraft carriers, supported by Admiral Moffett, soon broadened its horizons. The first tests of torpedo-carrying aircraft were undertaken in 1922.

One development near the end of the First World War that was to have a long-term impact on the development of commercial aviation was scheduled airmail, inaugurated on May 15, 1918, under the direction of Major Reuben H. Fleet. Standard Aircraft provided the first specialty mailplanes. Airmail routes expanded rapidly postwar, facilitated by the availability of large numbers of surplus DH-4s. Transcontinental airmail service, initially taking a day and a half, began on February 22, 1921. Aging equipment made the job of airmail pilot a dangerous one, but significant upgrading did not occur until the delivery of a fleet of new Douglas M-2s in 1926. The major long-term benefit to the industry was that airmail operations established the airplane's capability to fly scheduled routes. The financial rewards of airmail contracts enabled the commercial carriers to offer passenger services as well, eventually leading to a market for commercial airliners.

New industry associations were in the long run extremely beneficial in the development of an infrastructure conducive to aeronautical progress and national security needs. Associations alone could not persuade the government to order more aircraft for the services and could not rescue a firm facing financial failure, but in providing a forum for the airing of mutual concerns and for mutual support they more than justified their existence. Their major failing, of course, was that they could not counteract often deleterious congressional regulation.

Chapter 3 Building an Industry
The 1920s

No industry ever fell so far so rapidly as did the young aircraft industry, and since the recent war was the War to End All War, there was little interest in new military aircraft. Orders for no fewer than sixty-one thousand aircraft were canceled with the Armistice. By the summer of 1919 some 90 percent of peak production capacity had been liquidated.[1] No new military aircraft at all were funded in 1920, and the world-leading Martin bomber did not enter service until 1921. Many firms either were liquidated or turned to other business pursuits. Compounding industry difficulties, quantities of surplus British airplanes were shipped to Atlantic City, New Jersey, in 1919 for sale on the American market. Industry resistance to this scheme was interpreted by some in Congress as another instance of an aircraft trust to limit competition and was enough to kill antidumping legislation introduced in response.[2] The MAA ended the threat by invoking the Wright patent ruling, in that the aircraft were unlicensed in the United States.[3]

The State of Aeronautical Progress
Wartime development and production experience and rapid international diffusion of aviation technology had spurred major progress in aeronautics. The Army Engineering Division made many contributions, among them the aircraft supercharger. On May 8, 1919, three NC (Navy-Curtiss) flying boats set out across the Atlantic, and *NC.4* reached Lisbon, Portugal, on May 27, by way of Newfoundland and the Azores, completing the first Atlantic air crossing. But the feat was soon eclipsed by the nonstop flight of Alcock and Brown, in a British Vickers Vimy, on June 14, 1919. Inglis Uppercu's Aeromarine West Indian Airways initiated the first international passenger service, between Key West and Havana, on November 1, 1919.

The National Air Races, begun in 1920, stimulated public interest in aviation and were a major inducement to aeronautical progress. These were the Pulitzer Trophy Race, the Bendix Trophy Race, the Thompson Trophy Race, and the international Schneider Cup Race for seaplanes. The National Aeronautic Association Trophy, for the outstanding aeronautical accomplishment of the year, became the most prestigious award in the aviation world. While referred to informally as the Collier Trophy from the outset, it did not officially acquire that name until 1944. The military services supported the air races throughout the 1920s, feeling that they advanced aeronautical technology, redounding to their benefit as well as to the industry's.

Speed, nonstop distance, and altitude records were steadily extended, many by American pilots flying American machines. By 1924, 54 of the 103 world records recognized by the FAI were held by the United States.[4] Lieutenant J. A. Macready, in a Packard-LePere powered by a supercharged Liberty engine, raised the altitude record to 34,508 feet on September 28, 1921. On December 30, 1921, Eddie Stinson and Lloyd Bertrand remained aloft over Long Island in their Larsen J.L.6 (actually a German Junkers all-metal aircraft renamed due to wartime animosities) for twenty-six hours and nineteen minutes.[5] On September 5, 1922, Lieutenant Jimmy Doolittle made the first coast-to coast flight in a single day. The flight took twenty-two hours and thirty minutes, with only a single refueling stop. Then during May 2–3, 1923, Macready and Lieutenant Oakley G. Kelly, on their third attempt, completed the first nonstop coast-to-coast flight, from New York to California, in a modified Fokker T-2 transport. On October 18, 1922, Billy Mitchell brought the air speed record to the United States by attaining 223 miles per hour in the Curtiss *R-6 Racer.* That record, incidentally, owed more to Clement Keys than to army support, as Keys was determined to demonstrate U.S. aviation capabilities, especially to the French. The first aerial-refueling experiment took place over San Diego on June 27, 1923. The flight around the world by the Douglas World Cruiser fleet was another dramatic demonstration of the capabilities of the airplane. The arduous flight, which took almost six months due to mishaps and delays, and involved 371 flying hours, was completed on September 24, 1924. The aircraft were fitted with Boeing-built floats in Seattle for the overwater stages of the flight.

Airframes were becoming stronger and instruments more sophisticated. In Germany, Dr. Adolf Rohrbach developed the stressed-skin construction method, in which the skin itself bore part of the weight of the aircraft. Engines became more powerful and more reliable, but power plant progress still tended to lag airframe and aerodynamic advances. Inadequate power prevented many aircraft from attaining their design potential, and engineers constantly struggled to raise the horsepower/weight ratio. The limitations imposed by engines nagged the industry well into the supersonic era.

In the early 1920s duralumin, an aluminum-copper alloy, began to be used in aircraft structures, replacing wood, and the variable-pitch propeller, greatly enhancing performance at high altitude, was developed.

Yet the major ingredient for progress and stability, continuing development contracts and production orders, remained elusive. Progress in design, production, and performance was nothing less than remarkable given the general dearth of government support in the 1919–24 period. Once again the persistence and dedication of the pioneer industrialists led to successful results even in the face of severe financial constraints. Martin designed and produced the world's most advanced bomber, and Boeing, Consolidated, and Vought produced excellent training planes, but a vast surplus of wartime aircraft had severely limited the new aircraft market, both military and civil. A design failure, albeit a spectacular one, was the Engineering Division's Barling XNBL-1 heavy bomber of 1923. Designed by Walter Barling and built by the Wittemann Aircraft Company, it was the heaviest (at twenty-one tons) airplane built to that time but was seriously underpowered, even with six Liberty engines, unable to fly any useful distance with a bombload. A short-lived venture, but of special interest, was the Lawson Airplane Company of Milwaukee, founded by the Englishman Alfred W. Lawson in 1919. He designed and built, with the assistance of Vincent Burnelli, the eighteen-seat Model C-2, possibly the first true airliner. The airliner was tested successfully but was not ordered for production. The single example then crashed in 1921. Lawson, incidentally, is credited with coining the term *aircraft,* which supplanted the original *aeroplane.* Lawson organized a new venture in 1924 at the old Standard Aero New Jersey plant, for a multiengined airliner with a capacity of one hundred passengers, but again could not attract support.[6]

In July 1921, after delays and controversy, Billy Mitchell, seeking to demonstrate dramatically the capabilities of air power, located and sank old warships, including the captured German battleship *Ostfriesland,* with bombs from his Martin bombers. The ships were anchored some sixty miles offshore, and locating them from aircraft was far from assured, given the primitive navigation equipment of the day. In addition to disturbing the naval hierarchy, his methods also angered his superior, General Menoher, who threatened to resign if Mitchell were not removed. Menoher did in fact resign and was succeeded by Patrick. The Mitchell episode should have resulted in major tactical lessons being learned but instead degenerated into another in a long series of controversies between sea and air power advocates. Mitchell continued as the nation's foremost, if abrasive, advocate of air power.

Amid the troubles of the industry in the early twenties the original Wright patents expired on May 22, 1923, bringing that chapter to a legal close. Never again would patent controversies overhang the entire industry.

The State of the Industry

The industry faced a sheer survival struggle. While wartime production failures would dog the industry for years, the major frustration was that manufacturers legally held no exclusive production rights to their designs. As a result of the patent-pooling arrangement, the government could ask for production bids from all manufacturers, with no compensation to the designer. Designs were regarded as public property, and despite industry objections the principle was upheld in court rulings. For example, the Martin MB-2, the production version of the MB-1, was opened for bids in 1919, and the NBS-1 (its designation from other makers) was produced by L-W-F, Aeromarine, and Curtiss, although they as well as Martin incurred financial losses. Martin received no compensation for production of its design by others.[7] Martin also received no further army orders, although Mitchell had strongly supported the MB program.

Another major obstacle to growth and stability in the industry came from Congress. In pursuit of "free enterprise" competition and of low bids, Congress focused on profit restrictions, frequently feeling that taxpayers benefited if a manufacturer suffered a loss on either a development or production contract. Congress also found itself increasingly at odds with the army and navy, more interested in supporting and preserving the industry. Since the civil market was insignificant at the time, the industry remained abjectly dependent on government funding.

With the return to peace Boeing faced severe challenges to its survival. Employment dropped to only thirty in 1919, forcing Boeing to turn to furniture production to supplement sparse aircraft business.[8] Aircraft activity returned with a contract for modification and rebuilding of DH-4s, and Boeing also pioneered airmail service and subsequently passenger transportation. Boeing Air Transport inaugurated the first international airmail service, from Victoria, British Columbia, to Seattle, on March 3, 1919. Boeing's B-1 flying boat first flew on December 27, 1919, and the GAX triplane armored attack aircraft flew in May 1921. The combination of progressive aircraft designs and mail contracts gave Boeing a basis for future prosperity.

Boeing produced ten GA-1 models, developed from the GAX, from 1921, but of far greater importance was its contract to produce two hundred Thomas-Morse MB-3A pursuit ships, one of the few large contracts awarded during the early postwar period.[9] The MB-3A order was completed ahead of schedule in 1922, putting Boeing on the road to financial stability but sending Thomas-Morse into decline. Boeing also subsequently developed its own pursuit designs, beginning with the XPW-9 in

1923. The later navy F4B and army P-12 enabled Boeing to extend its position of leading fighter producer over some fifteen years. Airmail interests also led directly to the first Boeing transport, the Model 40 of 1925. William Boeing became chairman of the board in 1922, and Edward N. Gott became president. Gott, however, left late in 1925 after a dispute with his cousin and was succeeded by Phillip Johnson in February 1926.

Grover Loening fared somewhat better at the hands of the government than had Martin, as the NAF produced fifty Loening M8O two-seat fighters, while Loening produced thirty-three.[10] Curtiss won a contract for the Orenco Model D pursuit ship, the loss of which contract forced Orenco to liquidate in 1919. A successor Baldwin Aircraft was attempted but also failed, in 1924.

In 1919 the most active remaining firms were Curtiss, Wright Aeronautical, Martin, and Boeing. Thomas-Morse, Lewis and Vought, Loening, and a few others also persisted, but at a low level of activity. Companies simply could not survive without production orders or the prospect thereof. Standard Aircraft, Standard Aero, Engel, Springfield, and eventually L-W-F and Aeromarine, exited the industry. Cost overruns on the Barling bomber, slow army payments, and loss of financial backing led to Wittemann's collapse.

Even Curtiss, the largest and strongest company during the war, fell on hard times. The giant North Elmwood plant became redundant and was purchased by the government for $1.8 million.[11] In addition, the Hammondsport motor plant was closed, and corporate headquarters moved to Garden City. For John N. Willys, the drop in aircraft orders coincided with a crisis in his automobile empire, leading him to withdraw his interest in August 1920, forcing Curtiss into receivership. Clement Keys took over as president, acquired Willys's interest, arranged financing to manage the company's $650,000 debt, and led it again to sound financial status.[12] Keys, still unsalaried, reorganized the company in 1923, separating Curtiss Aeroplane and Motor from other activities. He received U.S. citizenship in 1924. Glenn Curtiss returned briefly as chief engineer of the company but again departed to develop Florida real estate. Keys closed the Garden City operation in October 1924 in order to keep Buffalo open. Theodore P. Wright joined Curtiss in 1925 as chief engineer, and Curtiss remained the largest American aircraft company through the 1920s, albeit with shaky profitability. The P-1, first in a long line of Hawk pursuits for the army, was ordered in September 1925, and Curtiss built a line of observation airplanes as well.

Chance Vought's first postwar aircraft was the successful VE-7 of May 1920, which entered production for the navy. Originally a trainer, the

VE-7 was adapted as a fighter and in 1922 made the first takeoff from the first American aircraft carrier, the USS *Langley*. Vought followed with his improved VE-11 design of 1921, regarded as the first true naval fighter. Vought reorganized his company as the Chance Vought Corporation in May 1922, with the participation of his family. His father became president, while Vought took the title of chief engineer. The first design to bear the famous Corsair name was the O2U-1 of 1926. The designer Rex B. Beisel, who had worked for the navy during the First World War, first joined Vought in 1921, but soon moved to the Bureau of Aeronautics.[13] Beisel then moved to Curtiss in February 1923, contributing to Curtiss designs throughout the 1920s.[14]

General Motors, in its first foray into aircraft production, purchased control of Dayton Wright along with the other Talbott interests in 1919, after having purchased Kettering's DELCO in 1916. General Motors paid $1 million for Dayton Wright as part of chief executive William C. Durant's postwar expansion program.[15] GM was attracted to aviation's growth potential, foreseeing an eventually saturated auto market, and feeling that the purchase would be an effective counter to Ford's interest in the field.[16]

The depressed postwar aircraft market led to continuing losses, however, leading General Motors to withdraw Dayton Wright from aircraft production on June 1, 1923. While unsuccessful as a manufacturer, the firm did possess the distinction of having tested the first retractable airplane landing gear, in 1920. General Motors sold design rights to the new Consolidated Aircraft of Reuben Fleet in 1923.

While Glenn Martin gained prominence with the MB bomber, he suffered a major blow with the departure of Donald Douglas in March 1920. Martin was neither an engineer nor an innovative designer and relied heavily on the expertise of others. After Douglas left, Martin made Larry Bell vice president and general manager and then gained control of his company from Cleveland financiers in 1921. Martin succeeded in naval aircraft and flying boats and delivered forty Curtiss-designed CS-2 torpedo bombers in 1926. Larry Bell began pressing Martin for part ownership of the firm, which Martin refused. Bell resigned on January 18, 1925, and left for the West Coast and other pursuits.[17]

There was a discernible tendency of promising young designers and engineers who began their careers with one of the pioneer firms to later strike off on their own. The rather close network of designers and manufacturers that existed facilitated such moves. The community was small enough that most company heads knew each other, often establishing strong personal ties even as they competed for orders. There was a strong sense among most leaders that they faced the same challenges and needed to work together closely to survive. This fraternal pattern was in fact more characteristic of the early industry than the Wright-Curtiss feud, which was an aberration. Grover Loening related the story of how in 1922, when banks had little interest in financing aircraft manufacturers, his friend Chance Vought was producing an order for ten VE-7H planes for the navy when Congress refused to appropriate funds for payment. Loening, realizing Vought was being starved financially, lent him sufficient funds from his then-profitable company so that Vought could survive until the government came through.[18] Loening later credited General Patrick with rescuing his company with a military order in 1924 after his commercial business had declined.

Clement Keys vocally blamed the government, with its lack of procurement support and severe contracting procedures, for the rather grim financial condition of the industry. Loening, however, took issue with Keys, stating that the "low bidding craze" by manufacturers created most of their problems.[19] Loening believed that firms were so desperate for contracts that they deliberately bid below costs, leading to later losses. The controversy would recur.

It remains noteworthy that, in common with other sectors of the economy then and now, more new aircraft ventures started up in hard times than in prosperity. The industry definitely faced hard times postwar, yet certain firms destined to endure in the industry were established. Accounts of the most significant efforts follow.

TABLE 3. U.S. Air Appropriations, 1920–27 (in thousands of dollars)

Year	Army	Navy	Post Office	NACA
1920	25,000	25,000	850	175
1921	33,000	20,000	1,250	200
1922	19,200	13,343	1,250	200
1923	12,895	14,684	1,900	210
1924	12,426	14,647	1,500	283
1925	13,476	15,250		470
1926	18,061	15,130	500	534
1927	18,257	19,065	2,000	513

Source: Aircraft Year Book, various editions.

Lawrence Sperry

Lawrence "Gyro" Sperry, son of Elmer Sperry of the Sperry Gyroscope Company, became fascinated with aviation in boyhood and learned to fly with Glenn Curtiss. At twenty he was the youngest pilot in America to hold an Aero Club certificate. He began experimenting with gyroscopic stabilizers in 1913, and on June 18, 1914, at age twenty-one, he successfully tested his stabilizer in France.[20] This led to a ten-thousand-dollar prize and to a practical automatic pilot. He was to be awarded twenty-six aviation patents.

Sperry built the first American amphibians, as opposed to floatplanes or flying boats, and in 1915 developed the first retractable landing gear.[21] Feeling confined as an officer of his father's company, Sperry formed Lawrence Sperry Aircraft Company on May 5, 1917, with a factory in Farmingdale, Long Island. He worked with the Curtiss aerial-torpedo design and experimented with his own design for an aerial torpedo, or flying bomb, using his gyroscope as the basis for automatic piloting of the vehicle. Although successful in tests, this concept did not win production orders and was too late for the war. But it may be regarded as a predecessor of later guided missiles.

Sperry designed and tested an innovative triplane amphibian in 1918 and achieved some postwar production success with amphibians. He won recognition with his Messenger, a light sportplane or "flivver" that won the support of Billy Mitchell and was produced for the Army Air Service. Sperry also collaborated with the designer Alfred Verville on a low-wing monoplane pursuit ship for the army, to the specifications of Billy Mitchell, who was eager to progress beyond the struts and wires of biplanes. Later Sperry won contracts from the army for development of the aerial torpedo using Standard E-1 biplanes. Sperry won wide attention on March 22, 1922, when he landed his Messenger on the grounds of the U.S. Capitol, in part to dramatize his difficulties in receiving payment for his contracts, but it was the Messenger's prospects as a mass-ownership flivver that primarily preoccupied him. Tragically, the young company died along with its founder when Sperry crashed in the English Channel at the end of 1923. He was thirty-one.

Donald Douglas

Donald Wills Douglas probably best typifies the men who made the aircraft industry what it was to become. He was possibly the first successful manufacturer who was an engineer and businessman rather than a pioneer aviator. In fact, Douglas's chief personal interest was sailing rather than flying, and he did not become a pilot until 1924. His interest in aviation was sparked by his witnessing of the Wright Flyer trials at Fort Myer in 1908.[22] Douglas won appointment to the U.S. Naval Academy in 1910 but showed more interest in designing aircraft and building models than in his naval studies. He resigned in 1912 and transferred to MIT, completing his mechanical-engineering degree in a remarkably brief two years.[23] His performance was such that he was offered a one-year appointment as an instructor, which he accepted over a more remunerative naval commission.

In that capacity Douglas came to the attention of Dr. Jerome C. Hunsaker, pioneer of MIT's program in aeronautical engineering, developer of the first large wind tunnel, and later a key founder of the Naval Aircraft Factory. Hunsaker appreciated Douglas's engineering potential and provided valuable contacts.[24] When Douglas became chief engineer for Martin in August 1915, he made major contributions in stress analysis, but his affiliation lasted little over a year, as Douglas was asked in November 1916 to join the Aviation Section of the Signal Corps.[25] Although holding the title of chief civilian engineer, Douglas soon became frustrated with army bureaucracy, and in December 1917 rejoined Martin in Cleveland. He headed design of the MB bomber, regarded as the best in the world at that time.[26] Douglas also became recognized as the best aeronautical engineer in the country.

By 1920 Douglas was again restless and ambitious to strike off on his own. Although perhaps not entirely comfortable with the somewhat eccentric Martin and the ambitious Larry Bell, Douglas was primarily motivated by the desire to run his own company. Although his personal assets were no more than one thousand dollars, Douglas left in March 1920 for Los Angeles, where he began organizing a company. Receiving only rejection from banks for such a dubious venture, Douglas eventually found a financial angel in the person of David R. Davis, a wealthy sportsman and aviation enthusiast. Davis was introduced to Douglas by the reporter Bill Henry, who had handled public relations for Martin earlier and was a friend of both men.[27] Davis was especially attracted by Douglas's plans for a plane that would fly coast-to-coast nonstop, and provided capital of forty thousand dollars. The Davis-Douglas Company was founded on July 22, 1920, in a storefront.

Bringing in Eric Stringer, formerly Martin's chief test pilot, and five other employees, Douglas set about designing and constructing his *Cloudster,* which flew successfully on February 24, 1921. Unfortunately, when Stringer and Davis began the transcontinental attempt, on June 27, 1921, the *Cloudster* became disabled in Texas and had to be returned for repairs.[28] After this setback David Davis's enthusiasm for the attempt, and for the company, waned. Davis withdrew his interest, taking the *Cloudster* with him. The *Cloudster,* incidentally, enjoyed a notable later career, eventually being sold to T. Claude Ryan of San Diego for six thousand dollars and converted to a passenger plane for his Ryan Airlines.[29]

Douglas in the meantime had become more oriented toward military contracts and began developing military designs. Douglas reorganized as the Douglas Company in July 1921. The navy concurrently had become interested in the potential of torpedo-carrying shipboard aircraft and formulated a specification. With the influence of Jerome Hunsaker, then with BuAer, Douglas secured an order for three DT torpedo bombers, based on his original *Cloudster* design.[30] The immediate problem was that of capital with which to complete the order. Once again Bill Henry came to the rescue, as he briefed his boss, the wealthy publisher Harry B. Chandler, on the venture. Chandler arranged to guarantee a loan of fifteen thousand dollars, which the frugal Douglas felt to be the minimum required to proceed.[31] Douglas also received a five-thousand-dollar loan from his banker father, William E. Douglas, but the total still proved to be inadequate. The company operated on a shoestring, with workers missing paydays, until the navy accepted and paid for the aircraft. More important, however, the order was profitable. The Douglas Company delivered six aircraft in 1922 with a total value of $130,890.[32] Major success followed as the navy ordered a further thirty-eight DTs, followed by orders for O-2 army observation planes and M-2 mailplanes, and by improved M-3 and M-4 models.

By 1924 Douglas was becoming established as a manufacturer. His company was strongly profitable. Such future industry leaders as John Northrop, Gerald Vultee, James H. Kindelberger, and Edward Heinemann had joined. Major recognition, however, came with production of the army World Cruiser fleet in 1924, which accomplished the first around-the-world flight. While the flight enhanced public support for aviation

TABLE 4. U.S. Aircraft Production, 1920–27

	Total	Military	Civil
1920	328	256	72
1921	437	389	48
1922	263	226	37
1923	743	687	56
1924	377	317	60
1925	789	447	342
1926	1,186	532	654
1927	1,995	621	1,374

Source: *Aerospace Facts and Figures,* 1959, 6–7.

and for the Army Air Service, the principal long-term beneficiary undoubtedly was Douglas, probably the most successful firm in the small industry of the 1920s. By the end of 1928 Douglas had delivered a total of 392 airplanes, most to the military, but a few to others.

While most early manufacturers located in the home areas of their founders, Douglas was the first to choose Southern California. He was to play a major role in developing that location as the center of the industry.

Huff Daland

Huff Daland Airplane Company was formed in 1920, in Ogdensburg, New York, by the engineers Thomas H. Huff and Elliott Daland, along with C. Talbot Porter. Huff, an MIT classmate of Donald Douglas, later served on the faculty of that institution. He had worked earlier for Sturtevant and Standard Aero. Despite a paucity of orders, Huff Daland produced a successful trainer and also pioneered aerial crop dusting. Huff Daland Dusters, locating first in Macon, Georgia, and then in Monroe, Louisiana, was the predecessor of Delta Air Lines. Later specializing in bombers, the Huff Daland/Keystone B-3 through B-6 served into the early 1930s.

In 1925 the firm moved to Bristol, Pennsylvania, to be nearer its major markets. Business accelerated in 1926 and 1927, helped in no small part by increasing Air Corps appropriations.[33] Huff, however, resigned as president in December 1926 and joined Fokker. On March 8, 1927, the firm was reorganized as Keystone Aircraft Corporation, with Edgar N. Gott, formerly of Boeing, as president. While predominantly military, the firm also produced crop dusters. Keystone was the recipient of a major investment by Wright Aeronautical in 1928, which led directly to its acquisition of Loening in that year.[34] Keystone operated independently until 1929 when it became a division of Curtiss-Wright. Elliott Daland departed in 1930.

Claude Ryan

The Kansan T. Claude Ryan moved to California early in life to pursue his interest in aviation and learned to fly in 1917. Putting together limited assets, he purchased his first airplane and started Ryan Flying Service in San Diego in September 1922, offering a combination of passenger and mail service along the coast, a flying school, and an excursion service.[35] He also rebuilt and resold airplanes for profit. Needing more capital, Ryan entered into a partnership on April 18, 1925, with the young Benjamin Franklin Mahoney, a reputed millionaire whom Ryan had taught to fly.[36] The company name was changed to Ryan Airlines, Inc., although not legally incorporated. Ryan developed his own designs, and his successful M-1 mailplane first flew on February 14, 1926.[37] The partnership was destined to be short-lived, however, as Ryan and the erratic playboy Mahoney proved to be incompatible. Unable to purchase Mahoney's interest, Ryan instead sold his interest to Mahoney and dissolved the partnership effective November 23, 1926, but continued to manage the company.[38]

In early 1927 came the telegram from a group of St. Louis businessmen asking Ryan if he could design and deliver in sixty days an airplane that would fly nonstop across the Atlantic. Ryan accepted the challenge, and the result, developed from his mailplane, became *The Spirit of St. Louis*. Although Ryan sought no personal glory, the company shared in Lindbergh's fame and briefly enjoyed enhanced prospects. But Ryan held no financial interest in the company by that time, and the business did not in fact benefit under the impractical Mahoney. With the favorable publicity, however, Mahoney changed the name of the company to B. F. Mahoney Aircraft Corporation, owner of Ryan Airlines, and prepared to take the company public. He successfully floated a stock offering of five hundred thousand dollars, of which he owned 55 percent, and Ryan and others the remainder. Soon the St. Louis investors who had backed Lindbergh bought control of the new corporation, and it became the Mahoney Aircraft Corporation of Missouri on December 31, 1927.[39] The factory remained in San Diego, but headquarters moved to St. Louis in 1928 under yet another name, Mahoney-Ryan Aircraft Corporation, to capitalize on the Ryan name. On May 24, 1929, the company was acquired by the ill-fated Detroit Aircraft Corporation. Detroit shut down Ryan's production in 1930, a year before the parent firm fell into bankruptcy. Thus the Ryan team that had helped spark the aviation boom failed to reap any long-term benefit.

Claude Ryan had in 1926 established a separate Ryan Aeronautical Company, primarily to import Siemens aircraft engines from Germany. Then Siemens, wishing to establish its own American distributor, bought Ryan out in 1928 for seventy-five thousand dollars, giving Ryan enough capital to act independently.[40] For the time he maintained his flying school, started on May 1, 1928. Mahoney, sadly, died in 1951, possibly destitute.

Reuben Fleet

Reuben H. Fleet, a native of Washington state, learned to fly at an early age and became fascinated with aviation. He left a promising political and business career to accept a commission as a pilot in the U.S. Army. After organizing the army airmail service in 1918, he was assigned to the Engineering Division as chief contracting officer. Learning the procurement business, he considered entering the aircraft industry in some capacity. Declining offers from Boeing and Curtiss, Fleet eventually accepted a position as general manager of Gallaudet in East Greenwich, Rhode Island, and resigned his army major's commission.[41]

Edson Gallaudet had already left active management of his firm, which again had become financially troubled, and Fleet saw an opportunity. On May 29, 1923, Fleet founded the Consolidated Aircraft Corporation by leasing the Gallaudet plant, taking over Gallaudet's activities, and also purchasing Dayton Wright design rights with the General Motors withdrawal from aircraft.[42] The Gallaudet firm itself was not acquired by Consolidated but had exited aviation, and Fleet left his position as general manager of Gallaudet in 1924. Edson Gallaudet later built a reputation as a producer of tennis rackets. Fleet had an authorized capital for his firm of sixty thousand dollars but began with only twenty-five thousand dollars, provided entirely by himself and his sister, Mrs. Lillian Fleet Bishop. Fleet, neither a designer nor engineer, was instead an aggressive salesman and used to advantage his Air Service experience. He won military contracts for existing and new designs. His PT-1 and NY trainers, from original Dayton Wright designs, were ordered by both the army and navy and were widely exported.

Facing a labor shortage in Rhode Island, Fleet moved to populous Buffalo, New York, in 1924, taking over the old Curtiss North Elmwood plant and enjoying access to a more plentiful labor supply.[43] Fleet experienced the glare of Congress on contract charges, a recurring situation in the industry, in 1927. Sensitive to the possibility that Consolidated had earned "excess profits" amounting to three hundred thousand dollars on trainer production, the Air Corps demanded a refund.[44] Fleet resolved the issue by supplying a further batch of fifty trainers to the army at a price of one dollar each, an alternative that pleased the army, since any refund would have disappeared into the Treasury.

Consolidated was one of the first aircraft firms to be formed by acquiring assets from existing firms, a pattern that would be seen frequently in the future. In 1927 the engineer I. M. Laddon, whom Fleet had met at Wright Field, joined the firm and led the design of naval flying boats.

Igor Sikorsky

On March 30, 1919, the accomplished Russian aircraft designer and manufacturer Igor I. Sikorsky arrived in New York. Sikorsky had gained international fame with large aircraft, pioneering their use as bombers and transports, but left in 1918 as a consequence of the Communist revolution.

First settling in Paris, he began the design of a bomber capable of carrying a 1,000 kg bomb, but the Armistice precluded its construction. Sikorsky thus was led to pursue his career in the United States.[45]

Having left most of his assets in Russia, Sikorsky arrived with very limited resources, but still determined to form his own aircraft firm. Frustrated in the virtually nonexistent market of 1919, he secured a temporary position with the Army Engineering Division as an engineer and consultant.[46] There he became acquainted with Major Reuben H. Fleet. Afterward he returned to New York, still without personal funds or financial backing for an aircraft firm. Supporting himself by teaching mathematics, Sikorsky soon made contact with the Russian émigré community.

With the moribund new aircraft market of 1922, Sikorsky was unable to find a position with an existing company. But he attracted modest investors, including Russian émigrés, to form his Sikorsky Aero Engineering Corporation on March 5, 1923, with limited facilities at Roosevelt Field, Long Island. His small staff often went without pay, and tools and working conditions were crude, but he persevered. Famed Russian pianist and composer Sergei Rachmaninoff eventually invested five thousand dollars.[47]

Sikorsky's first American design, the S-29-A twin-engined transport, first flew on May 4, 1924, from Roosevelt Field to Mitchel Field. Overloaded with several of its builders aboard, however, the plane crash-landed and was extensively damaged. Sikorsky in desperation raised a further twenty-five hundred dollars from his stockholders to keep the project going, including fitting it with improved engines. The reengined S-29 flew on September 25, 1924, and began service as a contract carrier of passengers and cargo.[48] But no government orders were forthcoming, leaving the company struggling. Major assistance came, however, from an investment group headed by businessman Arnold Dickinson, a friend of Sikorsky, which provided $1 million and enabled the firm to pursue new developments. The new Sikorsky Manufacturing Corporation was formed on July 25, 1925, with Dickinson as president and Sikorsky as vice president.

Sikorsky's next design, the trimotor S-35, was intended to win the Orteig Prize for the first nonstop New York–Paris flight. Sikorsky raised $80,000 on his own for the project, which eventually cost $150,000. The S-35 flew successfully early in 1926, but the Atlantic attempt, on September 20, 1926, by the famed French ace René Fonck, ended tragically as the S-35 crashed on takeoff, killing two of the crew.[49] Before the S-35 could be rebuilt, Lindbergh had succeeded, and again the survival of the Sikorsky firm was in doubt.

Shortly after that episode Reuben Fleet, seeking the army contract for a new bomber, collaborated with Sikorsky on a development of his S-35, the S-37, to compete for the contract.[50] While they lost out to Curtiss, this marked possibly the first instance of two unaffiliated companies teaming to pursue a major contract, in the manner in which major contemporary aerospace firms team to compete for increasingly rare development contacts.

Sikorsky finally achieved commercial success with amphibians, especially the large S-38, and again reorganized in October 1928 as Sikorsky Aviation Corporation, with a capitalization of $5 million, and moved to a large new plant in Stratford, Connecticut, in 1929.[51] Major investors, including Dickinson, were from New England and favored that location. The new firm took over all assets of the Sikorsky Manufacturing and Sikorsky Aero Engineering Corporations.

Sherman Fairchild

The young Sherman M. Fairchild, son of a founder of International Business Machines (IBM), had pioneered aerial photography during the First World War and was inventive in several fields. Fairchild Aerial Camera Corporation was formed in 1920, to exploit the development and application of the aerial camera, followed by Fairchild Aerial Survey Company in 1921. Fairchild attracted extensive survey work but soon became unsatisfied with his two Huff Daland aircraft and began developing his own. Forming the autonomous Fairchild Aviation Corporation in New York in 1925, Fairchild built his first design with the assistance of Igor Sikorsky. The firm grew to two production arms, the Fairchild Airplane Manufacturing Corporation of Farmingdale, Long Island, in the old Lawrence Sperry factory then being used for storage, and by 1928 the Kreider-Reisner Aircraft Company of Hagerstown, Maryland.[52] Kreider-Reisner was incorporated on September 5, 1925, by Ammon H. Kreider and Lewis Reisner primarily as an airplane dealership, but had produced original designs. Acquisition of Kreider-Reisner enabled Fairchild to broaden his product line. The Fairchild FC-1 photographic airplane first flew on June 14, 1926. Curtiss Flying Service ordered six, and Fairchild became a leading producer of light cabin aircraft.

Sherman Fairchild made further aeronautical contributions in the 1920s, developing wing slats and flaps, enclosed cabins, and folding wings.[53] The Caminez Engine Company was acquired in 1925 as Fairchild Engine Company. To control his growing operations, Fairchild incorporated Fairchild Aviation Corporation in Delaware in 1927, holding the

Aerial Camera, Aerial Surveys, Airplane Manufacturing, and Flying Corporations, plus subsidiaries in Canada and Mexico, thus evolving into a vertically integrated aviation company.[54] Kreider-Reisner was still a separate subsidiary but became a division in April 1929.

The Loughead Brothers and John Northrop

The brothers Allan and Malcolm Loughead of San Francisco were early flying enthusiasts further influenced by their older half-brother Victor, who had written a book on aeronautics. With the backing of businessman Max Mamlock they formed their Alco Hydro-Aeroplane Company in 1912.[55] Their first effort, the Model G floatplane, first flew on June 15, 1913.[56] The brothers earned money flying passengers, especially at the 1915 San Francisco World's Fair, but were unsuccessful in selling the Model G to others. Moving to Santa Barbara with the backing of businessman Burton Rodman, they formed the Loughead Aircraft Manufacturing Company in the summer of 1916.[57] Their first effort was the ambitious ten-passenger F-1 seaplane, which flew on March 28, 1918, and was an instant success. But the end of the war precluded any navy orders.[58] The brothers survived by building two Curtiss HS-2L flying boats and with subcontracting but continued to pursue original designs.

Neither brother was a trained designer or engineer, but Allan was a skilled pilot and salesman, while Malcolm was a gifted mechanic. Among their significant early moves was the hiring in 1916 of the twenty-year-old John K. Northrop as a draftsman and designer. Northrop, a New Jersey native who had arrived in Santa Barbara with his family at age nine, received a sound technical education at Santa Barbara High School. He became especially valuable in stress analysis. Allan Loughead and Northrop designed the S-1 sportplane, which featured among many innovations a molded plywood fuselage, later patented.[59] Unfortunately the S-1 did not sell in the face of huge numbers of cheap surplus planes on the market, and the firm became another victim of the market collapse, going bankrupt in 1921.[60] Malcolm Loughead had left for Detroit in 1919 to form the successful Lockheed Hydraulic Brake Company, and Allan became the West Coast distributor. Allan Loughead, who had made the acquaintance of Donald Douglas, then helped Northrop secure a position with the fledgling Douglas Company in 1923.

Northrop had begun design studies in 1925 while at Douglas for what became the Vega, and approached Allan Loughead about producing the design. Northrop believed his streamlined Vega, with monocoque fuselage, metal structure, and internally braced wings, would be an advance over any other transport flying.⁶¹ With Loughead's renewed interest, a new Lockheed Aircraft Corporation was formed on December 13, 1926, in Hollywood. Allan Loughead, having tired of the frequent mispronunciation of the original, spelled the name phonetically. The spelling was not changed legally until 1934. Of total capital of $25,000, businessman Fred S. Keeler provided $22,500, with Allan Loughead supplying the balance.⁶² Earlier Allan Loughead had hired the accountant Kenneth Jay to handle the firm's finances, and Jay had introduced him to Keeler.⁶³ The prototype Vega first flew on July 4, 1927, and Lockheed enjoyed considerable success in the boom of the late 1920s.⁶⁴

In March 1928 Lockheed relocated to a factory in Burbank, where it would remain for sixty years. Vega sales rose steadily in the boom years of 1928 and 1929. Total Vega production amounted to 141, good for the day, and was also profitable. Largely on the strength of the Vega, Lockheed's 1928 stock offering soon soared. The Vega also began the Lockheed tradition, suggested by Northrop, of astronomical names for its aircraft, and led to the later Sirius, Altair, and Orion. Northrop resigned again on June 28, 1928, to form his own firm but was succeeded by another capable engineer, Gerald Vultee. Kenneth Jay joined Northrop, and Whitley C. Collins succeeded him at Lockheed.

Anthony Fokker

Aircraft of the Dutch pioneer Anthony Fokker set notable flight records after the First World War, and sales increased in the United States, in part a reflection of an American design lag in the early 1920s. The Fokker import company of 1921, Netherlands Aircraft Corporation, led to an American factory: laws of the day required new aircraft sold in the American market to be manufactured there. Netherlands Aircraft was managed by Bob Noorduyn, later to became a manufacturer in his own right in Canada. Business began in December 1923, in the vacant Hasbrouck Heights Wittemann factory. Fokker acquired the remaining facilities of Wittemann in 1925 and was joined by the designer A. Francis Arcier, formerly with Wittemann. Originally named Atlantic Aircraft Corporation, possibly to avoid any stigma from Fokker's association with the German cause in the recent war, the company began by modifying surplus DH-4s. Renamed Fokker Aircraft Corporation on September 16, 1925, it attained success with airliners, most importantly the F10 trimotor of 1927. The successor Fokker Aircraft Corporation of America of December 1927 became a major airliner manufacturer in the late 1920s. A second factory opened in Wheeling, West Virginia, and total employment reached nine hundred.

Aeromarine and Successors

Aeromarine went through a series of reorganizations in an attempt to survive. Flying-boat production largely ended by 1923, and for diversification Aeromarine merged with the Healy Bus Company, a manufacturer of bus bodies, as the Healy-Aeromarine Bus Company in 1927. Remaining in Keyport, the company continued to bid on military aircraft contracts, but with little success. In 1928 the Aeromarine-Klemm Corporation was formed, with Inglis M. Uppercu as president and the designer Vincent J. Burnelli as vice president, although it is unclear whether this venture succeeded Healy-Aeromarine or was autonomous. Its primary objective was licensed production of the German-designed Klemm low-wing monoplane, but the poor market led to receivership in 1931.

Vincent Burnelli had struggled for years, with different backers, to develop his innovative lifting-fuselage RB-1 design, which he felt maximized aerodynamic efficiency. Locating in Keyport in 1928, he set up Burnelli Aircraft, Ltd., with 60 percent foreign ownership and apparently independent of his Aeromarine affiliation, to develop the more advanced RB-2. With support from Uppercu the firm was succeeded in 1930 by the Uppercu-Burnelli Corporation. Subsequently this became the Burnelli Aircraft Corporation, still housed in the old Aeromarine facilities in Keyport. Uppercu departed in 1933, ending his long but unremunerative involvement in aviation, and the Aeromarine factory closed finally in 1936. Burnelli continued his lifting-body developments, but never attained production.

The Emergence of National Aviation Policy

Recognizing the crisis at the time, Dr. Charles D. Walcott, chairman of NACA, attempted to formulate the first national aviation policy in 1920.[65] His initial report supported the creation of a Bureau of Aeronautics in the Department of Commerce and emphasized that commercial aviation should be regulated and supported for its economic potential as well as for its contribution to national security.[66] A more formal NACA proposal was transmitted through the president to Congress on December 5, 1922. The statement expressed concern over the contraction of the military sector and observed that the civil sector had not developed as hoped, so that the industry had become too small to be capable of rapid expansion in an emergency. It recommended that the industry specialize in designs with continuous-production prospects, and that the government plan for future needs and place orders with a view toward continuity of production so that special facilities and skills could be maintained.[67]

Total employment in the industry was only four thousand in 1921 and declining. In 1922 output was just 263 aircraft, almost none of which were civilian, contrasted with an output of 411 aircraft in 1916, the last prewar year.[68] There was little prospect of increased demand. Accident and fatality rates were high, related to a lack of government licensing, and much of the public viewed flying as dangerous. By 1923 there were rising concerns about the lack of aviation policy, even as the role of aviation grew, especially in such fields as agriculture, forestry, photography, and mapping. The struggling industry wanted a policy that would set the rules of competition, help develop a commercial market, and develop air transport. Further, there was a need for a body of air law. But the industry still felt that, even with its importance to defense, its future lay primarily as a commercial industry.

The Lassiter board, established by the War Department under the direction of Major General William Lassiter to analyze the aviation field, issued its report on October 11, 1923. It concluded,

> The aircraft industry in the United States at present is entirely inadequate to meet peace and war time requirements, it is rapidly diminishing and under present conditions will soon practically disappear. It depends for its existence almost wholly upon orders placed by Governmental services. . . . The development of commercial aviation will stimulate the aircraft industry, but orders from the military services must be depended upon, at least for the immediate future, if the industry is to be kept alive.[69]

Calling attention to declining military air strength, the board developed a ten-year plan for procurement of aircraft in the amount of $35 million per year, but this produced no supporting legislation or anything tangible for the services. Principal equipment remained the DH-4, named with some justification the "Flying Coffin." Nevertheless, the report represented a real plan for army aviation.[70] The board also advanced the idea of a General Headquarters Air Force, which was not to reach fruition for twelve years. A unified Department of National Defense also was first seriously studied as early as 1923, but President Coolidge was said to be opposed both to an independent air arm and to Billy Mitchell personally.

Eventually two independent investigations of long-term significance were undertaken: the Congressional Select Committee of Inquiry into the Operations of the United States Air Services, known as the Lampert-Perkins committee, and the President's Aircraft Board, known as the Morrow board after its chairman. Both reported in 1925, otherwise a nadir for aviation, with remarkably consistent findings. Lampert-Perkins, originally to investigate the MAA, changed its focus to the Air Service and addressed the numerous concerns of the industry, of which three were paramount:

1. Firms held no proprietary rights to their designs, military contracts were awarded solely on the basis of price, and there was no recovery of development costs by the designer/builder.
2. There was a lack of production continuity due to sporadic orders.
3. The physical capital of the industry at the end of the First World War was largely destroyed.

Lampert-Perkins held hearings from October 1924 to March 1925. The findings included the following:

1. There was no evidence of an aircraft trust.
2. Only about 10 percent of government air expenditures went for the purchase and overhaul of aircraft and engines.
3. The industry was essential to national defense.
4. Production capacity of the industry was only about twelve hundred aircraft annually, or about 10 percent of the peak war capacity.[71]

The recommendations were favorable to the industry in that they called for a single agency for civil aviation, proprietary design rights to manufacturers, and elimination of government competition with the industry. The industry complaint of competition from the military actually was less about new aircraft than with repair and rebuilding work. In fact, the Naval Aircraft Factory had largely withdrawn from aircraft production in 1923. Admiral Moffett had made the decision after protests from industry.

The Morrow board was appointed by President Coolidge on September 12, 1925, following Billy Mitchell's attacks on the War and Navy Departments that resulted in his court martial. Coolidge was not known as a strong supporter of the industry, but establishment of the board constituted an act of statesmanship on his part. Hearings were held throughout September and October 1925. Findings were strongly supportive of the congressional report, but perhaps were more significant in that they represented views of the president. Its final report, *Aircraft in National Defense,* endorsed a private, economically healthy industry.[72]

The major specific recommendations were the following:

1. Secure continuity of production by five-year plans and appropriations that would span a longer period than the "fits and starts" business had been subjected to.
2. Stop direct competition from government plants like the Naval Aircraft Factory in Philadelphia and McCook Field in Dayton.
3. Recognize and honor proprietary rights of design.
4. Limit competitive bidding so as to stop ruinous price competition.
5. Establish an Aviation Bureau in the Department of Commerce to develop aids to navigation, as had been done in marine commerce, by lighting, airfields, mapping, radio, and so forth.
6. Require the licensing of airplane pilots and the certification of aircraft on a safe but reasonable basis.[73]

Lampert-Perkins was supportive of an independent air force, but the Morrow board favored retaining aviation in the Department of War, while supporting more autonomy for the Air Corps as well as authorization of the army and navy five-year procurement programs.

In 1925, even before the issuance of the two reports, Secretary of Commerce Herbert Hoover responded to the industry's concerns and appointed a Joint Committee on Civil Aviation in collaboration with the American Engineering Council. His report, issued on November 26, 1925, among other things recommended the creation of an Aeronautics Branch. While carrying no immediate benefit, the actions of Hoover did recognize the vital nature of the industry.

Legislation followed. The Air Commerce Act of 1926 was a major step toward federal recognition of aviation as a critical industry for the United States and provided major encouragement to civil aeronautics. It recognized aviation as primarily a commercial industry and did much to promote air safety. A Bureau of Air Commerce was created, and the position of assistant secretary of commerce for aeronautics was assumed by William P. MacCracken on August 10, 1926. Major functions of the bureau were licensing of pilots and of aircraft, air rules and safety, establishment and maintenance of air navigation facilities and airports, and provision of weather information. A further activity was in the control of national airspace. Foreign aircraft needed clearance to fly over the United States.[74]

While hardly the answer to all the industry's needs, these bodies at least acknowledged the critical situation. The Morrow board, which issued its report on December 2, 1925, in the middle of the Billy Mitchell court martial, may also be credited with passage of the Air Commerce Act. Mitchell, incidentally, had been a key and vocal witness for the Lampert-Perkins committee.[75]

The lofty goals of the two bodies actually were not fulfilled until the Second World War, but at least they had been established. Further actions in the 1924–26 period with portents for the future of the industry included the Air Mail (Kelly) Act of 1925, which opened the already active airmail service to private competition. The act spurred development of new aircraft for the specific purpose of carrying mail, but also with potential for carrying passengers, and thus eventually to the development of commercial airlines. By 1926 the government had reduced its airmail activities to only the transcontinental and New York–Chicago routes. All others were contracted. On August 21, 1927, the government announced that its airmail flights were to end and awarded all activity to commercial contractors.[76] A national air express system, using private facilities, was inaugurated on September 1, 1927, and the last government airmail flight was on September 9, 1927.[77]

The aircraft procurement law also established the office of assistant secretary of war for air, upgraded the Army Air Service to corps status as the Army Air Corps, and authorized an increase in strength to 1,650 officers and fifteen thousand enlisted men. The program was passed in an atmosphere of controversy, however, coming on the heels of the sensational Mitchell court martial and vigorous opposition by Admiral Moffett of the BuAer. While not having any immediate impact on the industry or the size of the military aviation establishment, the reorganization enabled the Air Corps to become roughly comparable to the Marine Corps in its semiindependent status under the Navy Department. Another significant change was upgrading the Engineering Division at McCook Field to the Materiel Division, which relocated in 1927 to nearby Wright Field. The Air Corps also benefited from the appointment of the proaviation F. Trubee Davison as assistant secretary of war for air, while Edward P. Warner was appointed assistant secretary of the navy for air. Both appointments became effective on July 2, 1926.

Mitchell's push for a separate air force definitely was a stimulant to the industry. The industry tended to tread lightly on the issue, since it did not wish to alienate either existing service, on whose good will it was so dependent.[78] While the five-year programs were boosts for both services, procurement rules could be confusing and ambiguous, and there would be numerous controversies in the future.[79] Competition for limited orders still was intense, with major contenders being Boeing, Consolidated, Curtiss, Douglas, Huff Daland, Loening, and Vought.

The procurement program began July 1, 1927, but was curtailed by the economic depression, leaving reequipment still slow and inadequate. But military aviation and the industry had gained much during the decade and were poised for further progress. Major General Mason M. Patrick, who learned to fly at age sixty, retired as chief of the Air Corps in 1927 with a record of great accomplishment and was succeeded by Major General James E. Fechet.

In the private sector, the Daniel Guggenheim Fund for the Promotion of Aeronautics, Inc., was established in 1925. The fund began operation in January 1926 with a bequest of $2.5 million, with emphasis in four areas, education, scientific research, commercial development, and educational information.[80] Earlier, in June 1925, Guggenheim had established the Guggenheim School of Aeronautics at New York University with an endowment of five hundred thousand dollars.[81] Professor Alexander Klemin headed the school from 1929 to 1945. The fund enabled several universities to establish schools of aeronautics. California Institute of Technology built an advanced wind tunnel with its $350,000 Guggenheim grant. Daniel Guggenheim believed, correctly, that public support rested on the safety of flying, and he sponsored a national Safe Aircraft Competition, which attracted no fewer than twenty-seven entrants. The winner of the one-hundred-thousand-dollar award was the Curtiss Tanager in 1930. Although the Tanager did not go into production, it may be regarded as the seminal short-takeoff-and-landing (STOL) aircraft, a critical technology for the future. The Guggenheim Medal for aeronautics was first awarded in 1929.

TABLE 5. U.S. Aircraft Exports, 1913–27

Year	Aircraft Exported		Value of All Aeronautical Exports ($)
	Number	Value ($)	
1913	29	81,750	107,552
1914	34	188,924	226,149
1915	152	958,019	1,581,446
1916	269	2,168,395	7,002,005
1917	135	1,001,542	4,135,445
1918	20	206,120	9,084,097
1919	85	777,900	13,166,907
1920	65	598,274	1,152,649
1921	48	314,940	472,548
1922	37	156,630	494,930
1923	48	309,051	433,558
1924	59	412,738	798,273
1925	80	511,282	783,659
1926	50	303,149	1,027,210
1927	63	848,568	1,903,560

Source: *Aircraft Facts and Figures,* various editions.

The Emergence of Professional Management

Certain common factors emerge from a review of how aircraft firms typically were established. Most individuals striking out on their own did so by capitalizing on their contacts and experience. Such leaders as Leroy Grumman and Frederick Rentschler were military officers during the First World War and made their initial industry contacts while in uniform. Others worked for pioneer manufacturers, such as Loening with the Wright Company and Sturtevant, and later Grumman with Loening. Douglas worked for Martin, while Northrop worked for both Lockheed and Douglas. Martin perhaps more than any other firm served as an incubator of future industrial leaders; Douglas, Bell, and Kindelberger all began with Martin, and McDonnell served with Martin in the 1930s. Vultee was a protege of Northrop who worked for Douglas and Lockheed. When Reuben Fleet moved Consolidated to San Diego in 1935, Larry Bell, his factory manager, remained behind to build his own firm. J. Leland Atwood, an associate of Kindelberger and eventually his successor at North American, began with Douglas in 1930 as an engineer. James S. McDonnell worked for several firms from 1925 to 1933, before joining Martin. In all cases, diligence and determination extended several years prior to firm success. No one began a career in aviation as an owner of an aircraft company. A consistent pattern was a strong streak of entrepreneurial independence: they simply wanted to do it their way. It is also noteworthy that those who were to become leaders in the industry were youthful when first starting their firms. Many were still in their twenties.

Many pioneers focused primarily on aeronautics and only secondarily on commercial prospects and profits. Recognition of the industry's commercial potential and initiatives for forming large companies tended to come more from the banking and securities industries. It is not that such men as the Wrights and Curtiss were oblivious to profit—the Wright patent suits pointedly demonstrated that—but that their first love was aviation and not building an industrial empire. Neither Orville Wright nor Glenn Curtiss had successfully transformed from aviation pioneer to corporate executive, although both profited when they sold to investors primarily motivated by profit potential.

Professional management could be considered to have gained prominence from the First World War, when a manyfold increase in production capacity was necessary. Wartime needs led to the involvement of the automobile industry, larger and more experienced, in production. Earlier, Clement M. Keys had rescued Curtiss and helped transform it into a public corporation. General Motors' acquisition of Dayton Wright in 1919 also brought with it management expertise. Donald Douglas and Reuben Fleet, who were very much businessmen, should be regarded as among those who not only developed the industry but were first and foremost firm-handed managers of their companies.

The gradual shift in the industry from dominance by pioneer aviators to professional managers accelerated in the late 1920s with the formation of holding companies with Wall Street backing. These contributed to the maturation of the industry, as it became less fragmented and better financed. It remained noteworthy, however, that many new entrants into the industry were visionaries who not only wanted individual success, but also believed that America should have a strong aircraft industry. C. M. Keys, Canadian by birth, stated patriotically that his motivation for "the foolish thing" of investing in the bankrupt Curtiss Aeroplane and Motor Company in 1920 ("and I have often been sorry since," he remarked) was "giving the United States an aircraft industry."[82] By most measures, major European nations led in aviation in this period, and only the determination of those early industrial leaders enabled America to regain its initial supremacy.

One dominant characteristic emerged about the men who were dedicated airmen and who had a head for business: aircraft companies must be built and guided by aviation men, not by financiers or conglomerateurs. That characteristic remains valid to the present. Their counterparts in the automobile industry were known as "auto men" or "car guys." Whenever executives have held a narrow focus on costs and financial returns in aircraft development decisions, they tend to withhold the necessary investments, facing an ultimate exit from the airframe sector. This likely was the major reason for the exit of Curtiss-Wright from aircraft in the late 1940s.[83] Among true aviation men, however, persistence in aircraft development and production also may help explain why the aircraft industry has seldom ranked among the most profitable industries. Entry into the field rarely could be justified by conventional projections of financial return on investment.

Industry progress in its first two decades, while impressive, was tempered by the fact that it remained a creature of government, highly vulnerable to a Congress that tended to view it with suspicion. A major achievement was the establishment of national aviation policy. Events of even greater influence on the industry lay just ahead.

Chapter 4 ✈ The Industry's Golden Age

A period associated with the Great Depression and consequent economic hardship might appear to present a dubious case for being called a golden age. The aircraft industry suffered perhaps more than most during this period and had few champions in Congress, which still held strong sentiments not only for the Naval Aircraft Factory but also for an army aircraft factory. Even after the Air Corps Act there was no cohesive structure for dealings between the industry and the government. Contracting and procurement were largely on an adversarial and ad hoc basis. Manufacturers received no aid with planning, being told to rely on past patterns.[1]

Yet many developments during 1927–35, including advances in aircraft design, performance, and mission capabilities, expansion of potential applications, a stronger industry structure, and noteworthy accomplishments of individual firms, make a persuasive case that those years were in fact the golden age for the industry. The broader golden age of aviation, which included airline development, private flying, record-setting flights, and air races, generally encompasses the 1919–39 interwar period.

During the golden age the industry became more specialized, with air transport largely separated from aircraft holding companies by 1935, and with engine and component manufacturers becoming independent companies or subsidiaries. While some overlap remained, a distinct division between the larger military and commercial manufacturers and smaller private aircraft firms also emerged, a pattern that was to remain. Thus the aircraft industry, while still small and struggling, developed a modern structure in addition to making major progress in design and engineering.

The Lindbergh Boom: Becoming a Major Industry

While aircraft manufacturing was well established by 1927, and the field of aviation had gained much in public acceptance, the industry's viability still depended largely on government procurement and airmail contracts. But air travel was beginning, and airlines began to grow rapidly.

Rising public interest did not, however, translate into consistent business. Many builders were forced to shift production from aircraft to other goods due to weak demand. Progress in both military and commercial aviation positioned aircraft manufacturing to become a major industry, but that status could be attained only by steady demand. In part due to the

Air Corps Act and other legislation, total output grew to 1,186 planes in 1926 and increased to 4,346 in 1928.[2] In that same year civilian aircraft sales by value exceeded military aircraft sales for the first time. Overall demand increased sharply and attracted many new producers to the field. But the apparent boom unfortunately turned out to be a bubble, which burst with the Great Crash.

The airplane's capabilities were expanded steadily. Richard E. Byrd's flights over the North Pole, on May 9, 1926, and the South Pole, on November 28, 1929, were dramatic events. The Air Corps Pan American Goodwill Flight, using Loening amphibians, beginning December 21, 1926, and completed May 2, 1927, covered twenty-two thousand air miles. In April 1927, Clarence Chamberlin and Bert Acosta stayed aloft in their Wright-Bellanca for fifty-one hours.

If any single event or development opened the eyes of the public to the potential of the airplane, however, it was the solo flight of Charles Lindbergh across the Atlantic on May 20–21, 1927. The accomplishment of Lindbergh, formerly an airmail pilot, made him a worldwide hero and made the public air-minded as never before. Although essentially a stunt, breaking no new technological ground, the feat served as the catalyst for the aviation boom. Less recognized, but no less significant, was the non-stop flight of Chamberlin from New York to Germany only two weeks later, in the same Wright-Bellanca used in the endurance record. Wealthy aviation enthusiast Charles Levine, accompanying Chamberlin, thus became the first transatlantic air passenger.

The *Question Mark,* commanded by Major Carl A. Spatz (later Spaatz) and piloted by Captain Ira C. Eaker, stayed aloft for 151 hours, from January 1 to January 7, 1929, with the aid of aerial refueling. The first successful blind flight was achieved by Jimmy Doolittle, on September 24, 1929, in a Consolidated trainer. Sponsored by the Guggenheim Fund, the program marked a major advance in the use of instruments, many originally developed by Lawrence Sperry, and paved the way for instrument and bad-weather flying.[3]

The Aviation Boom was well under way by the end of 1927, and Wall Street began to pay attention. The public became interested in aeronautical stocks, and banks became eager to finance new and expanded ventures. In addition to new entrants, larger and financially stronger firms were organized from a base of smaller, formerly independent firms. Developments in established firms in the boom years of 1928 and 1929 also led to a stronger industry structure. On November 30, 1928, Douglas went public as the Douglas Aircraft Company, Inc., with an offering of three hundred thousand shares. Donald Douglas received two hundred thousand shares in exchange for the previous Douglas Company, which to that time had been owned solely by Douglas and his father.[4] Also in 1928, Douglas completed the large Clover Field plant at Santa Monica, to serve as its headquarters for four decades. Lockheed's new Burbank plant, initially at only twenty thousand square feet, was eventually to expand manyfold.

Glenn Martin, in sole control of his company since the early 1920s, decided to relocate in 1928. He sold the Cleveland factory and chose Baltimore, influenced by better weather for testing his flying boats, availability of open land for expansion, and proximity to Washington and the government. Civic concessions involved with Martin's acquisition of factory land even became a local political issue at the time. While others expanded with stock issues, Martin borrowed $3 million to finance the new factory. But the new venture did not proceed smoothly in the depression, and poor business led to default and bankruptcy proceedings in 1934, which Martin did not fully overcome until 1939.[5] Although strongly conservative, antiunion, and distrustful of government, Martin was forced to gain a $1.5 million loan from the Roosevelt administration's Reconstruction Finance Corporation (RFC) to survive. Martin's experience served to caution others in the industry about debt and expansion.[6]

In 1928 Reuben Fleet secured a contract to develop his XPY-1, the first of a long line of Consolidated flying boats, and in 1929 was able to take his firm public.[7] To concentrate his resources on new military development and production, Fleet in February 1929 organized Fleet Aircraft, Inc., as a 100 percent–owned subsidiary to manufacture the popular Fleet biplane trainer for the civil market. He brought in Larry Bell, then out of the industry, as vice president of Consolidated on July 1, 1928.[8] Bell was assigned added duty as president of Fleet Aircraft and later as president of Fleet Aircraft of Canada, formed on March 25, 1930, and located across the Niagara River in Fort Erie, Ontario.[9] The Fleetster eight-seat all-metal transport of 1929 with its NACA-designed cowling marked another advance, and the Fleet trainer continued its popularity into the mid-1930s.

Another major development for Consolidated was the acquisition of Thomas-Morse on August 5, 1929. Frank L. Morse previously had sold his company to Borg-Warner for $18 million, which then sold Thomas-Morse to Fleet for an exchange of stock. Thomas-Morse had returned to production in 1927 with an army order for O-19 observation aircraft, and Consolidated completed that order. The able B. Douglas Thomas was retained as general manager but entered early retirement in 1933.[10] Then a major manufacturer, Fleet was fiercely independent, showing no interest in joining one of the large holding companies being formed. Financially, however, he still struggled, losing $177,000 in 1931.

The number of aircraft manufacturers expanded steadily, although the failure rate remained high. The *Aircraft Year Book* for 1929 reported a total of 96 firms in the aircraft-manufacturing business.[11] The U.S. Department of Commerce listed a total of 157 aircraft manufacturers in 1929, although it is doubtful that most were active. With the exception of the Wichita personal-aircraft firms, the industry still was centered on the East Coast. In 1927 only two large firms, Boeing and Douglas, with six hundred and five hundred employees respectively, were on the West Coast.

The Design Revolution, 1929–35

Many seemingly established aircraft firms as well as new ventures failed during the depression of the early 1930s. Yet economic reverses did not cause an immediate industry collapse, as after the First World War. Instead, it shrank in stages, with such primarily military contractors as Douglas, Martin, and Consolidated continuing to fill orders through 1931. By 1932 the industry was extremely weak, however, since the procurement programs were complete, the civil-aviation market remained very small, and new airliner designs had not yet appeared. Private aircraft demand virtually disappeared. Yet the basic industry structure that was to carry the nation to victory in the Second World War rose from this weak market and was largely in place by 1935. Curtiss-Wright, Douglas, Boeing, Lockheed, Consolidated, Bell, North American, Fairchild, Martin, Grumman, and the Sikorsky and Chance Vought divisions of United Aircraft all were in operation.

It was also during this period that aircraft design progressed to the modern era of all-metal stressed-skin construction, enclosed cockpits, and retractable landing gear. Monoplanes gradually superseded biplanes, and engine power steadily increased. The Ryan NYP and John Northrop's Vega and the later Alpha, the first design under his own nameplate, with their advanced structures, were major steps. On September 29, 1931, the streamlined Supermarine S.6B floatplane, powered by a special "Sprint" engine, raised the world speed record to 407 mph, presaging Second World War fighter speeds, and Wiley Post completed his eight-day around-the-world flight on July 22, 1933. The period coincidentally marked the transition to the modern automobile, which achieved a more streamlined appearance.

In 1927 anodized aluminum structures were developed for naval aircraft, helping solve the persistent problem of saltwater corrosion. The variable-pitch propeller was another major advance. NACA research was highly active, and its streamlined cowling, enabling higher performance with no increase in engine power, won the Collier Trophy for 1929. NACA also demonstrated the advantages of fairing engine nacelles into the wings of multiengine aircraft rather than suspending them below, which manufacturers quickly adopted. Research by NACA applied equally to both military and civil aircraft; its airfoil advances and progress in understanding lift and drag greatly modernized aircraft design.

The design revolution accelerated with the streamlined, all-metal Boeing Monomail, which first flew on May 6, 1930. The Monomail was not produced in quantity, but led to the twin-engined B-9 bomber and then to the 247, the first modern airliner. These developments, however, were quickly eclipsed by the Martin B-10, the first modern bomber, and the Douglas DC-1 airliner. Further advances included such famous aircraft as the Douglas DST, which became the DC-3, and the Boeing 299, which became the B-17. Both first flew in 1935 and were to play major wartime roles. The Seversky P-35, the first modern fighter, and the Consolidated XP3Y flying boat also appeared in 1935. Design progress of the early 1930s laid the foundation for the industry's technological supremacy in the Second World War.

Geographical location also shifted significantly. Southern California would become the major manufacturing center before the Second World War, but in earlier years Long Island claimed that distinction. Early industry development centered on the Hempstead Plains of Nassau County, an area with several natural airfields. By 1931, however, Curtiss had withdrawn from Long Island to Buffalo. Loening left the area after a change of ownership, and Fairchild moved to Hagerstown, Maryland, in 1932. Chance Vought moved to Hartford and Sikorsky to Bridgeport, Connecticut, after their affiliation with United Aircraft and Transport Corporation (UATC). Long Island would continue as an important center, however, with the Brewster, Grumman, and Seversky (Republic) firms.

The international character of the industry continued to grow. Boeing, Consolidated (Fleet), and Curtiss-Wright all established subsidiaries in Canada. Canadian Pratt and Whitney was established on November 29, 1928, and Fairchild of Canada was established in May 1929.

During 1930 the industry lost two of its pioneering lights. Glenn H. Curtiss died at age fifty-two, and Chance M. Vought, very much active in design and production, was only forty when he died unexpectedly. Their deaths occurred only two days apart, on July 23 and July 25, respectively. Also in 1930, the aviation world lost Elmer Sperry and Daniel Guggenheim, whose support led to rapid aviation progress.

Across the Atlantic, twenty-two-year-old Pilot Officer Frank Whittle of the Royal Air Force applied for a patent on his design for an aircraft jet engine in 1930. In an episode of extraordinary myopia, no work would proceed on Whittle's design for several years. Also in Great Britain, the first report on the potential of radio detection and ranging was published on July 23, 1935. The development would later become known by the acronym *radar*.

New Military Firms

Grumman

In the summer of 1928, Grover Loening observed that the major public firms in the industry were Curtiss and Wright. Smaller firms like his were ignored by Wall Street and had difficulty obtaining financing despite their potential. Loening thereupon determined to sell if he could realize an advantageous price and contacted Richard Hoyt, chairman of both Wright Aeronautical and the investment banking firm Hayden, Stone and Company. Hayden, Stone purchased Loening Aeronautical Engineering for a reported $3 million, giving a price/earnings ratio of ten to one, a strong price for a smaller firm, then merged it into Keystone Aircraft.[12] Hoyt also became chairman of Keystone.

In 1929, after the Loening firm had moved to Pennsylvania with the Keystone merger, Leroy Grumman, who had resigned his naval commission in 1920 to become Loening's chief engineer, declined to move and remained on Long Island to found his own company.[13] Grumman retained two Loening colleagues who were to make major contributions to his company's success, Leon A. "Jake" Swirbul and William T. Schwendler. Having prospered from the sale of his company, Grover Loening and his brother Albert became the largest investors in the new venture, and other key employees from the Loening company moved to Grumman.[14] Among them was Clinton Towl, later to succeed Grumman himself. Albert P. Loening served as a Grumman director.

Despite its dubious timing, just after the Great Crash, Grumman Aircraft Engineering Corporation was incorporated on December 6, 1929, with total capital of $77,250 raised entirely from close associates.[15] The company opened for business on January 2, 1930, in a rented garage in Baldwin, Long Island, with twenty-one employees.[16] With his naval-aviation background, Grumman shared Loening's interest in seaplanes and dedicated his company to aircraft for the U.S. Navy.

Grumman managed to survive by repairing Loening amphibians and constructing duralumin seaplane floats. But the major success in the early years was a navy contract for its new FF-1 shipboard fighter, which was followed by orders for utility amphibian and scout versions of the FF-1. The FF-1 was a major advance over earlier shipboard fighters, featuring a NACA cowling, stressed-skin construction, and an aluminum propeller.[17] It became the first production combat aircraft with retractable landing gear. The company moved from Baldwin to Valley Stream in 1931, and then to the vacant Fairchild Farmingdale factory in 1932. Grumman's employment expanded to 207 by 1934.[18] Production was under Jake Swirbul and engineering under Bill Schwendler, and sales exceeded one million dollars in that year.

Orders followed for the improved F2F, and eventually for the F3F series, on which Brewster subcontracted, and which became the navy's last biplane fighters.[19] The venture paid off brilliantly during the Second World War, as Grumman became the leading supplier of aircraft to the navy.

Brewster

Brewster represents an instance of a long-established company, in this case since 1810, that moved into aircraft manufacturing. The Brewster Corporation, a manufacturer of automobile bodies, first entered aviation in 1920 with duralumin seaplane floats. Brewster Aeronautical Corporation of Long Island City, New York, was formed in 1932 when the engineer James Work, formerly with the Naval Aircraft Factory and briefly with Detroit Aircraft, purchased the name, equipment, and goodwill of the then-dormant Brewster aircraft division for a total of thirty thousand dollars. He began operations in older factory space in Brooklyn. While continuing the manufacture of floats and components, and subcontracting to Vought and Grumman, Brewster undertook original military aircraft.[20] The designer Dayton Brown led design of its first aircraft, the Buffalo and the Buccaneer (named the Bermuda for export), which were to provide strong competition for Grumman. The first production contract, awarded in 1936 for the XSB2A, later named Buccaneer, positioned Brewster to become a major producer.

Seversky

The Russian émigré Major Alexander P. de Seversky, a pilot, designer, and engineer, came to the United States in 1918. He developed a successful bombsight, then founded Seversky Aero Corporation in 1923 for aircraft parts and instruments, but not aircraft. While this first venture did not survive, Seversky established a national reputation as a pilot, became a U.S. citizen in 1927, and determined to enter aircraft manufacturing. He attracted the backing of the millionaire banker Paul Moore and others to found Seversky Aircraft Company on Long Island, on February 16, 1931, specifically for military aircraft. Initially leasing production space in the Edo float factory, Seversky won attention with his SEV-3 float amphibian and subsequent developments, including the BT-8 trainer.[21] The firm later shared factory space with Grumman in the former Fairchild facility in Farmingdale.[22] Seversky brought in the talented Alexander Kartveli, also a Russian émigré, as chief engineer in 1934, and Kartveli was primarily responsible for the P-35 pursuit. Seversky attracted attention with advanced designs, but production orders were sparse and profits elusive.

Others

The engineer Charles Ward Hall formed Hall-Aluminum Aircraft Corporation in 1928. He had begun aircraft aluminum research with Aeromarine in 1916 and built his first airplane in 1922. First locating in Buffalo and sharing space with Consolidated, in 1934 Hall moved to Bristol, Pennsylvania, occupying the then-closed Keystone factory. He eventually attracted navy interest with his PH-2 aluminum-hulled flying boat of 1929. Hall advanced flying-boat development, and naval orders kept his small company operating during the 1930s, as it became a potential competitor to Consolidated and Martin in the field.[23]

A second flying-boat developer of the period was Fleetwings, Inc., organized in Buffalo, New York, in 1929, as an outgrowth of an earlier nonaviation company started in 1926. It advanced the shot-welded stainless-steel aircraft structure originally developed by the Budd Company. It moved first to Roosevelt Field, Long Island, and then to Bristol, Pennsylvania, in 1933, sharing space with Hall-Aluminum. Its first aircraft, the small Sea Bird amphibian constructed of stainless steel, was not tested until 1936. Significant orders never materialized, and the firm turned to subcontracting in order to survive, but it remained an active researcher in steel and aluminum alloys for aircraft.

The Great Lakes Aircraft Corporation was incorporated on October 22, 1928, by W. R. Wilson and Colonel B. F. Castle, in the Cleveland factory vacated by the Glenn L. Martin Company. From 1929 to 1932 it produced the 2T-1 biplane trainer that won some recognition, and it developed army dive bombers and navy amphibians. But its XBTG-1 torpedo bomber lost out to the Douglas XTBD for naval orders in 1936, forcing the firm out of business.

The Involvement of the Automobile Industry

An especially significant factor in the development of the aircraft industry during the golden age was the automobile industry. Ford, Fisher Body, and Packard had been involved with production of aircraft and engines during the First World War. Then General Motors, whose involvement extended to the Dayton Wright Company in 1919, renewed its interest by acquiring a 40 percent interest in Fokker in May 1929.[24] Anthony Fokker felt that GM's backing would convey a competitive advantage over Ford and its tri-motor airliner. While intending to remain a passive investor, General Motors soon became concerned over growing financial losses at Fokker and at what it considered deficient management.[25] Also, the four-engined F32 airliner, first flying in September 1929, proved too large for the market. Becoming more active, GM changed the name of the Fokker operation to General Aviation Manufacturing Company (GAMC), then in May 1930 placed the firm under the newly formed General Aviation Corporation (GAC) with its other aviation interests, including air transport. After the board meeting of July 10, 1931, Anthony Fokker departed in disagreement. By the summer of 1931 GM had gained total ownership of Fokker/GAMC, but after completion of existing orders largely shut down aircraft production. Then in April 1933 GAC was merged into the new North American Aviation holding company by a direct exchange of GAC stock for 30 percent of NAA stock, with GM taking effective control of NAA in the process.[26] Aircraft production by various GM divisions during the Second World War, which was of high importance, is covered in chapter 8.

Henry Ford also renewed his early interest in the aircraft industry, and on the recommendation of his son Edsel acquired control of the Stout Metal Airplane Company in August 1925, along with the design for what became the legendary Ford "Tin Goose" trimotor. William Bushnell Stout, chief engineer of the aircraft division of Packard during the First World War, built his first airplane in 1919 and had formed his Stout Metal Airplane Company in November 1922, with considerable assistance from the auto industry.[27] Stout was particularly attracted to duralumin and all-metal corrugated construction rather than the stressed-skin technique. After merger, the operation became the Stout Metal Airplane Division of Ford Motor Company, and later simply the Airplane Division. The Ford 4-AT trimotor first flew on June 11, 1926, and was the first successful American passenger airliner, with more than two hundred produced. But Stout, unhappy with Henry Ford's interference in design, left in 1930 to develop new designs independently. The Ford Airplane Division remained active until 1933, with the shutdown more a matter of market conditions rather than a loss of interest by Ford. Ford, like GM, was to become a major wartime factor in aircraft, especially with its huge Willow Run plant.

The automobile manufacturer E. L. Cord pursued aviation involvement for several years. He became interested in air transport and particularly airmail in the 1920s and moved into manufacturing in 1929 by acquiring both Lycoming Manufacturing, already a supplier for his automobiles, for engines, and Stinson Airplane, a manufacturer of personal aircraft and small airliners.[28] Cord then sponsored a third venture, backing Gerald Vultee in the formation of the Airplane Development Corporation on January 26, 1932. Vultee had departed as chief engineer of Lockheed in 1930, before the Detroit bankruptcy, and Cord provided him with fifty thousand dollars to develop a new airliner design.

Thus the aircraft industry had become strongly linked to the auto industry. It is noteworthy that the auto industry typically did not establish or create new aircraft operations or subsidiaries but tended to invest in existing firms. Despite a superficial resemblance between the two industries, it eventually was realized that they possessed little in common. Technology, market applications, and production processes were vastly different. The major area of commonality was transportation.

The Aviation Holding Companies

Of even greater long-term significance than auto involvement to the development of the aircraft industry was the formation of holding companies, a structure still in existence for many major industries. The broad economic boom of the 1920s highlighted such growth industries as oil, automobiles, rubber, and radio, and many forecast that aviation would be the next growth industry. Consequently, large aviation-related corporations, financed by public stock offerings underwritten by investment-banking firms, were established.

The Wall Street interest led to speculation, perhaps to a greater degree in aviation stocks than in stocks overall. Grover Loening reported the following financial figures during the Wall Street boom of 1928–29, when many aircraft firms, including holding companies, issued public stock:

- AVCO: 2.8 million shares @ 20 = $56 million
- Curtiss-Wright: 1,150,000 shares class A; 6.4 million shares common @ 31 = $244 million
- Douglas: 360,000 shares @ 45 = $16.2 million

- North American: 3.4 million @ 19 = $64.6 million
- UATC: 7.5 million preferred; 2 million common @ 162 = $331.5 million[29]

These, combined with stocks of smaller firms traded on the New York Curb Exchange (over-the-counter), having a total value of some $150 million, gave the industry a total market value in excess of $850 million. This was at a time when total net earnings of the firms involved were no more than $8 million, resulting in a price/earnings ratio of more than one hundred to one.[30] While not greatly different in degree from the speculative excesses in other industries in the stock market of the period, these values were obviously unsustainable, and aircraft industry shares fell further than most after the Great Crash.

The initial structures of early holding companies were so convoluted as to almost defy rationality, but market processes and government restrictions eventually transformed them into more durable structures. Some divisions or subsidiaries were spun off from holding companies as independent firms, adding to the complexity of the industry. The new holding companies were Curtiss-Wright; United Aircraft and Transport (now United Technologies); Aviation Corporation (later AVCO); North American Aviation (now Rockwell International), founded by Clement Keys primarily as an investment trust; and Detroit Aircraft Corporation, which was to survive only four years. The giants were United and Curtiss-Wright, as North American was years away from becoming a major manufacturer. Fairchild also emerged as a smaller holding company in aviation.

Detroit Aircraft Corporation

With the aviation boom of 1927–29, efforts were made to restructure the industry along the lines of the automobile industry, consolidating smaller firms into larger and financially stronger entities. The earliest, and possibly most grandiose, such effort was in the home of the auto industry. A group of prominent businessmen, some of whom had been involved with lighter-than-air craft in the Aviation Development Corporation (not to be confused with Jerry Vultee's later Airplane Development Corporation), organized the Detroit Aircraft Corporation in 1927. Edward S. Evans was president, and influential investors included the banker Harold S. Emmons, Charles S. Mott of GM, and William B. Mayo of Ford. Openly aspiring to become the General Motors of the aircraft industry, Detroit acquired a number of existing firms in aviation. By 1929 it had acquired no fewer than eleven such firms, most prominently Ryan and Lockheed.

The Lockheed board in July 1929 made the fateful decision, against the wishes of Allan Lockheed, to merge the company into Detroit.[31] The sale later revealed astuteness on the part of majority owner Fred Keeler, however, in that the market crashed three months later. Lockheed in fact was the largest and most profitable of the firms held by Detroit, with its noted Vega and later Star series, and with a capable manager in Carl Squier.[32] Richard von Hake, formerly with Douglas, replaced the departed Jerry Vultee. Problems ensued when the Detroit board milked Lockheed earnings for dividends and provided no new funding for research and development, a fatal policy given rapidly changing aircraft technology.[33] But Lockheed still managed to develop its first fighter, the two-place XP-900, designed by Robert J. Woods and given the military designation YP-24. The prototype first flew in 1931 and was delivered to the Air Corps, but after a crash no further development was pursued at the time. With the civil market in a state of collapse by 1931 and no military orders, the Detroit venture was destined to be another victim of the depression. The company fell into receivership and was liquidated on October 26, 1931. Most affiliates, including Ryan, died with it.

Lockheed attempted to operate in bankruptcy but eventually was forced to close. The P-24 pursuit design was taken over by Consolidated and developed by Woods, who had moved to Consolidated, into the YP-25, eventually produced as the P-30. Lockheed's assets were valued by the bankruptcy court at only $130,000, but a rescuer was at hand.

United Aircraft and Transport

Frederick B. Rentschler, an Ohio native and former Air Service pilot, organized a new engine venture in 1925 within the Pratt and Whitney subsidiary of the Niles-Bement-Pond Company of Hartford, Connecticut, in which he had family connections.[34] He obtained a ninety-thousand-dollar development contract from Admiral Moffett, who favored a second source for naval-aircraft engines.[35] Rentschler was well qualified, having helped found Wright Aeronautical in 1919 and serving as its president. In 1923, prodded by Moffett, he had merged Lawrance Aeronautical into Wright, bringing the radial engine expertise of Charles L. Lawrance to the company. But Rentschler, advocating greater R&D funding, subsequently resigned over policy disagreements.[36]

Rentschler became interested in Pratt and Whitney, which specialized in tools, primarily for its name recognition, and his new Pratt and Whitney Aircraft Company operated autonomously. With a new board of directors including Edward A. Deeds, then president of Niles-Bement-Pond, and a production team under George L. Mead, he developed the new engine, later named the Wasp. Production was in the old Pope-Hartford auto factory. Success of the Wasp was a bright spot in an otherwise depressing period for the aircraft industry, and orders flowed as army and navy procurement increased after 1927. The company presented a strong challenge to Wright domination in the field.[37]

The civil market also expanded after Air Mail Act of 1925 and the Air Commerce Act of 1926. The air transport and communications infrastructure was strengthened, especially with respect to air safety, facilitating airline formation and growth and benefiting all engine producers.[38] One of the most successful airmail companies was Boeing Air Transport, founded by William Boeing and Elbert Hubbard on February 17, 1927, whose transports were powered by Wasp engines. Boeing then formed the public Boeing Airplane and Transport Corporation late in 1928, which controlled all production, Boeing Air Transport, and Pacific Air Transport. Boeing's young legal counsel William M. Allen handled the incorporation work.

Both Boeing and Rentschler recognized the potential for passenger transport built onto an airmail contract base, and negotiated an affiliation. Rentschler also induced his friend Chance M. Vought, whom he had known since the First World War and who had adopted the Wasp for his designs, to bring his firm into the new organization. Legally a Boeing Airplane and Transport acquisition of Pratt and Whitney Aircraft, the United Aircraft and Transport Corporation was incorporated on February 1, 1929.

Still in the era when commercial banks could underwrite stock offerings, financial backing was provided by National City Bank of New York, of which Frederick Rentschler's brother George was president, and by the Daytonians Deeds and Kettering. Boeing became chairman and Rentschler president of UATC. Fifty percent of the stock was provided by Pratt and Whitney Aircraft, which was then separated from the original Pratt and Whitney, and the remaining 50 percent was provided by Boeing, Hamilton Aero, and Vought.[39] The combination raised antitrust questions, but the Justice Department ruled that the firms were noncompetitive and did not threaten restraint of trade. Both Boeing and Rentschler gained millions when UATC stock boomed before the crash, raising later questions in Congress, but neither was ever charged with manipulation or fraud.

Major operating subsidiaries were Pratt and Whitney, Chance Vought Aircraft, Boeing Airplane and Transport, and Hamilton Metalplane of Milwaukee, founded in 1921, which had become a division of Boeing in February 1929.[40] Hamilton originated in Seattle, where it first came to the attention of Boeing. On July 30, 1929, Sikorsky Aircraft joined UATC, bringing the extraordinary engineering expertise of Igor Sikorsky. Stearman Aircraft of Wichita also joined, giving UATC a position in the personal-airplane market. Standard Steel Propellers of Pittsburgh, dating from 1919, was acquired in September 1929 and merged with Hamilton to become the Hamilton Standard Division. By 1931 Standard's operations were transferred to East Hartford to consolidate production. Thus UATC became the leading propeller producer.

The naval aviator Eugene E. Wilson, who had worked on the Wasp at the BuAer, retired and joined UATC as president of the propeller division and then became president of Sikorsky. He was to serve as president of the successor United Aircraft. Rex Beisel, who had joined the new Spartan Aircraft of Tulsa in 1929 only to become unemployed with the depression, rejoined Chance Vought Aircraft in September 1931 as chief engineer, effectively replacing Vought himself.[41] The transport group consisting of Boeing Air Transport, National Air Transport, and others, including Varney, evolved into United Air Lines.[42] Elbert Hubbard died on December 18, 1928, and was succeeded by W. A. "Pat" Patterson, to head UAL for three decades.

John K. Northrop also was to have a brief affiliation with UATC. Northrop had formed the Avion Corporation in 1928 in Burbank, in partnership with Kenneth Jay. He received some support from the publisher George Randolph Hearst, a Vega customer, for his venture, and Hearst served as the company's president. But persistent financial difficulties led to the merger into UATC in October 1929.[43] William Boeing, impressed with the Alpha design, initiated the purchase from Northrop's financiers. The new Northrop Aircraft Corporation was formed on January 1, 1930, as a subsidiary of UATC. Boeing became chairman of the board of Northrop, Jay president, and Northrop vice president. John Northrop began developing the flying wing, which he first conceived in 1923. Always seeking higher performance and greater aerodynamic efficiency, Northrop strongly believed that the flying wing would achieve that optimal efficiency. His first such design, known by its registration number X216H, while not a true flying wing, tested the concept.

Even with the support of UATC, the Northrop company still experienced financial and market problems, leading UATC to consolidate Northrop with Stearman on September 1, 1931. Lloyd Stearman had left earlier that year over disagreements with certain UATC policies. All production was to be in Wichita, with the Stearman factory to take over Northrop's designs. Unwilling to move to Wichita, Northrop resigned from UATC in January 1932, seeking again to establish his own firm. He gained backing from Donald Douglas, and no Northrop designs were built in Wichita.[44]

The United Aircraft and Transport Corporation became one of the strongest aviation companies in the world. Unlike others, UATC exercised active control of its subsidiaries.[45] UATC initially prospered to the point that it became a particular target of congressional investigations into both airmail and military procurement contracts.[46]

By 1934, however, UATC was in serious trouble. Boeing production was running down with the loss of the new bomber contract to Martin, and with the DC-2 threatening the 247 airliner. The four-engined Sikorsky S-40 was impressive, but only three were built, and the more advanced S-42 proved expensive to develop. Loss of airmail contracts and a strike in Hartford were added complications. The Air Mail Act of 1934 forced major organizational changes, including creation of United Air Lines. UATC was in fact the first of the holding companies to separate airline operations. Boeing also separated, but retained Stearman as a wholly owned subsidiary. Bill Boeing, only fifty-two, retired in 1934 from the company he founded and sold his stock, bitter over the investigations and the forced breakup of the organization.[47] Phillip Johnson also was forced out as president, moving to Canada, where he helped found Trans-Canada Airlines.

The Boeing Airplane Company was reincorporated in Delaware on June 19, 1934. Under the parent were the Boeing Aircraft Company of Seattle, which controlled Boeing Aircraft Company of Canada, Ltd., and Stearman Aircraft Company of Wichita.[48] Claire Egtvedt became president and pointed the firm toward development of large aircraft.

United Aircraft and Transport finally dissolved legally on September 26, 1934. By June 1935, Chance Vought, Pratt and Whitney, Sikorsky, and Hamilton Standard were consolidated into the United Aircraft Manufacturing Company. A separate United Aircraft Exports Corporation was organized. All in turn were subsidiaries of the United Aircraft Corporation.

The Aviation Corporation

Among the most complex of the new holding companies was the Aviation Corporation, later known as AVCO, a so-called aviation investment trust. Aviation Corporation was formed on March 1, 1929, underwritten by a combination of Harriman, Lehman Brothers, and Mellon banking interests. Primarily oriented toward air transport, the Aviation Corporation soon consisted of eight subsidiaries, including a manufacturing arm in Fairchild. While established as a manufacturer, Sherman Fairchild realized that to survive in the depression he needed a smaller, cheaper airplane. Partially in pursuit of that development he had formed Fairchild

Aviation Corporation as a holding company. When AVCO acquired 55 percent of the stock in 1929, it gained control of Fairchild operations along with the Farmingdale plant.[49] The purchase was paid in AVCO stock, thereby making Sherman Fairchild, who became vice president, the largest stockholder in AVCO. Fairchild thus indirectly was instrumental in the founding of AVCO but soon became disenchanted and withdrew his interest on April 1, 1931, regaining his 55 percent share of Fairchild Aviation Corporation. In payment he transferred to AVCO the Fairchild Airplane Manufacturing Corporation and Fairchild Engine Corporation subsidiaries.[50] The Fairchild stock was retired. Thus AVCO secured total control of the original Fairchild companies in April 1931, combining them as American Airplane and Engine Company.[51] Aircraft were marketed under the name Pilgrim and engines as Ranger.

While large enough to provide a challenge to the Curtiss-Wright and UATC giants, Aviation Corporation never established the clear objectives and strong organization of others, including the later General Motors–North American Aviation organization. With further acquisitions it developed an unwieldy structure, at one time possessing a total of fifty-six directors, and its activities were constantly fluctuating even apart from government-imposed changes that affected all aviation holding companies. At one point AVCO even proposed to acquire control of North American, with a view to combining their air transport interests principally as a defense against the looming threat of E. L. Cord.

Cord controlled over 30 percent of AVCO stock by November 1931.[52] Cord opposed, among other things, the North American merger and, after a bitter and protracted public proxy battle, gained board representation on November 13, 1932. By March 1933 Cord had gained complete control, forcing out the Harriman-Lehman Brothers interests.[53] AVCO then became a subsidiary of Cord Corporation. The major attraction for Cord was airmail contracts, a field he had been pursuing for several years. He brought in the financier La Motte T. Cohu as president of AVCO, replacing Richard Hoyt, earlier with Curtiss-Wright, and manufacturing within AVCO was soon discontinued. Cord also hired C. R. Smith to run American Airways, founded in 1930 under AVCO. It became independent in 1934 as American Airlines.

Fairchild, having left his Farmingdale plant and airplane and engine operations with AVCO, relocated his headquarters to Hagerstown in 1932. From 1931 the Kreider-Reisner Aircraft Company was the only aircraft-manufacturing unit of Fairchild but remained operationally autonomous. Kreider-Reisner added to its accomplishments with the XC-31, the first purpose-designed military cargo aircraft. The largest single-engined airplane to that time, the XC-31 first flew on September 22, 1934. While a notable effort, the C-31 experienced stability problems and was not ordered by the Air Corps.[54] It was to lead, however, to further specialized military cargo aircraft developments. Also in 1934, Fairchild Aviation Corporation regained the aerial camera, survey, and engine operations.[55] All operations were in Hagerstown except for the renamed Ranger Engine Division on Long Island. The Fairchild F-24 cabin monoplane of 1932 was the major aircraft production program and achieved success in both civil and military markets.[56]

Aviation Corporation reentered aircraft production in 1934 when it established Aviation Manufacturing Corporation (AMC), based in Chicago. AMC was built on AVCO's acquisition of Jerry Vultee's Airplane Development Corporation, a Cord subsidiary located in Glendale, California.[57] Vultee's V-1 eight-passenger transport, designed in cooperation with Vance Breese, was a successful design. But the day of the single-engine passenger transport was largely over, and the V-1 won few orders, only some twenty-five being produced. Vultee then redeveloped the design into the V-11 attack aircraft for the Air Corps, but it also won few orders.

After a 1936 reorganization AMC consisted of three divisions, Vultee's Airplane Development, then of Downey, California, Stinson Aircraft of Wayne, Michigan, and Lycoming Aero-Engine of Williamsport, Pennsylvania. American Airplane and Engine activities were phased out, and with American Airways out, airmail activities also were lost.[58] Cord's aviation interests now were fully integrated within AVCO. Further changes loomed, however, as Cord suffered financial reverses both in automobiles and aviation due to the depression and began to divest his AVCO interests.

Curtiss-Wright

The Curtiss-Wright merger, a transforming event in the development of the aircraft industry, was conceived by Clement Keys and completed on August 29, 1929, with Curtiss A and M merging with Wright Aeronautical. Keys regarded the combination as an effective counter to the United Aircraft and Transport threat. Along with UATC and later Fairchild, Curtiss-Wright was the most vertically integrated of the aviation combines. Subsidiaries included Curtiss-Robertson, Curtiss-Caproni, Keystone, Loening (through Keystone), Moth, and the personal-aircraft builder Travel Air, through affiliation with Wright Aeronautical. Guy A. Vaughan, though opposed to the merger, remained to succeed Richard Hoyt as president of Wright in 1930, but Charles A. Lawrance resigned for a new venture. Curtiss engine operations moved to Paterson, combined with those of Wright. Keys became chairman and Hoyt president of Curtiss-Wright. By 1929 Clement Keys was the most prominent figure in both the aircraft and airline industries, although enjoying that status only briefly.

Neither Glenn Curtiss nor Orville Wright had any involvement with the firm that carried their names. Bitterness remained, but more by Wright than Curtiss. Near the end of his life Curtiss wrote Orville Wright to attempt a reconciliation; Wright never acknowledged the letter.[59]

Curtiss-Wright organization continued to evolve, and the number of divisions declined steadily, but the company remained the largest aircraft firm during most of the 1930s. The Airplane Division designed the A-8 attack aircraft in 1930, the first Curtiss monoplane combat aircraft, which led to the production A-12 Shrike of 1934. The popular Robertson-developed Curtiss Robin ended production in 1932, and the Curtiss-Caproni subsidiary closed. Keystone, having lost its bomber line to Martin, shut down its Bristol factory, transferred remaining work to Buffalo on September 1, 1932, and was absorbed completely into the parent company in 1934. The Condor airliner, developed from the B-2 bomber, was designed by the St. Louis Division and first flew on January 30, 1933. Only twenty Condor airliners were built, but their production kept the division, intended for the civil market, open.[60]

A particular emphasis of Curtiss-Wright as well as others during this period was exports to China, regarded as the most lucrative export market.[61] The Intercontinent Corporation was set up in China in 1934, originally as a subsidiary of Curtiss-Wright, for aircraft manufacturing.[62]

North American Aviation

North American Aviation was another bold conception of Clement Keys. He began its formation in 1928, before implementing the Curtiss-Wright merger. North American represented one of the first instances of a major aircraft firm organized by financial interests rather than by a pioneer aviator or engineer. Keys, however, was as committed to the industry as any aviator or engineer and was determined to develop a strong structure. His plan was that North American should not become another aircraft manufacturer but would operate as a holding company investing in a range of aviation interests. In that respect it was similar to AVCO. Incorporated in Delaware on December 6, 1928, North American Aviation was owned 70 percent by Curtiss, 12 percent each by Transcontinental Air Transport (later TWA) and Curtiss Flying Service, and 6 percent by Douglas and had a capitalization of $25 million.[63] Keys and Douglas served on each other's boards.

North American Aviation's first acquisitions in 1929 were Sperry Gyroscope, founded in 1910, and the passenger and airmail operations of Pitcairn Aviation Company, including flying schools and associated services, founded by autogiro pioneer Harold Pitcairn. Pitcairn felt that the airline industry eventually would be concentrated, and he decided to sell when most advantageous. The deal, completed on June 12, 1929, for $2.5 million, provided a major profit for Pitcairn.[64] Keys then resold the operations to NAA. Pitcairn's manufacturing arm was not included. The flying services became Eastern Air Transport on January 1, 1930.

In 1930 NAA also acquired the Ford Instrument Company and Berliner-Joyce Aircraft, originally founded by Henry A. Berliner in 1926. Temple N. Joyce, an Air Service pilot earlier with Chance Vought, invested in Berliner in 1929 and became president. Also in 1930, NAA expanded its holdings into air transport, with control of Transcontinental Air Transport and Western Air Express (to be combined as Transcontinental and Western Air with Clement Keys as chairman). The holding company was located in Dundalk, Maryland, at the site of the Berliner-Joyce factory, but production activity was very limited, the major output being the P-16 pursuit.[65] Other investments included Curtiss, Curtiss-Wright Flying Service, and even United Aircraft and Transport. By the end of 1930 NAA's investments were valued at $5,850,000, but for which it had originally paid some $15 million.[66] Encountering increasingly serious managerial disputes after the TAT/WAE merger, Keys resigned as chairman of the airlines in January 1931. Harold E. Talbott Jr. replaced Keys but resigned himself in October 1932. Under GM pressure, and with his health concerns, Keys then resigned as chairman of North American on December 31, 1931.

Meanwhile General Motors' GAC expanded its aviation interests, spurred by Charles F. Kettering, a director of both GM and UATC. That expansion led to the 30 percent interest in NAA, thereby giving GM interests in Berliner-Joyce and Fokker, and in the transport companies TAT, WAE, and Eastern. Operating units of North American thus became Berliner-Joyce (B/J), and General Aviation Manufacturing Corporation (GAMC). GM earlier closed the Hasbrouck Heights factory and returned the license to Fokker. Both units' products were traded under the designation GA, for General Aviation, and GM executive Ernest R. Breech served as chairman, succeeding Keys. The original NAA structure was not to be long lived, however, as the Air Mail Act of 1934 forced, among other things, the separation of manufacturers from air transport companies.[67] Although thinking that the holding-company organization might exempt it, and reluctant to lose its lucrative airmail subsidies, GM nonetheless disposed of all its aviation interests except for Eastern and its NAA stock. GM did not dispose of its interest in Eastern until March 31, 1938. The transport units eventually became Trans World Airlines (TWA) and Western Airlines.

By 1934 the initially ambitious North American Aviation was not faring well. The combination of NAA with GAC was accomplished on April 28, 1933, with NAA purchasing GAMC from GAC by means of a direct stock swap.[68] Then in June 1933 GAMC was merged with B/J. With Fokker production suspended, GAMC was reduced to only two hundred employees. North American now had to determine if it had a future in either transport or manufacturing.

In a move with a lasting impact on North American, General Motors hired James H. "Dutch" Kindelberger from Douglas on July 13, 1934, as president and general manager of GAMC. Kindelberger in turn hired his Douglas engineering colleagues J. L. "Lee" Atwood and J. S. "Stan" Smithson. In December 1934, GM separated North American from GAC, which was liquidated. Kindelberger was made president and managing director, while Breech continued as chairman. NAA sold its Douglas stock in 1934, while its holdings of Curtiss-Wright, Sperry Gyroscope, and Ford Instrument were exchanged for stock in the newly formed Sperry Corporation.[69] Temple N. Joyce departed as president of GAC, becoming president of Bellanca in 1935. But General Motors continued its aviation involvement by acquiring Allison Engineering, an engine manufacturer, in 1929, and also investing in Bendix Aviation. Allison was to challenge Pratt and Whitney and Wright in engines.

North American Aviation, largely in reaction to the Air Mail Act of 1934, decided to concentrate on manufacturing. Kindelberger also decided to move the corporation from Maryland to Inglewood, California, in March 1935, along with some seventy-five employees, completing the six-year transition from holding company to manufacturer. North American had little active production left from former B/J and Fokker/GAMC operations, but it did have a trained and highly skilled labor force. It occupied leased space while formulating a market strategy. Kindelberger felt that North American could not compete with such established giants as Douglas, United, Martin, and Boeing in large aircraft and decided to concentrate on small trainers to win military business.[70] NAA earlier had dropped a trimotor airliner project to succeed the Fokker models but completed design of the GA-15 observation aircraft and the GA-16 open-cockpit trainer before the move. Each led to further developments, as the

GA-15 became the Air Corps O-47, and the GA-16 led to a series of successful trainers including the famed wartime AT-6. Thus were the modest beginnings of what would become a brilliant thirty-year record in military aircraft.

New Independent Firms

Lockheed Redux

A young San Francisco investor with aviation interests named Robert E. Gross, originally from Boston, was made aware of aware of the Lockheed opportunity by Carl Squier. Gross had already demonstrated considerable financial acumen, making a substantial fortune in the stock market in the 1920s. His first involvement with aviation came in 1927 with a trip to Wichita for his employer, Lee, Higginson and Company, to evaluate Lloyd Stearman as a financing client. Lee, Higginson rejected Stearman and his small company, but Gross then resigned and personally invested twenty thousand dollars in Stearman. With his younger brother Courtlandt he also formed the Viking Flying Boat Company in Connecticut in 1928, to manufacture and market sport flying boats from a French design. This venture declined in the depression, however, with Gross suffering severe losses. In San Francisco, Gross also took a position in Varney Speed Lines and assisted in the sale of Stearman to UATC in 1929.

Gross lacked available cash personally but assembled a small group of investors, including Varney, who purchased Lockheed's assets for forty thousand dollars on June 16, 1932, directly from bankruptcy court and immediately set about to revive the company.[71] Staking his future on the venture, Gross chartered the new Lockheed Aircraft Corporation with himself as treasurer, Carl Squier as vice president for sales, and Cyril Chappellet as secretary.[72] Stearman, having earlier left UATC, was appointed president, largely because of his name recognition in the industry. Initial employment was fifteen persons, and cash on hand amounted to eight hundred dollars. Gross retained Richard von Hake and soon added the gifted Hall Hibbard, previously with Viking. Then in 1933 he hired a young engineer who would, more than any other, make Lockheed's design reputation: Clarence L. "Kelly" Johnson. Courtlandt Gross became Lockheed's eastern contact as well as liquidating remaining activities of Viking.

The Lockheed nameplate possessed wide recognition, primarily with the Vega, but that meant little in the market of 1932. The revived company reestablished existing designs and turned a profit of $25,000 on sales of

$355,000 in 1933. Gross realized, however, that he could not rely on the Vega and the Star series much longer, with more advanced designs coming from competitors. He then decided in 1933 to develop an advanced, small airliner to fill a market niche below the larger Boeing 247 and Douglas DC-1 then being developed. Lloyd Stearman had begun such a design before associating with Lockheed. The company also foresaw that the future lay with multiengined designs and ended single-engined production in 1934. The Model 10 Electra made its first flight on February 23, 1934, although suffering minor damage on landing. Market success of the Electra led to a long line of Lockheed civil aircraft.[73]

Development costs of the Electra were of such magnitude that Gross was forced to obtain an RFC working capital loan, and began to consider military business as a potential offset. Capital was also raised by stock sales, initially through Sutro and Company, but Brashears and Company was employed from June 1933. Varney eventually sold his interest, and Stearman, somewhat isolated from the other Lockheed executives, resigned as president on December 15, 1934, moving to the Bureau of Air Commerce. Robert Gross, although a minority stockholder, then assumed the office of president, and Courtlandt Gross became general manager.[74]

Industry Veterans Reenter
Grover Loening's 1928 sale agreement prohibited him from entering direct competition in aircraft, and he formed Loening Laboratories in 1929 to specialize in research. He then founded Grover Loening Aircraft Company in Garden City, Long Island, in 1932, and remained active until 1938, producing more amphibians and subcontracting for others. Claude Ryan, flush with the seventy-five thousand received from sale of his company to Siemens, remained in business with his flying school.[75] He eventually formed Ryan School of Aeronautics on June 5, 1931, followed on May 26, 1934, by a new Ryan Aeronautical Company, with the school eventually as a subsidiary.[76] With this organization he was well positioned for the war period. His successful S-T sportplane and trainer first flew on June 8, 1934. Ryan was among the first to recognize the value of subcontracting to maintain stability during slack periods. Later all aircraft manufacturers would subcontract for others to some degree.

In 1930 Allan Lockheed, having departed his company after the Detroit acquisition, formed the Loughead Brothers Aircraft Corporation, reverting to the original spelling to avoid confusion with the earlier firm. His brother Malcolm lent his name but did not actively participate. The company lasted only until 1934, as the innovative Duo-6 transport crashed. Ironically, Allan Lockheed had attempted to raise capital to repurchase his original firm from bankruptcy court in 1932 but felt that it would require at least one hundred thousand dollars, and so gave up the attempt.[77] Malcolm Lockheed sold his company to Bendix in 1932, while Allan reentered aviation with the Alhambra Airport and Air Transport Company. Then in February 1937 he formed ALCOR (for Allan Lockheed Corporation). He developed the C.6.1 twin-engined transport from the earlier Duo-6 concept, but the crash of the prototype brought the venture to an end in June 1938. Allan Lockheed emerged as general manager of a small aircraft parts plant in Michigan during the Second World War but essentially was out of the industry afterward.[78]

John Northrop launched his new Northrop Corporation in January 1932, first locating in the old Moreland Aircraft plant in Inglewood. Douglas provided major financial backing by taking a 51 percent ownership position. Ken Jay again joined Northrop as vice president, dealing with finance and administration. Northrop made major contributions to Douglas designs, most notably the DC-1, and his own Alpha, Delta, and Gamma mailplanes and transports were particularly noteworthy. The engineer Don R. Berlin joined, contributing to the series. Of greater significance for both Northrop and Douglas, however, was the affiliation of Ed Heinemann. Heinemann, originally hired by Douglas in 1926 only to be laid off in 1927 during a slack period, was working for Standard Oil of California when hired by Northrop. He became chief engineer in 1936, and went on to become one of the greatest aircraft designers.

Glenn H. Curtiss with his "hydroaeroplane" prior to the first successful flight made from the water, San Diego, California, January 25, 1911

H-16 and F-5L seaplanes under construction at Naval Aircraft Factory, Final Assembly Plant No. 2, Philadelphia, October 21, 1918

Eddie Hubbard and William Boeing with a Model C floatplane, 1919. (Reprinted with permission of The Boeing Company Archives.)

Left to right: William Hawley Bowlus, a Ryan Company factory manager; Benjamin Franklin Mahoney, a Ryan Company partner; Captain Charles A. Lindbergh, Missouri Air National Guard; and Donald Hall, Ryan Company's chief engineer, prior to Lindbergh's transatlantic flight in the Ryan Company's *The Spirit of St. Louis*, circa 1927

Lockheed Orion fuselage production, Burbank, California, circa 1930. (Reprinted with permission of Lockheed Corporation.)

Announcement

ON JUNE 16, 1932, a group of men with wide experience in the aviation field purchased all assets, good will, trade name, and manufacturing facilities of Lockheed Aircraft Company. A new company has been formed with ample resources to carry on the development and production of the famous Lockheed line of airplanes and to insure leadership in the field of high speed air transportation.

• The active management of the company will be in the hands of Lloyd Stearman, President; Carl B. Squier, Vice President; Robert E. Gross, Treasurer; and Cyril Chappellet, Secretary. The board of directors includes Walter T. Varney, Thomas F. Ryan III, Lloyd Stearman, Robert E. Gross, Robert C. Walker, and Cyril Chappellet.

• The new company solicits your interest in new orders, spare part and repair work on all Lockheed models.

LOCKHEED AIRCRAFT CORPORATION

July 1, 1932   Burbank, Calif.

Announcement of the formation of Lockheed Aircraft Corporation, July 1, 1932

Boeing 247 airliners, the first low-wing, multiengine transports, in assembly, circa 1933. (Reprinted with permission of The Boeing Company Archives.)

Lockheed Aircraft Corporation executive personnel, Burbank, California, July 26, 1934. *Left to right:* Assistant Treasurer Ronald King; Vice President for Sales Carl Squier; President Lloyd Stearman; Chairman of the Board and Treasurer Robert E. Gross; Secretary Cyril Chappellet; Vice President and Chief Engineer Hall Hibbard. (Reprinted with permission of Lockheed Corporation.)

Grumman Corporation's Bethpage, Long Island, New York factory, 1938. (Reprinted with permission of Grumman Corporation.)

Leroy Grumman and Leon A. "Jake" Swirbul, circa 1940. (Reprinted with permission of Grumman Corporation.)

B-25 production at North American Aviation's Inglewood, California, plant, circa 1941. (Reprinted with permission of Rockwell.)

Boeing B-29 Superfortresses on the Wichita, Kansas, flight line, each waiting for a crew and a mission during the Second World War. (Reprinted with permission of The Boeing Company Archives.)

Edgar Schmued, James H. "Dutch" Kindleberger, and J. L. "Lee" Atwood of North American Aviation, circa late 1940s. (Reprinted with permission of Rockwell.)

David S. Lewis, James S. McDonnell, Donald Douglas Sr., and Donald Douglas Jr. at the McDonnell Douglas merger, April 28, 1967. (Courtesy of McDonnell Douglas.)

William M. Allen boarding a Boeing B-47 jet bomber, circa 1950s. (Reprinted with permission of The Boeing Company Archives.)

James S. McDonnell with McDonnell Douglas's 5,000th Phantom jet fighter, circa 1978. (Courtesy of McDonnell Douglas.)

McDonnell Douglas F-4s in production at the St. Louis, Missouri, factory, circa 1970s. (Courtesy of McDonnell Douglas.)

Grumman F-14 production, 1986. (Reprinted with permission of Grumman Corporation.)

C-130H final assembly, Lockheed Martin Marietta plant, 1994. Note P-3C final assembly in background. (Reprinted with permission of Lockheed Corporation.)

Chapter 5 Progress amid Struggle
The Mid-1930s

The design revolution, a more stable structure, and a trickle of orders gave the industry hope even in the face of continuing economic troubles. Three further, concurrent developments, in airmail, government investigations, and airliners, were of particular significance for the future.

The Airmail Uproar
On February 9, 1934, after almost seven years of exclusively commercial airmail operations, President Roosevelt revoked all airmail contracts due to charges, primarily by smaller operators, of widespread collusion in contract awards. By that time three firms, UATC, AVCO, and GM/NAA, controlled some 90 percent of airmail contracts.[1] The relationship between the Hoover administration postmaster general, Walter F. Brown, and the air transport industry was highly suspect. The postmaster general, with authority over airmail contracts, effectively controlled commercial aviation, and the Air Mail (Watres) Act of 1930, which encouraged carrying passengers by airmail lines, extended that authority. The new act had been opposed by Representative Kelly, author of the earlier act, as centering too much authority on the postmaster general. The Senate Special Committee to Investigate Air Mail Contracts, under Senator Hugo Black of Alabama, reported at the end of 1933 and provided Roosevelt with the basis for contract cancellation.

Roosevelt subsequently ordered the Army Air Corps Mail Operation (AACMO). Major General Benjamin Foulois, chief of the Air Corps since 1931 and a tenacious fighter for its interests, sought the assignment, feeling it would support Air Corps budget requests, but that proved to be a major error as well as a political crisis for Roosevelt. The AACMO experience was tragic, due to severe weather, lack of specialized training, and obsolete aircraft that did not even possess the Sperry turn and bank indicator. Twelve pilots lost their lives before commercial service was restored. Roosevelt appointed a Federal Aviation Commission, chaired by the publisher Clark Howell, with Edward P. Warner as vice chairman, to resolve the situation. A new airmail act resulting placed civil aviation under three agencies. The U.S. Post Office awarded all mail contracts and set routes and schedules, the Interstate Commerce Commission fixed rates and payments, and the reorganized Bureau of Air Commerce controlled airways and licensing.[2]

The airmail act became effective on June 12, 1934, and commercial airmail was back in full operation by July 1. The market for new airliners capable of carrying the mail in addition to passengers, rather than the other way around as in the 1920s, expanded. Mail service contracts also provided a sounder financial base for the development of passenger transports as well as markets for the manufacturers. Perhaps the most far-reaching provision of the act, however, was that a single corporation could no longer control both aircraft production and air transport.[3] That arrangement had been held responsible for excess concentration and insufficient competition in the industry, and forced the reorganization of UATC and others. Almost lost in the flurry of new activity was that the original charges of collusion between the Post Office and the industry were dismissed.

Further Investigations

Rear Admiral W. A. Moffett, chief of the Bureau of Aeronautics, was killed in the crash of the airship Macon in 1933. His successor was Rear Admiral Ernest J. King, highly regarded by President Roosevelt, a former undersecretary of the navy. King was to serve as chief of naval operations in the coming war. Efforts of Foulois and King notwithstanding, industry hopes of greater support from Roosevelt soon were dashed. Severe financial constraints curtailed procurement, and the overall economic outlook for the industry was grim. The lack of proprietary design rights remained among the most pressing issues.

The airmail fiasco, despite the attention focused on the Air Corps, did not bring it any immediate benefit. Instead, General Foulois was criticized for its state of preparedness, and the situation did little to enhance congressional support for its equipment and training needs.[4]

Economic conditions had curtailed reequipment of the army and navy air arms during 1929–32 to far less than had been envisioned. But Boeing enjoyed a particular distinction during this period in that its P-12/F4B fighter series was produced in larger numbers, a total of 586, than any basic type of military aircraft since the First World War.[5] More advanced military types also found their way into the inventories of the army and navy. Grumman's FF-1 shipboard fighter entered service in 1933, and Boeing delivered its P-26 pursuit monoplane also in 1933. The Martin B-10 of 1934, more than one hundred of which were delivered, was an advanced design that brought bomber speeds to in excess of 200 mph, challenging the best fighters of the period.

General Douglas MacArthur, army chief of staff, earlier negotiated an agreement in January 1931 with Admiral W. V. Pratt, chief of naval operations, in which the Air Corps would have responsibility for air and coastal defenses of the United States.[6] The navy thus was free to undertake offensive operations, using the capabilities of the new aircraft carriers, while the Air Corps could use its capable new Martin B-10.[7] The air arms of the respective services still defined their separate missions, but the agreement led to an Air Corps requirement for longer-range aircraft, and to a role beyond support of army field forces. Thus the MacArthur-Pratt agreement did not end conflicts, and new controversies over aviation and coastal defense erupted.

Foulois regularly visited the major manufacturers as chief of the Air Corps, encouraging them to continue development even when production funding was nonexistent.[8] He later expressed satisfaction that despite budgetary cutbacks of the time, new developments from 1933 had led to the B-17 and B-24 heavy bombers of the Second World War.[9]

Concurrently with investigations into airmail contracts and the Air Corps fatalities in flying the routes, the Baker board was constituted by the War Department in April 1934 under the chairmanship of former secretary of war Newton D. Baker to examine all phases of aviation. General Foulois was a member. Reporting in July 1934, its findings were consistent with those of earlier investigations, and while they set back again the idea of an independent air force, still led to the official establishment of the GHQ Air Force, on March 1, 1935. For the first time field control of air units was removed from army corps commanders, leading to the independent employment of air forces rather than their restriction to support for the field forces. The Baker board gave particular attention to Air Corps aviation and national defense and recommended a new five-year procurement plan for the Air Corps, to begin in 1935, so that new orders began to flow.

Concurrently, the Rogers subcommittee of the House Military Affairs Committee began to investigate Air Corps procurement practices and relationships with the industry. Always suspicious of fraud and conspiracy on the part of the industry, the subcommittee subjected Foulois to criticism for the practice of direct negotiation with manufacturers for new aircraft. Although the Air Corps Act of 1926 authorized the secretary of war to purchase aircraft with or without competition, by contract or otherwise, Congress insisted on competitive bidding and acceptance of the low bid. The environment remained unchanged from the 1920s. Foulois adamantly maintained that a low bid from a manufacturer did not mean competence, that he knew the capabilities of the various companies, and that direct negotiation on a cost-plus basis would result in greater value to the taxpayer.[10] Further, he emphasized that the Air Corps only recommended approval, and that only the secretary of war could officially enter into a contract. Foulois eventually was cleared of any illegalities or conflicts of interest, but acrimoniousness with Congress eventually led to his forced retirement at the end of 1935, a step he accepted in order not to jeopardize future appropriations.[11]

The 1934 Delaney hearings into naval affairs also affected the industry. These hearings focused on the apparent lack of competition within the industry for naval procurement contracts for both aircraft and engines.[12] The engine industry consisted only of Wright and Pratt and Whitney at the time. United Aircraft and Transport, the parent of P & W, was a major target. The term "excess profits" was heard frequently. The fact that airframe contracts were less profitable than engine contracts was credited to the restraining influence of the Naval Aircraft Factory.[13] Overall, however, the Delaney hearings largely vindicated the navy and its procurement practices and recognized that military aircraft were unlike other commodities.[14] The old proposal for an army aircraft factory was also revived.

There were in fact some procurement reforms in 1934 to ensure more competition, and accordingly no new legislation resulted.[15] The Federal Aviation (Howell) Commission, also reporting in July 1934, presented little contradiction to the Baker board, but said little about strengthening the Air Corps, which still benefited from the added publicity.[16]

The Modern Airliner

The first successful American multiplace airliner was the Ford, originally Stout, "Tin Goose," and principal air transports from the late 1920s were Ford, Fokker, and Stinson trimotors. The single-engined Boeing 40-A of 1927 was the first combined mail and passenger carrier, offering coast-to-coast service in twenty-seven hours. The Boeing Model 80 trimotor of 1928 carried up to eighteen passengers, but the passenger market still was very small. In 1930 the civil air transport industry employed a total of 640 aircraft, of which four hundred were single-engined.[17]

A major advance came in 1933 with the Boeing 247, a twin-engined all-metal ten-passenger craft. United, then affiliated with Boeing under UATC, ordered a fleet in 1932, conveying a major competitive edge over other airlines.[18] The Model 247, a streamlined design with retractable landing gear, had enough power to climb on one engine with a full load and was also the first airliner to use wing flaps. The Model 247 drew on the technology of Boeing's B-9 bomber of 1931, which while experimental, possessed such advanced features as all-metal construction and retractable landing gear. This linkage between military and commercial design would serve Boeing spectacularly well twenty years later when it developed the 707 jet airliner.

Transcontinental and Western Air (TWA) in the meantime suffered a major tragedy on March 31, 1931, when the crash of a Fokker F10 killed all aboard, including the legendary football coach Knute Rockne. Investigation by the Bureau of Air Commerce pointed to structural deficiencies with the Fokker, and its reputation was irretrievably damaged.[19] Additionally, Ford suffered major losses on its trimotor and ended production in 1932.[20] With that background, Jack Frye, vice president of TWA, developed specifications for a new trimotor airliner and solicited bids by Martin, General Aviation, Consolidated, Curtiss-Wright, and Douglas.[21] Frye felt enormous pressure to regain the lead from United and the Boeing 247.

Donald Douglas studied the Frye letter, dated August 2, 1932, which contained the airliner specifications. At that time he was motivated toward commercial business largely as an offset to lean military orders: Douglas delivered only seventy airplanes in 1932.[22] Douglas had not devoted major effort to the commercial market to that point but had developed the twin-engined Dolphin flying boat in 1929, which found a market with millionaires of the day. Even William E. Boeing became an owner. Harold E. Talbott Jr., then with North American Aviation and a friend of Douglas, also encouraged him to enter the commercial market, even though he was potentially an indirect competitor through the former Fokker operations within GM/NAA.[23] North American, incidentally, had not yet sold its eighty-nine thousand Douglas shares.

Douglas was awarded the contract, and the company was also rewarded with a vote of confidence by the stock market, as the stock more than doubled on the news.[24] The Douglas team was confident that it could meet the specifications, but Dutch Kindelberger insisted on a twin-engined design rather than a trimotor, to that point the preferred arrangement.[25] The design for the Douglas Commercial model 1, or DC-1, met with the approval of TWA. But the first flight of the DC-1 prototype, on July 1, 1933, very nearly ended in disaster as the engines began to sputter and lose power during climbs. Investigation showed that the problem was in the carburetor floats of the Wright Cyclone engines, which were placed to the rear and tended to cut off fuel flow during a climb.[26] After modification, the performance of the DC-1 met or exceeded all expectations.

The production model was the improved fourteen-seat DC-2, which appeared in May 1934. TWA ordered twenty-five DC-2s that it first placed into service on the Newark–Los Angeles route, advertising coast-to-coast in eighteen hours. The DC-2's new controllable-pitch propellers enhanced speed and altitude performance, and its direct operating costs were one-third lower than those of the trimotors. Critical for the future of Douglas, the DC-2 also was profitable. The first DC-1 prototype had cost Douglas $307,000, a financial strain at that time, and the TWA order was fulfilled at a price of sixty-five thousand dollars each, excluding engines, but the program quickly reached breakeven at 75 aircraft.[27] Production amounted to 138, an impressive number for the time.[28] Boeing, by comparison, produced only seventy-five Model 247s, as the DC-2 literally drove it from the market. Most major U.S. airlines equipped with DC-2s, as did those of twelve other countries. It served in the Air Corps as the C-32, C-33 and C-39, and a further development of the basic design was the B-18. By the time the DC-2 entered production, the main plant at Clover Field had expanded to 350,000 square feet and nine hundred employees, while the Northrop subsidiary, in nearby El Segundo, had 140,000 square feet and some one thousand employees.[29] Thus there was more than ample production capacity to meet market demand. Arthur Raymond, who had succeeded Kindelberger as vice president of engineering, led development and production of the DC-2. Douglas Aircraft Company, Inc., enjoyed record sales of $5.3 million in 1934.[30]

Meanwhile, American Airlines had on May 12, 1934, inaugurated its first "sleeper" transport service, using Curtiss Condors. The Condor, a biplane design that entered service in 1933, was hardly as advanced as the DC-2 or the Boeing 247, but the sleeper version had seats capable of being converted into six berths for night flights. C. R. Smith of American approached Douglas with a proposal to develop a sleeper version of the DC-2 to succeed the Condors.[31] Douglas thereupon developed the Douglas Sleeper Transport (DST), an enlarged and improved DC-2. The day transport version was designated DC-3 and first rolled out of the factory on December 17, 1935, exactly thirty-two years after Kitty Hawk. The DC-3 perhaps more than any other airplane was to have the adjective *legendary* applied, figuratively teaching an entire industry to fly. Airline progress in the mid-1930s was speeded by the Boeing 247, the DC-2 and DC-3, and the Lockheed L-10 Electra, which collectively replaced old trimotors and the Curtiss Condor.

Douglas, for fourteen years almost exclusively dependent on the government, became the major commercial-aircraft manufacturer. Donald Douglas by that time also was a millionaire and possibly the most prominent individual in the industry. He had assembled a superb technical staff. He and his staff received the 1935 Collier Trophy from President

Roosevelt for the DC-2.[32] But Boeing, Douglas's chief competitor, had fallen on lean times. Loss of the airliner market to Douglas, the bomber order to Martin, and the looming completion of the P-26 pursuit program led to a decline in employment from more than two thousand to only six hundred by September 1934.[33] While Boeing pursued advanced airliner developments, its future appeared to depend on the then-dubious four-engined B-17 bomber.

The State of the Industry
The years 1933 and 1934 were very difficult for the military and the industry, yet certain developments still carried a positive long-term impact, much as developments in the lean years of 1924 and 1925 led to the eventual strengthening of the industry, military aviation, and air transport. The political ferment of 1933 and 1934, combined with rapid aeronautical progress and ominous world developments, marked a major turning point.

By the close of the 1927–35 period the aircraft industry, although small, financially shaky, and operating in the Great Depression, had transformed remarkably. The short-lived aviation boom ended as rapidly as it had begun, and investment interest had largely eroded, but the industry was now recognized as important and vital to security. In management, it had made the transition from the pioneer days to boardroom decision making. Pioneer aviators remaining active in the industry now were more likely to be corporate chief executives. The design revolution produced many world-leading aircraft, yet the industry remained the object of considerable suspicion and controversy regarding profits on military contracts. Such companies as Curtiss-Wright, Douglas, and United were indeed profitable, leading to imposition of profit limitations.

In foreign policy, the United States remained largely isolationist, and proposals for a major military buildup found little political support. President Roosevelt still was not regarded as a strong supporter of the air arm, given his naval background. Procurement and R&D support lagged behind industry and military needs, yet potential threats to the peace were emerging: Italy had invaded Ethiopia, Germany and Japan were arming rapidly, and in March 1935 Hermann Göring revealed the rebirth of the Luftwaffe.

While the industry was starved for military orders, commercial-airliner progress was heartening, and new military firms formed by end of the 1930s would have a major impact on war production and after. Bell in 1935 was followed by Vultee and McDonnell in 1939. An independent Northrop and a new Republic reorganized from Seversky also were incorporated in 1939. The new firms, joining the established Boeing, Consolidated (later merged with Vultee), Curtiss-Wright, Douglas, Fairchild, Grumman, Lockheed, Martin, North American, Ryan, and United Aircraft, constituted the engineering and production base on which the war industry would build its record.

Despite inadequate support for the industry in the mid-1930s, a remarkable cadre of military leaders fought to keep the production lines open and the companies healthy. Arnold, MacArthur, and Foulois were principal among them. The Air Corps "Educational Order" procurement policy, in which available work was spread among several firms to keep their facilities active and updated, began in the 1930s.[34] The government issued letter contracts to the industry to start production, and to make money available.

Thirty years after its beginning with the first army order, after participation in a world war, after rapid progress in aeronautics, and after gaining general public acceptance, the aircraft industry still had little impact on the overall economy. Total employment in 1938 was only thirty-six thousand, and sales were $131 million.[35] The value of annual output was less than that for the bicycle industry. The reality, before the start of the military buildup, was that the typical firm in the industry was wholly dependent upon the next order for survival. Martin and Boeing especially experienced periodic losses and even had difficulty meeting payrolls.[36] By the end of 1938 the cumulative production total of the American industry was about fifty-five thousand aircraft.[37] Soon that figure would be exceeded in a single year.

Nineteen thirty-eight was also the last year in which aircraft would be

TABLE 6. Prewar Production and Exports of Aircraft

Year	Aircraft Produced	Aircraft Exported	Value of Exports ($)
1931	2,552	140	1,812,809
1932	1,422	280	4,358,967
1933	1,353	396	5,389,739
1934	1,699	490	8,258,484
1935	1,902	334	6,638,515
1936	3,200	500	11,299,451
1937	3,859	629	21,036,361
1938[a]	2,873	616	30,936,642

Source: Army Air Corps, *Aircraft Production*.
a. January 3–September 30.

considered a small industry. The industry expanded rapidly from 1939 and, except for the brief demobilization period following the Second World War, would remain one of the largest manufacturing industries. With expansion and increasing exports, the aircraft industry also became a component of the international munitions industry, enmeshing it in political controversy of a different nature from that it had experienced heretofore. Disputes between the industry and the government over contracts, costs, and profits continued. There were also those who labeled military-aircraft manufacturers as "merchants of death," due to exports to unstable or warring countries, although critics tended to overlook the fact that no exports could take place without a government license.

Strikes and unionization drives in the aircraft industry were rare before 1937.[38] But legislation enacted in 1935 spurred widespread industrial organization, and from 1937 successful efforts were led primarily by the United Auto Workers (UAW) of the Congress of Industrial Organizations and the International Association of Machinists (IAM) of the American Federation of Labor. Compounding the generally unfavorable situation for workers at the time was that—as the Congress and the military were unsympathetic to the industry—the industry tended to be unsympathetic to workers and the unions. Major organizing drives still proceeded, centering in the Los Angeles area. Donald Douglas was especially noted for his hard line on unionization, although few firms welcomed unions. Aircraft industry wages actually tended to be higher than those in manufacturing overall, but that was a function of its structure, in that the major proportion of the labor force consisted of skilled and semiskilled workers.[39] The view still persisted in Congress that the industry, small size notwithstanding, was subsidized by taxpayers.[40] The industry, on the other hand, held that it was supporting the government with its research and design efforts, to which it held no proprietary rights. Robert E. Gross of Lockheed complained about government procurement policies to the Howell commission, advocating that experimental contracts should be separate from production contracts, since manufacturers otherwise could not recover their capital outlays. The industry thereby operated in an environment of profit limitation but unlimited potential losses.[41]

Those running the industry in the prewar period retained a major dedication to aviation even as they struggled to earn profits. Most felt a strong common bond in the adversity they all faced.[42] The industry was highly uncertain as an investment vehicle. Richard Barclay Harding, surveying the industry in 1937 for its investment attractiveness, formed eight general conclusions.

1. Air transportation companies are highly speculative.
2. Manufacturing companies are more predictable.
3. Engine manufacturing is relatively stable.
4. Military procurement policies are a vital factor.
5. Transoceanic transportation has a promising outlook.
6. Companies catering to private flying are generally profitable.
7. Export business is currently most profitable.
8. A short-term investment viewpoint is recommended.[43]

Grover Loening, in his survey of the industry at the end of 1938, emphasized that aircraft production as an industry was still very small, citing sales of some $115 million in 1937 and $140 million in 1938, of which half, $70 million, were from exports.[44] Loening also pointed out that the industry was not and never was comparable to the automobile industry, despite many efforts to compare the two. He further emphasized that the aircraft industry was a contract industry in terms of production, with large sums involved in an order to a single customer. Consequently it was subject to wide and sudden fluctuations.[45] Those characteristics were to remain. Operationally the industry was more comparable to construction than to automobiles, in that activity swung from contract to contract.

The Vinson-Trammell Treaty Navy Bill of 1934, enacted amid the flurry of investigations into the industry, established a firm policy that at least 10 percent of naval aircraft and engines must be government manufactured.[46] While primarily an appropriations bill, Vinson-Trammell had a major impact on the industry, including enabling the Naval Aircraft Factory to reenter production, in which it had been inactive for more than a decade.[47] The act specifically authorized production of both aircraft and engines.[48] Vinson-Trammell also authorized the *Yorktown* and *Enterprise* aircraft carriers and a five- to seven-year naval-aircraft procurement program, which strengthened naval aviation as war approached. At the time there were no statutory profit limitations on army aircraft, but a limitation of 12 percent was imposed in 1939, at which time the navy contract limit also was raised to 12 percent.[49] This limitation, reflecting official suspicion of the industry, remained troublesome.[50] Among other things it imposed a major audit burden on the government but essentially was inoperative since most development contracts resulted in losses, while production contracts were narrowly profitable, if at all. A further difficulty was that losses on development contracts could not be recovered by crediting them against any production profits. The two remained entirely separate. Whatever their flaws, the Air Corps Act and the Vinson-

Trammell act largely constituted the prewar operating environment for the industry. Further legislative efforts in the late 1930s to restore manufacturers' design rights were unsuccessful.

Just as export orders began to rise rapidly, especially from Europe, the outbreak of war in September 1939 led to a conflict between such exports and the American position of neutrality. The Neutrality Act of August 1935 authorized the president to prohibit sales of weaponry to belligerents. Consequently, military exports to combatant nations, even Canada, were embargoed, illustrative of how political positions could adversely affect the industry. During the 1937–39 period, the industry's output consisted of 40 percent military, 23 percent air transport, 10 percent civil, and 27 percent exports.[51] Largely consistent with Loening's estimates, the total value of aircraft output from January 1 to September 30, 1938, was $102,053,922, of which $30,936,642 arose from exports. This corresponded to 2,873 complete aircraft produced, of which 616 were exported.[52] Then the Neutrality Act was repealed in January 1939, followed by Roosevelt's executive order of November 4 that enabled aircraft to be allocated to friendly nations.[53] It is no exaggeration to say that exports saved the aircraft industry during this period.

Military Developments

Brigadier General Billy Mitchell died on February 19, 1936, only fifty-six but largely forgotten. The Second World War would validate his views, and he would gain posthumous recognition for influencing both air power and the industry.

On September 21, 1938, Major General Oscar B. Westover, chief of the Air Corps, was killed in a flying accident. Major General Henry H. Arnold was appointed to succeed him on September 29. Arnold, an unflinching supporter of the aircraft industry whose career closely paralleled the development of military and civil aviation, was to lead the air arm throughout the Second World War.

United States Naval Aviation developed a comprehensive structure; organization paralleled that of the army, extending even to basic research, flight training, and procurement. In addition to a civilian assistant secretary of the navy for air, there was a deputy chief of naval operations for air. The Bureau of Aeronautics reported to the deputy CNO for air. In 1939 Rear Admiral John Towers, the navy's most senior aviator, became chief, BuAer. Marine Corps Aviation remained an integral part of Naval Aviation.

A pronounced interservice rivalry, especially involving the air arms, continued. Conflicts over air missions and strategic responsibilities would persist well into the aerospace age. Although resulting in obvious duplication and overlap in design, procurement, and missions, the rivalry probably benefited the industry in that it resulted in a larger overall aircraft market than would have been the case with more rationalization or consolidation. The independence of naval aviation undoubtedly enhanced naval-aircraft performance and mission capabilities as well.[54]

The Italian air minister Italo Balbo, in another dramatic demonstration of the long-range potential of the airplane, led a flight of Savoia-Marchetti bombers from Italy to Chicago in June 1933. In recognition, the Air Corps formulated a specification for a Very Long Range Bomber (VLRB) in May 1934. This resulted in the experimental Boeing XB-15 and later the Douglas XB-19 but again sparked a protracted battle both within the army and with the navy over strategic missions. Roughly concurrently, a 1934 competition for an eventual replacement for the B-10 in the multiengined bomber mission resulted in Boeing's development of the B-17, the army's first four-engined bomber. Other competitors were the Martin B-12, an improvement of the B-10, and the Douglas B-18. The four-engined bomber project, spearheaded by Claire Egtvedt, represented a considerable risk to Boeing, since the prevailing preference was for twin-engined bombers; many in the Air Corps supported the smaller B-18 on the grounds that greater numbers could be had for the same price as smaller numbers of the larger model. In addition, the navy was protective of its long-range mission and did not look with favor on any potential encroachment by land-based aircraft with the capability of the B-17.[55] To add to Boeing's difficulties, the prototype XB-17 crashed on takeoff on October 30, 1935. It was determined, however, that the crash was not due to any design flaw, and the army continued the program. Smaller and less expensive bombers won the day, however, with the B-18 selected for quantity production. But General Foulois also managed to secure an order for thirteen YB-17s for service test, which kept the program alive.[56]

All thirteen YB-17s were delivered by the end of 1937, and successful tests led to a production order for thirty-nine B-17Bs in 1938, for operational service from 1939.[57] In February 1938 a flight of six B-17s flew from Miami to Buenos Aires and back in a dramatic demonstration of long-range aviation. Lead navigator for the mission was First Lieutenant Curtis E. LeMay. Then on May 12, 1938, a flight of B-17s located and intercepted the Italian liner *Rex* some seven hundred miles out in the Atlantic.[58] But once again a land-based long-range air demonstration upset the navy, which pushed through an agreement restricting Air Corps operations to within one hundred miles of the coastline. Still, the B-17 gave the United States its only true long-range aerial capability at the beginning of the war.

As military aircraft development intensified, the industry pressed its view that it subsidized the military and the government, rather than the reverse. Indeed, for the twenty-year interwar period Air Corps annual R&D spending averaged only $4 million.[59] The 1939 Air Corps R&D budget was only $5 million, and NACA also remained very small, with only some five hundred employees in 1939.[60]

While the need for a military expansion became increasingly recognized, the Air Corps simply had no experience with mobilization, and procurement restrictions further limited its ability to act quickly. A major step came in September 1938, when the Office of the Chief of the Air Corps (OCAC) held a Procurement Planning Conference, the main objective of which was to address the capability of the industry to meet aircraft production requirements in an emergency.[61] In common with many such efforts conducted lacking a clear emergency, nothing happened in direct result. The rapidly changing world situation would soon lead to definite measures, however. After the conference General Arnold ordered a survey of the industry regarding its capabilities to support an order buildup and was deluged with responses from firms eager for business.[62] There was a further OCAC conference in July 1939, in which procedures for industry cooperation and subcontracting were addressed.[63] The manufacturers again stressed the need for more information in advance of production commitments.

Military production in 1939 still lagged behind operating requirements. Only twenty-eight hundred military aircraft were delivered that year, of which twenty-two hundred were trainers, leaving only six hundred combat aircraft. Although the industry had grown steadily since 1935, by the end of 1939 the thirteen leading aircraft manufacturers had a net worth of only about $138 million and net working capital of some $60 million, still minuscule compared to the automobile and other major industries.[64] But increases in output and additions to production capacity were heartening. The industry produced 5,856 aircraft in 1939, although the majority were light personal planes, while potential capacity was estimated at fifteen thousand per year.[65] Production then began to build rapidly as the War Department placed record orders for aircraft on August 10, 1939, allocating $41 million for bombers alone.

Aeronautical Progress

Progress accelerated with the primacy of the monoplane and all-metal, stressed-skin construction. Features such as retractable landing gear and more powerful engines became commonplace, leading to major performance advances. The Allison V-1710 of 1935, the first large liquid-cooled aircraft engine, powered the new P-38, P-39, and P-40 pursuits, and Pratt and Whitney and Wright developed more powerful radial engines.

The wealthy young aviation enthusiast Howard Hughes, already an experienced pilot, determined to set a new world speed record for landplanes and commissioned a new aircraft in 1934 expressly for that purpose. Called the *H-1 Racer* and designed by Richard W. Palmer, earlier with Vultee, the aircraft raised the world record to 352 mph on September 13, 1935.[66] Modifying the *Racer* with extended-span wings and extra fuel tanks, Hughes established a new cross-country record of seven hours, twenty-eight minutes on January 19, 1937. The advanced configuration of the *Racer* was reflected in several wartime fighter designs, including the Focke-Wulf FW 190. Hughes, with a crew, also made an around-the-world flight in a modified Lockheed Model 14, from July 10 to July 14, 1938. The flight, entirely in the northern hemisphere, covered about fifteen thousand miles. The Douglas World Cruiser flight fourteen years earlier, while of greater distance, had required almost six months. Hughes found himself welcomed on his return as the greatest aviation hero since Lindbergh and was awarded the Collier Trophy.

Lockheed, following on its success with its Model 10 Electra airliner, developed the XC-35 military transport version featuring cabin pressurization, which was test flown successfully on May 9, 1937.[67] Development extended to Dr. Sanford Moss's turbosupercharger experiments of 1918, but the XC-35 was the first full airborne test and led the way to more

TABLE 7. U.S. Aircraft Exports, 1928–39

Year	Aircraft Exported		Value of All Aeronautical Exports ($)
	Number	Value ($)	
1928	162	1,759,653	3,664,723
1929	348	5,484,600	9,125,345
1930	321	4,819,669	8,818,110
1931	140	1,812,809	4,867,687
1932	280	4,358,967	7,946,533
1933	406	5,391,493	9,180,328
1934	490	8,195,484	17,662,938
1935	333	6,598,515	14,290,843
1936	527	11,601,893	23,143,203
1937	628	21,076,170	39,404,469
1938	875	37,977,324	68,227,689
1939	1,220	67,112,736	117,807,212

Source: *Aircraft Facts and Figures*, various editions.

efficient and higher-flying airliners and bombers. Then Boeing's system, designed by Nathan C. Price and patented in 1937, found a production application on the Model 307 Stratoliner.[68] Production was licensed to the new Garrett firm, later Garrett AiResearch.

By the late 1930s true intercontinental or transoceanic air transport had arrived, but due to safety concerns long-range overwater flights were made only by flying boats.[69] Sikorsky, Martin, and Boeing developed flying boats capable of ocean crossings, and which of course were capable of emergency water landings. Glenn Martin had responded to a specification from Juan Trippe of Pan American for a transoceanic flying boat in 1931, and the Martin M-130 China Clipper first began transpacific mail service on November 22, 1935, following with passenger service on October 21, 1936. Sikorsky produced 10 S-42 models, but the line ended with the advanced S-44 Excalibur, with only three built. The advanced Boeing 314 Clipper, designed by the young engineer Wellwood Beall and the apex of flying-boat passenger service, began service with Pan Am on March 26, 1939. Unfortunately the large flying boats were all money losers for their producers.[70] Only Consolidated was profitable with flying boats, and its output was entirely military. Major carriers such as Pan Am and TWA continued for a time to sponsor advanced flying-boat developments, but the Second World War was to demonstrate the superiority of the long-range landplane, bringing to an end the romantic era of ocean-spanning flying-boat airliners.[71]

Civil air regulation, especially of scheduled airline service, continued to make great strides in the late 1930s. Air Traffic Control, first established by provisions of the Air Commerce Act of 1926, led to the establishment of the first Air Traffic Control Center at Newark Airport in 1935. Then the Federal Aviation Commission of 1934 led to the Civil Aeronautics Act of June 23, 1938, which established the Civil Aeronautics Authority and the first comprehensive regulation of civil aviation. The Authority began operating on August 22, 1938, as an independent agency, removed from the Department of Commerce. The Authority had the power to issue air carrier route certificates and regulate airline fares. In a further reorganization on June 11, 1940, the Authority was split into two agencies, the Civil Aeronautics Board (CAB), and the Civil Aeronautics Administration (CAA). The CAB dealt directly with airline regulation and safety matters, while the CAA was responsible for ATC, airman and aircraft certifications, and airway development. The CAA expanded ATC centers from the time of its establishment, while the CAB, by facilitating long-range air routes, encouraged the development of four-engined airliners, to become a significant segment of the aircraft industry.

The Air Transport Association was founded on January 3, 1936, to advance the interests of scheduled carriers and to cooperate in noncompetitive areas. The association's most significant contribution to the war effort came in 1941, when its members collectively decided to cooperate and offer their services in military transport needs. Planes, pilots, and technicians were pooled into a worldwide transport system for the armed forces, precluding the industry from being commandeered or taken over directly.[72]

Boeing, Douglas, and Lockheed led the development of large four-engined airliners. Boeing's efforts, as earlier with the Model 247, benefited from military development, as B-17 technology was applied directly to the Model 307 Stratoliner.[73] The Stratoliner first flew on December 31, 1938, and entered service with TWA and Pan American in April 1940. But a fatal crash of the prototype handicapped the Stratoliner in the market, and only ten production models were built. Five were converted in wartime as C-75 military transports, with pressurization removed.

Douglas dominated the airliner market up to the Second World War. The DC-3 was the first airliner with which airlines could earn a profit on passenger carrying alone, and some 430 DC-3s had been produced for American companies by the time of Pearl Harbor.[74] As the DC-3 became standard equipment, Pat Patterson of United, on behalf of five major airlines, informed manufacturers in 1936 of their desire for a large, four-engined "dream airliner," for which development they would finance up to three hundred thousand dollars.[75] Even with that support, most were too shaky financially to undertake the risk, and only Douglas ventured into the field. The result was the DC-4E (experimental), which flew on June 7, 1938, but which turned out to be too large for airline needs at the time. The design then was redeveloped into the somewhat smaller but more efficient DC-4 of 1940. War prevented its entry into airline service, but like the DC-3, the DC-4 was to render outstanding military service as the C-54 and supported Douglas's postwar airliner developments.

Lockheed, the last to enter the four-engined landplane market, conceived the Excalibur project in 1937, which featured the signature triple tail. It was to carry thirty-two passengers but did not move beyond the design study stage.[76] The firm then began the Model 049 in 1939 to the requirements of Jack Frye of TWA. TWA at the time could not finance the new airliner, but its major stockholder, Howard Hughes, agreed to finance the first order through his prosperous Hughes Tool Company.[77] Thus the progenitor of the famed Constellation series moved to the hardware stage. The wing of the Model 049 was virtually a scaled-up version of the wing Lockheed had designed for its new P-38 pursuit.

The United States unquestionably led the world in commercial aviation. Due in no small part to long domestic routes and separation from other major markets by two oceans, air travel became more of a necessity in the United States than in, for example, Europe. Airliners such as the Douglas DC-3 and Boeing 307 Stratoliner and 314 Clipper were easily the most advanced in the world. Although the 307 and 314 were built only in small numbers, their aeronautical advances strengthened Boeing's reputation in large aircraft.

In military aviation, the Seversky P-35 and Curtiss P-36 monoplane pursuits, capable of speeds approaching 300 mph, entered service in 1937 and 1938, respectively. The development of one-hundred-octane gasoline further enhanced engine performance. At the other end of the performance spectrum, the German Fieseler Fi 156 Storch, the first true short-takeoff-and-landing (STOL) aircraft, first flew in 1936.

Many published sources alleged that American military designs of the immediate prewar period were below world standards.[78] This simply does not hold up under scrutiny. While U.S. air strength was very small, ranking perhaps sixth in the world, many designs were quite advanced and were often better than their counterparts in other major air powers.[79] The standard pursuit, the Curtiss P-36, did not equal the speed of the Supermarine Spitfire or the Messerschmitt Bf 109 when those fighters entered service in the late 1930s, but it outperformed most others. The Northrop A-17 was a capable attack aircraft that also won export orders, and the Douglas DB-7 of 1939, which became the A-20, was the first attack aircraft to exceed 300 mph. No European power possessed the equal of the B-17 or DC-3. In particular American aircraft possessed longer range capability than most European aircraft and were also noted for their rugged construction, invaluable in combat.[80] Further, American military aircraft were exported to Europe in large numbers from the late 1930s, hardly a confirmation of design inferiority. British and French purchasing commissions first visited the United States in 1938, and the British Military Commission ordered new designs from 1940.[81]

The questionable contentions of American design inferiority of this period may have been influenced by the fact that American air doctrine was focused more on the bomber than the fighter, given the low likelihood that enemy aircraft would invade American airspace, and that the coastal-defense mission depended more on bombers than fighters. Also, despite the official U.S. policy of continental defense, Air Corps strategy was oriented toward offensive operations, again involving bombers. The U.S. military situation thus was markedly different than that of the British, French, or Germans, who possessed few long-range aircraft but emphasized fast short-range fighters.

Both Britain and Germany were well in advance of the United States in jet development, however. Power Jets Ltd., formed for the specific purpose of developing the jet engine, made its first test run of the Whittle design on April 12, 1937, and Hans von Ohain pursued parallel development in Germany.

As fixed-wing aircraft advanced rapidly in the 1935–39 period, rotary-wing development was in its embryonic stage. Autogiro pioneer Juan de la Cierva died in an airliner crash in December 1936, but development of autogiros by Harold Pitcairn and the Kellett brothers, his American licensees, continued. While successful as flying machines, autogiros still could not attract significant orders, and both developers turned to the helicopter. The concept of the helicopter, rising and landing vertically, with hovering capability, extended to Leonardo da Vinci, but operation presented bedeviling problems of power, stability, and control. The French Breguet-Dorand *Gyroplane Laboratoire,* a coaxial rotor design, flew successfully on December 22, 1935, but the German Focke-Wulf FW 61 generally is credited as the first aerodynamically practical helicopter. It first flew on June 26, 1936, but its structure was so heavy that it could lift only its pilot.

In the United States, Henry Berliner had experimented with the helicopter as early as 1920, but it fell to Igor Sikorsky to achieve full success. His major opportunity came in 1938. Directed by United Aircraft to cease development of money-losing flying boats, Sikorsky was instead authorized $250,000 for helicopter development, enabling him to focus on design of a practical model. His VS-300, featuring a single lifting rotor with a small tail rotor to offset torque, was successfully tested under tether on September 14, 1939. It achieved free flight on May 13, 1940, but, while maneuverable, still had difficulty in forward flight. Its potential, however, led to military support and to a permanent American lead in the field.

Chapter 6 Industry Survey, 1935–40

Reuben Fleet's Consolidated Aircraft encountered recurring problems with testing flying boats in the icy weather of the Buffalo area. Fleet determined to move his company to the sunnier, warmer weather of San Diego, where he had learned to fly.[1] He accomplished the move in late 1935, a prodigious effort for the day, involving quantities of production tooling, components, and assemblies as well as personnel.[2] The move also enabled Fleet to benefit from lower California labor costs and a union-free environment. He continued development of flying boats for the navy and production of the P-30 two-place pursuit for the army. The XP3Y-1, directed by I. M. Laddon, first flew on March 31, 1935, and led to the PBY Catalina of wartime fame. The navy order in 1937 for sixty P3Y-1 (later PBY-1), worth some $22 million, was the largest single military aircraft order placed since the First World War.[3] The larger XPB2Y-1 Coronado flying boat first flew on December 17, 1937. It also saw wartime service, but not in the numbers of the PBY.

Charles Ward Hall died in a crash on August 21, 1936. He was succeeded by his son Archibald, but Hall-Aluminum Aircraft Corporation declined thereafter. The company had produced only twenty-nine aircraft but had contributed significantly to aluminum flying-boat development. In September 1940 Consolidated acquired HAAC, which, although declining, was removed as a potential competitor.[4] Archibald Hall worked for Fleet for a time afterward. Fleet earlier had sold the rights to the Fleet Trainer to Canadian interests, and the former Fleet Division became Fleet Aircraft, Ltd.

In 1939, even with the threat of war increasing, Consolidated was facing a crisis. The PBY order was completed, and no further production contracts were in prospect. When Reuben Fleet was approached by the Air Corps about becoming a second source of production for the B-17, then enjoying a higher priority, he countered with a proposal to develop a new, more advanced bomber. A contract for a prototype B-24 was awarded in March 1939. While of similar dimensions and power to the B-17, the B-24 was a radically different design, with a long-span, high-aspect-ratio cantilevered wing, a deep fuselage offering large bomb-carrying capacity, and tricycle landing gear, the first on a large aircraft. The B-24 program was directed by I. M. Laddon, but the Davis Wing of the B-24 was the contribution of the same David R. Davis who had supported Donald Douglas in his first venture.[5] Davis in fact received royalties from Consolidated for his design. The B-24, after a remarkably brief development time of nine months, flew on December 29, 1939, and became Consolidated's major production program.

Manufacturing activities of the Aviation Corporation evolved primarily through the efforts of the investment banker Victor Emanuel, a native of Dayton, Ohio, with strong aviation interests. His firm, Emanuel and Company, took an equity position in AVCO in 1934, the same year in which Gerald Vultee formed Aviation Manufacturing Corporation as a subsidiary, enabling AVCO to remain active in aircraft.

In what amounted to a reverse takeover, the Aviation Manufacturing Corporation was liquidated on January 1, 1936. Vultee Aircraft Division then was formed as an autonomous subsidiary of AVCO, acquiring the assets of the previous AMC, including Lycoming and Stinson.[6] Jerry Vultee turned his full attention to developing military aircraft. Then in August 1937 Victor Emanuel and Tom Girdler, president of Republic Steel, purchased all of E. L. Cord's stock in the Cord Corporation for $2.6 million.[7] The Cord Corporation held 29.25 percent of AVCO stock, and with the sale Cord completely severed his interest in AVCO and aviation and moved to Europe. The Aviation Corporation had lost money in seven of its nine years of existence, and Emanuel set about to transform it into a more stable firm.[8]

The Vultee Aircraft Division was again reorganized on November 14, 1939, as the independent Vultee Aircraft, Inc., though still owned 5 percent by AVCO.[9] Jerry Vultee did not live to see this development, however. He and his wife died in a plane crash in the California mountains on January 29, 1938, when returning from a sales trip to the Air Corps, marking the premature passing of an outstanding designer and engineer. Vultee was succeeded by Richard Millar.[10] Vultee undertook advanced military designs, but its major production programs were the BT-13 Valiant trainer and the V-72 Vengeance, serving as the A-31 and A-35.

Continuing to rationalize its organization, Curtiss-Wright by 1936 consisted of only three major divisions, Curtiss Aeroplane and Motor, Curtiss-Wright Airplane (for civil aviation), and Wright Aeronautical. In 1937 Curtiss-Wright Airplane became the St. Louis Airplane Division.[11] Ralph Damon became president of the St. Louis operation. The CW-25 Coupe business aircraft, later the AT-9 Jeep in wartime service, was developed at St. Louis under a Bureau of Air Commerce contract. St. Louis also developed the CW-20 twin-engined airliner in 1937, to succeed the Condor as the Curtiss airliner.[12] Although never to see airline service, it became the famous C-46 Commando of the Second World War.

Thomas A. Morgan, earlier with Sperry, was elected board chairman in 1935, succeeding the late Richard Hoyt, and Guy W. Vaughan, who had risen through Wright Aeronautical, became president. Vaughan was to direct the company through the war. The P-36 designed by Don R. Berlin was selected as the standard Air Corps fighter in July 1937, and from 1938 also was ordered in quantity by the British and French. The P-40, developed from the P-36, first flew on October 14, 1938, and became the standard fighter from 1940. Curtiss followed a long series of biplane naval fighters with the SB2C Helldiver, which first flew on December 18, 1940, and which became a standard dive bomber for the navy. Although a major firm in the industry, both in military and civil markets, Curtiss-Wright was never a design leader as such firms as Boeing, Douglas, Lockheed, and North American became, clouding its future.

Fairchild maintained aircraft production, largely extending from Kreider-Reisner, in the Fairchild Aircraft Corporation formed in 1934. Then on November 10, 1936, the Fairchild Engine and Airplane Corporation was formed to acquire from the senior firm all manufacturing operations, including Fairchild Canada.[13] Sherman Fairchild's friend Howard Hughes was a major investor. Fairchild Engine and Airplane enjoyed a profitable year in 1937.[14] A separate Fairchild Aviation Corporation controlled the camera and survey operations and reacquired the Ranger engine division from AVCO in 1934. Fairchild Aviation eventually became Fairchild Camera, and engine operations were transferred into Fairchild E & A upon its formation. In 1939 the former Fairchild Aircraft entity became the Aircraft Division of Fairchild E & A. Fairchild's established F-24 cabin aircraft became a long-term success, and the firm began development of military trainers. The low-wing M-62 trainer, which first flew on May 15, 1939, was ordered on September 22, 1939, after winning the Air Corps competition. It became a major war program under the PT-19 and PT-26 designations.

By 1940, with military orders expanding, Sherman Fairchild sought eight hundred thousand dollars in working capital. Smith, Barney and Co., concerned by the firm's losses in 1938 and 1939 and by Fairchild's reputation as a nonchalant manager, required as a condition of the loan that he bring in an experienced professional to handle daily operations.[15] Fairchild then hired J. Carlton Ward, a vice president of United Aircraft, as president and general manager, while he remained chairman as well as the major stockholder. Ward was to be a conspicuously successful manager during the war but would also encounter conflicts with Sherman Fairchild.

Robert Gross, firmly in control of Lockheed by 1935, found the company the subject of acquisition interest but determined to remain independent. Despite its success with the Electra, Lockheed still struggled to raise capital, supplementing its two-hundred-thousand-dollar RFC loan with continued stock sales in the Los Angeles area.[16] Concerned over its market vulnerability as single-product company, Gross proposed a bomber development of the Electra but did not win support. More successful was the larger Model 14 Super Electra of 1937. Howard Hughes's around-the-world flight in a Model 14 had immediate benefits for Lockheed in that it spurred orders from several of the world's airlines.[17] The Model 14 also enabled production to be maintained as the original Electra suffered a market decline after several crashes. The larger Model 18 Lodestar of 1938 became an even greater success. Having little military business at the time and available capacity, Lockheed sent Courtlandt Gross to London to negotiate a contract with the British. Lockheed received an initial British order on June 23, 1938, for two hundred modified bomber versions of the Model 18 named the Hudson.[18] At $25 million the order was the largest in the industry to date.[19]

Lockheed had already turned to completely new military designs. In February 1937 the Air Corps specification for an advanced, high-altitude pursuit capable of 360 mph was issued. Lockheed responded with the twin-engined, twin-boom, tricycle-geared Model 322, designated XP-38, its first military design under Kelly Johnson.[20] Despite its then-radical features, the XP-38 won a development contract on June 2, 1937. Not incidentally, the British Hudson order provided Lockheed with the financing to proceed. The XP-38 first flew on January 27, 1939, and exceeded specifications, in fact becoming the first service aircraft to exceed 400 mph and the first to encounter compressibility problems as it approached the speed of sound.[21] With heavy orders, the production P-38D Lightning was in service at the time of Pearl Harbor. Lockheed won respect as possessing one of the strongest management and design teams in the industry.

While fighter and airliner development, plus the Hudson order, gave Lockheed a sound basis for the future, Robert Gross still pursued diversification. He briefly considered acquiring Northrop after Douglas closed the division during a labor dispute.[22] Then Victor Emanuel approached Gross in 1937 with a proposal to sell his AVCO manufacturing operations, Stinson, Vultee, and Lycoming, with Gross then merging his Menasco with Lycoming into a single engine subsidiary. Also involved was a Stinson move from Michigan to the El Segundo factory, which Gross would then buy from Douglas; Vultee would remain exclusively military.[23] This proposal likewise did not reach fruition.

Lockheed, while profitable in 1937, still lacked the financing for major acquisitions and instead established a new AiRover Aircraft subsidiary on August 17, 1937. The affiliate was intended to give Lockheed a position in personal aviation and developed the innovative StarLiner business airplane. The AiRover name was changed to Vega Airplane Company effective June 1, 1938, with Mac Short, formerly manager of Boeing's Stearman Division, as president and Robert Gross as treasurer. The StarLiner first flew on April 22, 1939, but no production was undertaken in the depressed market at the time. The Vega subsidiary converted to military activity in wartime. Courtlandt Gross managed Vega from June 1940, while Mac Short moved to the parent Lockheed as vice president of engineering.

Boeing continued to struggle financially in the late 1930s. The initial contract for thirteen B-17s resulted in a financial loss, as did the first production contract, and transport and flying-boat production also were unprofitable. The company sustained a loss of more than $3 million in 1939 before returning to modest profitability in 1940.[24] Phillip G. Johnson returned as president in August 1939 to lead the military production buildup, while Claire Egtvedt became chairman of the board. The company continued to emphasize large, long-range aircraft. The XB-15 for the army VLRB requirement first flew on October 15, 1937. While seriously underpowered for the bombing role, it enjoyed later accomplishments in long-range flights and as a transport.

Drawing from its B-15 and B-17 experience, Boeing considered even larger bombers in advance of any Air Corps requirement. War in Europe also brought awareness that the United States would need bombers even longer ranged than the B-17 and B-24. When the Air Corps issued a specification for an advanced bomber with a five-thousand-mile range on February 5, 1940, as B-17 production was slowly building up, Boeing accelerated design. After several revisions, Boeing settled on its Model 345, which was ordered as the XB-29 on August 24, 1940.[25] The B-29 program was led by Wellwood E. Beall, vice president of engineering, and would become the largest and most important wartime aircraft program, solidifying Boeing's leadership in the bomber field.

Boeing's Stearman Company in Wichita developed the Model 75 biplane trainer, which entered production in 1936 as the PT-13. Although popularly known as the Stearman trainer, the aircraft did not owe any direct design influence to Lloyd Stearman. Stearman fully merged into Boeing on April 30, 1938, as the Stearman Aircraft Division.

North American opened its new factory in January 1936 and produced the BT-9 series of trainers and the O-47 observation aircraft for the

Air Corps.²⁶ Production of the BT-9 totaled 266, and 239 O-47s were built. A major success was the order by the British Purchasing Commission for two hundred NA-49 Harvard advanced trainers, developed from the new BC-1 design. As the AT-6 it became the most famous allied trainer of the war. It also served with the navy as the SNJ. North American, operationally autonomous despite the 30 percent GM stock holding, still reflected in the AT-6 General Motors advances in standardization and mass production.²⁷

While continuing his initial strategy of specializing in trainers, Dutch Kindelberger also decided to enter the bomber field. His first design, the twin-engined XB-21, lost to the Douglas B-18A for production orders, but Kindelberger persisted and developed the Model NA-40, a more advanced design with tricycle landing gear. The NA-40 first flew on February 10, 1939, and became the famous B-25 Mitchell.

Chance Vought struggled as a low-volume supplier of naval aircraft. The SBU biplane scout bomber delivered from 1935 led to the more advanced low-winged SB2U Vindicator, which first flew on January 4, 1936. The SB2U also was exported to Britain and France as the V-156F for wartime service but was already obsolete when entering European combat in 1940 and was quickly relegated to training in the United States. More successful was the OS2U Kingfisher scout/observation aircraft, which entered service in 1940 and also was produced by the NAF. The most conspicuous production lack for Vought was a naval fighter. Earlier designs had failed to win orders, but the advanced, powerful XF4U project begun in 1938 under Rex Beisel was to thrust Vought into the front rank of fighter producers.

United Aircraft, having ended flying-boat development and having lost the military flying-boat contract to Consolidated, combined Sikorsky and Chance Vought into the Vought-Sikorsky Division on April 1, 1939. Aircraft manufacturing by United now was under a single division, and Chance Vought was moved from East Hartford to Stratford.

While Seversky had attracted notice with its P-35 fighter, designed by Alexander Kartveli, and with more advanced developments, the firm became increasingly troubled. The naval version of the P-35, the NF-1, was beaten in competition by the Brewster F2A in 1938.²⁸ Alexander de Seversky himself was in some official disfavor by 1938, as the Air Corps was increasingly concerned about management and P-35 production delays.²⁹ The firm had incurred continuous losses, leading to further doubts about its production capabilities to support the military buildup. Major de Seversky devoted increasing personal attention to record-breaking flight attempts and to sales trips for his aircraft, especially pursuing exports to China and Japan.

With financial problems mounting, including losses on P-35 production, the board undertook a reorganization in November 1938, after Paul Moore had secured added financing from White, Weld and Company.[30] Reporting the changes to the War Department, board member W. Wallace Kellett referred to Seversky's efforts to attract export business as a potential offset to its lack of production orders after the P-35 contract had been completed.[31] Then on October 12, 1939, the board, possibly with the influence of General Arnold, ousted Seversky as president.[32] Kellett, whom Seversky had hired as vice president due to his record of profitability in the autogiro field, became president.[33] The company was reorganized as Republic Aviation Corporation on October 13, 1939, with Paul Moore remaining the largest stockholder. Seversky's departure was acrimonious, and he blamed the Air Corps' preference for the Curtiss P-36 over his P-35 as due not to merit but to Curtiss-Wright political connections.[34] While Republic was small, with only some five hundred employees, new management pared the payroll further, to 189 at year end, although this number would expand manyfold with war production.[35] Seversky then was removed from the board in May 1940 but sued for compensation for his contributions.[36]

Douglas continued to develop its airliner business but by no means to the exclusion of military contracts, and produced the O-38 through O-46 series of observation planes from 1931 to 1937. Production of the DC-3 and its military variants remained strong and was joined by the DC-4. Douglas had produced its first naval-fighter prototype, the XFD-1, in 1933, but more important were the TBD-1 Devastator torpedo and SBD-1 Dauntless scout bombers. Ed Heinemann, chief engineer for Douglas El Segundo after the dissolution of Northrop in 1937, led development of the naval designs. The SBD-1 extended from the Northrop BT-1 of 1935, designed by Heinemann, in turn developed from the Northrop A-17.

The Model 7B, the advanced attack bomber prototype, first flew on October 26, 1938. The developed DB-7, ordered by the British as the Boston and by the United States as the A-20 Havoc, first flew on August 13, 1939. The design contract for the giant XBLR-2 was awarded on October 31, 1935, and a construction contract followed on September 29, 1936, but development of the redesignated XB-19 was delayed until 1937. Douglas employment, seven thousand in 1938, was the largest in the industry, and prospects appeared strong. The first of what would become some ten thousand military versions of the DC-3, most as the C-47, was ordered in 1940.

Martin, after struggling during the decade, finally emerged from bankruptcy. The B-10 program had enabled the company to become profitable, and the RFC loan was repaid in January 1937. Still, Martin faced a definite crisis after delivering the last B-10 on August 8, 1936, delivering three flying boats for Pan Am at a loss, and losing the new bomber order to the Douglas B-18. Glenn Martin's dream of giant flying boats crossing the Pacific was a victim of financial realities but led to bitterness by Martin toward Juan Trippe. Martin's prospects soon improved, however, with exports of the B-10 and with large orders from the British and French, initially for the Maryland bomber, which had been rejected by the Air Corps. Export profits provided the resources to develop the B-26, making Martin a major aircraft supplier to U.S. forces once more.

Grumman, with a stable management team, a force of loyal employees, and a clearly defined product line for essentially a single customer, expanded through the 1930s. Its XF3F-1 first flew on March 20, 1935, and more than 160 were delivered in three series. Several significant events for the company occurred in 1937. First, the new Bethpage, Long Island, plant opened on April 8. Second, Grumman issued its first public stock offering, of one hundred thousand shares, providing a major profit for the Loening brothers among others, and giving Grumman a strong capital base for expansion. Leroy Grumman continued to hold 40 percent of the outstanding stock.[37] Third, Grumman diversified into the commercial market with the successful Goose. Fourth, Grumman advanced carrier fighter design with the new XF4F-2.

Both Grumman and the navy realized that the F3F series represented the ultimate development of the biplane fighter, and Grumman turned to a monoplane development of the F3F as its successor. The XF4F-2, forerunner of the wartime Wildcat, first flew on September 2, 1937, but lost the production order to the Brewster F2A Buffalo. But Grumman continued development, resulting in the improved F4F-3 version, which then entered production in 1939.[38] The F4F Wildcat also marked the beginning of the Grumman tradition of feline fighter names. Also in 1939, Grumman began development of the large TBF torpedo bomber, which would play a major role in the coming war, and pursued further amphibian designs.

Goodyear organized Goodyear Aircraft Corporation on December 5, 1939, from its aircraft division, primarily for the development of lighter-than-air craft, but also to fulfill subcontracts on the Martin B-26. With the approach of war Goodyear moved strongly into subcontracting and eventually manufactured complete aircraft under license.

The Naval Aircraft Factory resumed active aircraft production in 1937 with the N3N-1 trainer, which served successfully as a wartime basic trainer.[39] The NAF produced 183 N3N-1s in 1937–38 and developed further designs.

Ryan received orders from both services and for export for trainer developments of its S-T and also developed the civil S-C cabin version. Also notable was the experimental YO-51 Dragonfly observation craft, which pioneered STOL techniques. Ryan progressed as a major subcontractor and as a flying-school operator.

Brewster's victory in the 1937 naval competition over Grumman and Seversky with its XF2A, after its contract for the SB2A Buccaneer, appeared to position the firm strongly in military aircraft. Temple Joyce, an industry veteran, joined the firm in 1937 as sales manager. The F2A Buffalo, flying in December 1937, actually preceded the Grumman Wildcat as the navy's first monoplane fighter, entering service in December 1939. But later difficulties with the Buffalo led the navy to return to Grumman, and design, management, and production problems loomed for Brewster.

Howard Hughes followed his success with the *Racer* by establishing Hughes Aircraft Company in 1936 as a division of Hughes Tool Company. Hughes intended to enter military aircraft but experienced numerous pitfalls, complicated by the manner in which he dealt with contracting officials. While neither an engineer nor a designer, Hughes could attract and exploit the talents of those who were and brought many capable men to his small company.[40] He could propose advanced designs and had visions of record-setting performance capabilities, but he simply never learned to do business the army's way, as had other, successful companies.

From the First World War, industry-government contracting depended heavily on personal relationships and collaboration. Rather than cultivating such relationships, Hughes instead managed to alienate such key individuals as General Arnold and Colonel Oliver P. Echols, chief of the Materiel Division. In 1937, after his record-breaking coast-to-coast flight, he failed to keep an appointment at Wright Field with Colonel Echols, who wanted to inspect the *Racer*.[41] In 1939, when Hughes proposed his D-2 advanced twin-engined bomber to the Air Corps, Hughes employees denied General Arnold access to the Hughes hangar in California when he attempted to inspect the facilities.[42] While Hughes claimed a case of mistaken identity, the incident did not enhance his standing as a military contractor. Wartime urgency would bring no relief.

New firms also were organized for military production. With Consolidated's departure, the city of Buffalo became concerned over the loss of a major industry. Lawrence D. Bell, Consolidated's general manager as well as president of the Fleet Aircraft subsidiary, had long desired to run his own company. He also received major civic encouragement to form a new aircraft company.⁴³ Consequently Bell and his associate Ray Whitman, retaining the engineer Robert J. Woods as well as many Consolidated production personnel, set about organizing the new company. Originally to be named Niagara Aircraft, Whitman argued the venture should be named after Bell, giving it a more individual identity. Thus Bell Aircraft Corporation was incorporated on July 10, 1935, with total capital of five hundred thousand dollars, to specialize in the military sector.⁴⁴ Investors were slow to commit funds, however, and the company teetered on the brink financially. Eventual acquisition of sufficient capital enabled the firm to expand to 145,000 square feet by 1936, including leased space.⁴⁵

Bell won a contract on May 12, 1936, to develop the pusher-engined, multiseat FM-1 Airacuda interceptor. The company's principal business was subcontracting, including making wings for Consolidated's flying boats, but the FM-1 entered flight testing in July 1937. The FM-1 was not ordered beyond the evaluation series of thirteen but gained Bell support for another innovative design, the P-39 Airacobra. The P-39 was Bell's entry into the competition for an advanced, high-altitude pursuit, also met by the Lockheed P-38, and first flew on April 6, 1938.⁴⁶ Bell survived 1940 with a British P-39 order accompanied by a $2 million advance payment.⁴⁷ Bell also received an Emergency Plant Facilities (EPF) contract for an additional plant at Niagara Falls and began rapid expansion.

Mounting labor tensions at Douglas, also affecting the Northrop Division, resulted in a strike on February 23, 1937. Douglas acquired the remaining 49 percent of Northrop stock on April 5, 1937, by exchanging its stock for that of Northrop.⁴⁸ Douglas first shut down, then legally dissolved Northrop on September 8, in the aftermath of another strike in which the labor force had disregarded an agreement negotiated earlier.⁴⁹ Almost fourteen hundred jobs were lost. The former Northrop operation then reopened on October 23, 1939, as the Douglas El Segundo Division. Douglas retained all designs and patents, and chief engineer Ed Heinemann remained with Douglas. Both the SBD Dauntless and the A-20 Havoc were El Segundo designs.

John Northrop resigned as a Douglas vice president on January 1, 1938, and announced that he would not reenter the industry. He was financially flush for the first time in his life through the stock sale, however, and soon began to explore organizing a new firm. Northrop decided to remain in the Los Angeles area due to the favorable climate, low construction costs, and availability of skilled labor.⁵⁰ After surveying locations, Northrop chartered his Northrop Aircraft, Inc., in August 1939, in suburban Hawthorne, and raised $1,250,000 by a stock sale.⁵¹ Public financing was difficult given the economic environment and the widespread view that there were already sufficient aircraft firms, but Northrop was determined that stock be widely held so that he would not be completely vulnerable to a single backer, as he had been in his earlier ventures. Investor Floyd Odlum's Atlas Corporation purchased the unsold 20 percent of the initial stock offering.[52] Northrop hired La Motte T. Cohu, formerly of AVCO and TWA, as vice president and general manager, while he served as president and chief engineer. Northrop broke ground for a new factory in September and opened in February 1940, primarily in subcontracting.

Jack Northrop initially aspired to return to flying-wing development but with the threat of war turned to military production.[53] The navy, seeking a second source for SBD-3 production, awarded a contract for two hundred to Northrop, but after Douglas lodged a protest the navy reneged, and the contract remained with Douglas, leaving Northrop outraged.[54] He angrily refused the navy offer to become principal subcontractor for the SBD-3. Jack Northrop and Donald Douglas, while never close personally, became somewhat estranged after dissolution of the Northrop subsidiary, the SBD-3 competition, and the move of many former El Segundo employees to Northrop.

Business began with a large PBY subcontract from Consolidated.[55] Sales reached $20 million in the first full year, but the major step was a production contract for the Vultee Vengeance for Great Britain (for which Vultee lacked adequate capacity), obtained with the assistance of Richard Millar.[56] John Northrop also developed the N-3 patrol bomber and with steadily increasing production finally was able to resume flying-wing development. He began the N-1M model in 1940, the first true flying wing, which would lead to the giant XB-35 flying-wing bomber. His reputation as an innovative designer would continue to grow.

James S. McDonnell, born in Denver and raised in Arkansas, moved east for his education. After his first airplane ride while a student at Princeton in 1917, he thereupon decided on a career in aviation. Graduating in 1921 and entering MIT, he concurrently undertook flying training through ROTC. After earning an M.S. in aeronautical engineering, he won a reserve commission and wings in August 1923, along with future chiefs of staff Hoyt S. Vandenberg and Nathan F. Twining. Leaving active duty in 1924, McDonnell entered an industry of only some two hundred engineers and twenty-eight hundred employees in total.[57] He worked briefly for Huff Daland and Consolidated, then moved to Stout to help develop the trimotor. After Ford acquired Stout, McDonnell moved to Ford also but suffered the dubious distinction of being fired personally by Henry Ford.[58]

After a short period with Hamilton Aero of Milwaukee in 1928, McDonnell left to begin his first independent venture, James S. McDonnell and Associates, to compete for the Guggenheim Safe Aircraft Competition prize. But McDonnell's design, the *Doodlebug,* crashed, and Curtiss won the competition. Unable to obtain production financing, the venture was dissolved in 1931. McDonnell then spent two years with Great Lakes and moved to Martin in March 1933, where he became chief engineer for landplanes.[59]

By 1939, with his extensive experience, McDonnell felt ready to establish his own company. He located in St. Louis, in part due to the government's concern over having the industry concentrated on both coasts, potentially more vulnerable to enemy attack, but also because he was impressed with the area's aviation facilities and record of support. McDonnell incorporated McDonnell Aircraft Corporation on July 6, 1939. He established offices in American Airlines space at Lambert Field, invested $35,000 on his own and raised further capital of $135,000, including $5,000 from Laurance Rockefeller.[60] He determined to seek Air Corps contracts, but from the outset the company struggled to survive, as first-year sales were zero.[61]

Chapter 7 The Wartime Buildup Begins
1940–42

Before 1940 a production order for one hundred aircraft would have been considered major for the American industry. Orders reflected the small size of both the Air Corps and Naval Aviation. While activity increased from 1938, the prewar environment was far from prosperous. Fixed-price contracts were standard, and working capital to fulfill new orders was invariably difficult to obtain. The procurement process still was such that aircraft were not paid for until delivered: progress payments were unknown. It remains remarkable that in the 1930s new competitors entered an industry that appeared so unpromising, but one favorable factor was that enhanced competition forestalled any charges of oligopoly in the industry. Throughout, the industry managed to avoid both becoming a complete captive of the government and being labeled a competition-reducing cartel. The industry giants of the late 1930s were Boeing, Consolidated, Curtiss-Wright, Douglas, and Martin, and Pratt and Whitney and Wright in engines, but others were mounting a challenge.

Military Orders Rise; Wartime Organization Evolves

The United States by 1940 still had no defense industry in the later sense of the term. There were the relatively few established aircraft manufacturers along with some shipbuilders and munitions makers, but there was no cohesive industry primarily oriented toward armaments and national defense. A defense industry would be a legacy of the wartime production effort and the Cold War tensions following. There was heavy military influence in the industry even earlier, of course, including direct intervention in the management of major firms, as with Republic and later Consolidated. Defense companies became too important for strategic business decisions to be left entirely in private hands.

A sidelight to the wartime success of the aircraft industry is that most of its leaders were contemporaries in age with the major wartime political and military leaders. A comparison of birth years illustrates the point: Martin 1886, Fleet and Rentschler 1887, Loening 1888, Sikorsky 1889, Douglas 1892, Bell 1894, Northrop and Kindelberger 1895, Fairchild 1896, and Gross 1897. For the war leaders: Marshall and MacArthur 1880, Roosevelt 1882, Arnold 1886, Eisenhower 1890, Spaatz 1891, Doolittle and Eaker 1896, and so on. Whether this enhanced the relationship of the industry with the military and political leadership is an intriguing speculation. There is no question that close personal relationships were carefully cultivated. Northrop and Arnold went on a California camping trip together September 20 to 27, 1939, as war was breaking out in Europe.[1] And Donald Douglas's daughter Barbara married General Arnold's son Bruce during the war.

The locational trend of aircraft makers toward Southern California was pronounced. By 1940 some 45 percent of all airframes were manufactured there, and the percentage was even higher when extended to the entire West Coast to include Boeing.[2] Engine producers were still concentrated in the East, resulting in the complication, though manageable, of cross-country shipments to Seattle, Los Angeles, and San Diego for installation. Labor relations in the industry, despite a strong unionization movement, were fairly smooth. Access to a plentiful labor supply remained the single most important locational consideration in the industry.[3]

NACA conducted research with both military and civil applications before the outbreak of war. Airfoil and cowling design were equally applicable to both. After 1940, however, its activity became almost exclusively military. NACA established the Ames Laboratory adjacent to Moffett Field near San Francisco in 1940 as the West Coast counterpart to the Langley facility. This establishment also recognized the increasing industry concentration on the West Coast and was to make major aeronautical contributions in the war years and consequently to the industry.[4] It was named for Joseph F. Ames, a chairman of NACA. The Lewis Laboratory in Cleveland, for engine research, was established soon after.

The United States began to respond, however slowly, to the rising military threat. Major expansions began for both the Air Corps and Naval Aviation from 1939. In January 1939 the Air Corps was approved for a strength of fifty-five hundred aircraft overall, with the intention that this strength would be reached by June 30, 1941.[5] That figure of course would be overtaken by events but represented an ambitious, yet still attainable, goal. Then President Roosevelt, a firm convert to airpower, called for 50,000 airplanes a year in his speech on May 16, 1940.[6] The 50,000-plane program was to be divided into 36,500 for the army and 13,500 for the navy. Actually, Roosevelt referred only to an expansion to that capacity rather than immediate production of that quantity, so that the major short-term impact of the goal was psychological.[7] Such capacity presented a formidable challenge, however, as it involved an expansion from little more than 2,000 planes per year to 4,000 per month.

Aviation expenditures remained small during this period. The combined budgets of the Air Corps and the Bureau of Aeronautics totaled only

$157,665,000 in 1940, 8.76 percent of the total military budget and only 1.75 percent of the total federal budget.[8] But procurement budgets began to rise sharply. Then on March 25, 1940, the industry was authorized to sell freely to friendly nations, providing a further stimulus to expansion. General Arnold had determined by 1939 that his most effective efforts should be in the training of massive numbers of pilots and maintenance personnel to operate the coming flood of airplanes and began to develop programs for those goals, including a large civilian flying-school program.[9] The program trained two hundred thousand pilots by 1944. Industry output rose rapidly even before Pearl Harbor. In the first half of 1941, 7,433 aircraft were produced, more than had been produced in all of 1940.[10]

A Joint Aircraft Committee was established, with representatives from the army, navy, and the British Purchasing Commission.[11] This committee helped allocate strained U.S. capacity between domestic and export production and emphasized standardization of types to be produced.[12] At the end of 1940 there were some fifty-five types of aircraft on order by the military, dramatically underscoring the need for rationalization.[13] Then the Lend-Lease Act of March 1941 eliminated competition with allies for orders, as the U.S. forces received all production, of which certain quantities were allocated to allied countries.

The Vinson-Trammell act was amended on June 28, 1940, to limit contract profits to 8 percent of the value of the contract.[14] The major authority needed by the government, however, was that for direct contract

TABLE 8. U.S. Air Appropriations, 1928–40 (in thousands of dollars)

Year	Army	Navy	Post Office	NACA
1928	25,612	20,100	4,650	550
1929	33,911	32,189	14,480	629
1930	34,910	31,430	19,300	1,508
1931	38,893	32,033	24,600	1,321
1932	31,850	31,145	27,000	1,051
1933	25,673	32,745	26,460	920
1934	34,038	21,957	22,000	695
1935	30,918	34,842	19,003	726
1936	50,287	40,732	18,700	1,178
1937	62,608	38,588	20,230	2,545
1938	67,308	49,500	24,406	1,734
1939	73,800	48,074	27,592	2,508
1940	320,448	111,459	31,163	6,543

Source: *Aircraft Year Book,* various editions.

negotiation, given the imminence of war, but this practice had been criticized by President Roosevelt back in the days of General Foulois.[15] Eventually, contract policy was changed from fixed-price to cost plus fixed fee (CPFF), easing somewhat the financial pressure on manufacturers.[16] Then the Second Revenue Act of October 8, 1940, suspended profit limits on contracts and authorized five-year write-offs of production facilities.[17]

In military organization, the GHQ Air Force was abolished and replaced by the Combat Air Command on November 19, 1940. The CAC enjoyed a separate status under Army Field Forces. The arrangement soon proved to be unsatisfactory, however, and led to further changes. Pursuant to the War Powers Act of 1941, the U.S. Army Air Forces were established by executive order on June 20, 1941. Major General Henry H. Arnold was appointed chief of the Army Air Forces. Under Arnold were the chief of Air Corps and the commanding general, Combat Command. The office of assistant secretary of war for air, vacant since F. Trubee Davison resigned in 1933, was revived under the able Robert A. Lovett.[18] In a further reorganization, on March 9, 1942, the Army Air Forces became an equal organization under the War Department with the Army Ground Forces and the Services of Supply. Arnold, then with three stars, became commanding general of the AAF as well as assistant chief of staff of the army. The Air Corps and Combat Command organizations were eliminated, resulting in the basic organization that would carry the air forces through the war. Then organic army aviation was established on June 6, 1942, involving primarily the artillery spotter function, controlled by field commanders.

The March 9 reorganization also resulted in formation of the AAF Materiel Command, which replaced the old Materiel Division and continued to coordinate procurement and production with the industry. The major divisions of the new command, the Production Division of April 3, the Engineering Division of December 3, and the Procurement Division of December 31, followed.[19] The new structure represented a massive change from the original Materiel Division of 1926.

Transformation to War Production

Several "alphabet soup" organizations were established to expedite the military buildup. The earlier Council on National Defense became the National Defense Advisory Commission (NDAC) in May 1940, with seven members, and was in reality advisory. It was joined by the Supplies, Priorities, and Allocations Board (SPAB). The NDAC then was succeeded on January 7, 1941, by the Office of Production Management (OPM), under the direction of William S. Knudsen, previously president of General

Motors.²⁰ An Aircraft Division of the OPM was established under Merrill C. Meigs, with a first task to involve the automobile industry in aircraft production. Once again the capabilities of that industry were regarded as transferrable to aircraft, and Knudsen, with his auto background, accorded the measure a high priority. Automobile production was suspended entirely in 1942, with the industry's total capacity devoted to the war effort, and a major part of that capacity to aircraft.

Unlike the situation in the First World War, however, the auto industry requested and received the help of the aircraft industry in transforming its production, with notably greater success than with the earlier effort. Preceding the cooperative organizations of the aircraft industry, the automobile industry formed the Automotive Council for War Production on December 31, 1941, to coordinate and foster cooperation within the industry on its role in aircraft production.

In January 1942, after American entry into the war, the War Production Board, succeeding the SPAB, was established under Donald Nelson.²¹ Theodore P. Wright of Curtiss-Wright headed the Aircraft Branch of the WPB, and Grover Loening served as aircraft consultant. Later Harold E. Talbott Jr. became director of aircraft production for the WPB. Despite the technical brilliance of many aircraft and the prodigious production achievements of the industry during the war, organizational measures involving the military, other government agencies, and the industry itself were of equal significance to the success of the effort. Organization for production proved to be critical to the production result.

The industry launched many initiatives to organize and coordinate itself, in part to preclude government intervention. First, the Senate Special Committee to Investigate the National Defense Program, which was established in March 1, 1941, and known as the Truman committee after its chairman, had originally recommended that an aircraft production "czar" be appointed.²² A czar, as Coffin and later Ryan had been in the First World War, never came about, largely due to effective industry initiatives. The United States also never found a separate Ministry of Aircraft Production to be necessary, as had Churchill had in the crisis of 1940.

Major firms recognized their common problems of personnel, materiel, organization, and subcontracting. The Aeronautical Chamber of Commerce (ACC), struggling through most of the decade, enjoyed a revival by 1939 with the military buildup. After disagreements about the nature of industry-government relationships, however, such West Coast manufacturers as Vultee, Northrop, North American, and Douglas withdrew from the ACC, leading to formation of a new specialized organization.²³ Eight major West Coast manufacturers set up the Aircraft War Production Council (AWPC) in April 1942 to coordinate production know-how, share materials, and deal with common problems.[24] Membership was reduced to seven with the Consolidated-Vultee merger, but Boeing joined in March 1943, bringing membership back to eight. The organization came about in part also because the West Coast firms felt that the ACC was too heavily influenced by the East Coast firms. In one example of the AWPC's activities, Northrop designed its advanced P-61 night fighter with tricycle landing gear, with which it had no engineering expertise. North American then provided Northrop with a workable system from its B-25 bomber, saving further development effort.[25]

The East Coast manufacturers followed with a similar council. Then, on April 14, 1943, the industry established a National Aircraft War Production Council, sharing technical reports between the two regional councils.[26] The National AWPC also carried out an effective liaison among the WPB and the War and Navy Departments. Glenn L. Martin served as its first president. One impact of the original West Coast–East Coast split, however, was that it largely destroyed the effectiveness of the ACC.[27] But with reorganization, the AAC won back by February 1943 most of those who had resigned.

The limited production capacity of the prewar industry soon was exceeded by war requirements, but manufacturers were reluctant to undertake major plant expansions on their own resources, mindful of the history of boom and bust extending to the first war.[28] They did not want to find themselves, after a frenzy of wartime production, without continuing business. It was understood, moreover, that the massive war expansion required was more than simply one of physical capacity: that was possibly the easiest area to expand. The most pressing difficulties lay in the areas of experienced management, quality control, and worker training. Resources were stretched in these critical areas. Advances in production tooling, however, enabled production to rise rapidly even with hastily trained workers, as production steps could be easily standardized. Hardly surprising, all did not progress smoothly in the wartime production effort. Worker turnover, for example, was extremely high, especially in the early years. There was a continuing tug-of-war between the industry and the armed forces, seeking to draft workers into uniform. One result was that women aircraft workers were employed on a large scale for the first time.

Beyond the labor force, several major developments combined to enable the industry to meet war demands. First, the British Purchasing Commission, increasingly desperate to augment strained British aircraft capacity, financed several American firms in developing the production capability to fulfill British orders.[29] Initially $74 million was provided to Bell, Curtiss-Wright, Douglas, Lockheed, Martin, and Vultee. The first British purchasing mission, to determine first the suitability of American designs for British service, and then to place orders, left for the United States on April 20, 1938. The first order, as already related, was for the Lockheed Hudson. Many American designs would see British service during the Second World War. By 1941 the British had ordered no fewer than twenty-six thousand American aircraft, necessitating the Joint Aircraft Committee.[30] French orders also were taken over by the British after France fell in June 1940. It could be fairly stated that the British and French governments largely financed the expansion of the American aircraft industry before the United States began its major buildup.

Second, the U.S. government developed programs to expand production capacity. The NDAC established the Emergency Plant Facilities (EPF) program, under which a manufacturer could build a new plant, with the government reimbursing the manufacturer over five years for his cost and assuming title. The manufacturer then had the option to buy after that time. Several major plant expansions were undertaken under this program from 1940, but the arrangement became regarded as unsatisfactory by both the government and the industry.[31] The EPF operation also took over the Douglas Long Beach plant, freeing Douglas of an eight-hundred-thousand-dollar tax liability in the process.[32] The more permanent policy was administered under the Reconstruction Finance Corporation. Legislation established the Defense Plant Corporation (DPC) on August 22, 1940, as an RFC subsidiary, to build, equip, and hold title to several major aircraft assembly plants.[33] These then were leased to the manufacturers. Also, the defense acts of June 28 and July 2, 1940, authorized the RFC to make loans to and buy the stock of firms to finance production or new plant construction.[34] Actual construction and outfitting of new plants was by the Corps of Engineers. The major early plants built in this manner were those in Cleveland, for Fisher Body; Chicago and Oklahoma City, for Douglas; and Marietta, Georgia, for Bell.[35] There was also an eightfold expansion of the Curtiss-Wright St. Louis plant, beginning in 1940 and completed in 1941.[36] Defense plants generally were in full production by the second half of 1943. All new plants were located inland due to concerns about vulnerability of both coasts to enemy attack.[37]

In a related activity, in August 1942 the Army Air Forces sought increased capacity for forthcoming B-29 production, then centered in Wichita. Boeing's commitment to B-17 production left no capacity for the B-29 in Seattle. The AAF then took over Boeing's new Renton plant on Lake Washington, intended for production of the canceled PBB-1 flying boat, transformed it to B-29 production, and built an adjacent airfield for the B-29s coming off the line.[38]

Engine plants were financed as well. An early example involved Wright Aeronautical, which won approval from the RFC on August 7, 1940, to build a large plant in Lockland, Ohio, a Cincinnati suburb. Wright secured $37 million for construction and $35 million for working capital, as well as pledges of $20 million for cooperating suppliers.[39] The DPC then bought the land, leasing it back to Wright.[40] Of $3,840,000,000 invested in aviation plants during 1940–1945, 89 percent was government funded, most through the DPC.[41]

A third factor was that, as in the First World War, the automobile industry and related industries would become directly involved in aircraft production. Already the largest supplier to the aircraft industry, the automobile industry accounted for a major portion of total aircraft production. General Motors, Ford, and Goodyear were the most prominent, but all auto-related companies would play a role. The mass-production experience of the auto industry also aided the adaptation of aircraft production to assembly line techniques. Despite numerous implementation problems, results overall were quite successful. There was also massive cross-licensing during the war, within the aircraft industry and between aircraft and nonaircraft firms. It developed, however, that comparatively few aircraft were produced in directly converted automobile plants: modifications required were too extensive.[42]

A fourth factor in the wartime production record was that most light aircraft producers shifted over to war production, either through subcontracting, license production, or production of military versions of their civil designs. Aeronca, Piper, Stinson, Beech, Taylorcraft, Bellanca, Cessna, and Waco were prominently involved.

A fifth factor was that new companies were established to contribute to war production, generally by licensing or subcontracting rather than production of original designs. Prominent among these were Columbia Aircraft Corporation, American Aviation Corporation, and Globe Aircraft Corporation, originally formed for light aircraft.[43] St. Louis, whose XPT-15 trainer of 1939 had lost to Fairchild for orders, also became a significant subcontractor and license producer during the war. Rohr Aircraft Corporation of San Diego, established in 1940 by former Ryan engineers, made a notable contribution with development of the "power-pack" engine subassemblies for large bombers. By December 1944, 38 percent of all airframes and 28 percent of engines were subcontracted.[44]

The government was concerned from the outset about the financial condition of major firms in the industry, since most had struggled merely to survive the 1930s. In wartime, financial and production problems affected both the industry and the military. Manufacturers did not know in advance what their costs would be, being vulnerable to uncontrollables such as the availability of materials and changes in priorities. The military services, on the other had, often were unsure of what they wanted. The companies' narrow capital base did not provide for a significant margin of error, requiring government guarantees of recovery of all costs. The cost-plus-fixed-fee system was used throughout the war but incurred hostility from the industry to the end. Record keeping was burdensome, as contracting officers had to approve all allowable costs. The greatest use of CPFF was by the Army Air Forces, with a majority of its contracts in excess of $10 million by this method.[45] Fees allowed generally ranged from 7 percent down to 4 percent and, while small, still could result in large profits in relation to a firm's net worth.[46] Many contracts were still awarded on a fixed-price basis, however.

In November 1941, shortly before the U.S. entry into the war, Artemis L. Gates, assistant secretary of the navy for air, ordered a study of the financial condition of those contractors with which the navy might do business.[47] The report, delayed somewhat by the outbreak of war, was submitted on May 14, 1942. Titled *An Investigation of the Financial Condition and Recent Earnings of Aircraft Contractors*, the study was under the direction of Captain S. J. Zeigler, a former head of the Naval Aircraft Factory, and John W. Meador, a contract employee. The investigators vis-

TABLE 9. U.S. Aircraft Production, 1928–41

	Total	Military	Civil
1928	4,346	1,219	3,127
1929	6,193	677	5,516
1930	3,437	747	2,690
1931	2,800	812	1,988
1932	1,396	593	803
1933	1,324	466	858
1934	1,615	437	1,178
1935	1,710	459	1,251
1936	3,010	1,141	1,869
1937	3,773	949	2,824
1938	3,623	1,800	1,823
1939	5,856	2,195	3,661
1940	12,813	6,028	6,785
1941	26,289	19,445	6,844

Source: *Aerospace Facts and Figures*, 1959, 6–7.

ited more than twenty plant sites of eleven major companies. Their findings confirmed the overall financial weakness of the firms investigated and stressed that some form of governmental assistance would be necessary. The report also recognized the political sensitivity of wartime "excess profits," which might occur.

Several changes in naval contracting and procurement policies were recommended. In particular the report viewed unfavorably the then-current CPFF system, recommending its removal and that any excess profits be recovered through taxation.[48] The report also noted that with some 70 percent of complete aircraft subcontracted, the practice should not go any further.[49] The authors recognized the increasingly dispersed ownership of major firms, and the trend away from total control by the founder and toward professional management. After Reuben Fleet left Consolidated, only Glenn Martin, Claude Ryan, and Leroy Grumman of the major firms represented effectively one-man ownership and control.[50] A point of particular difficulty in any financial analysis of the industry at the time was that of adjustments for DPC and AAF plant expenditures in assets, and for the amount of government-furnished equipment (GFE) in the final output.[51]

Certain recommendations were reflected in the Renegotiation Act of 1942, which provided for recovery of excess profits, imposed the Excess Profits Tax, and established the Renegotiation Board to implement. Working-capital shortages continued, and profits tended to lag the astronomical sales increases of the wartime expansion. Wartime hardly ended financial risks in the industry.

Chapter 8 The Industry in Wartime

While the history of the American aircraft industry may have received inadequate attention, that deficiency does not apply to the industry's wartime record. That period has been exhaustively researched and reported, and from several perspectives. Accordingly, it will suffice to summarize the period primarily with regard to technical, organizational, and managerial developments in individual firms, along with production. It is meaningful, however, to consider both government and industry plans developed during wartime for the postwar industry and the role of aviation.

Design and Production Achievements

The war achievements of the industry were due to a combination of factors, including labor, management, government support, scientific and technical efforts, and continuing supplies of vital resources. Support was provided not only through the military services and NACA, but also through cooperation with other institutions. For example, the joint research organization of Guggenheim Aeronautical Laboratories and the California Institute of Technology, known as GALCIT, was a pioneer in rocket research under the direction of Theodor von Karman. Von Karman, who had taken a permanent position at Cal Tech in April 1930, was one of the world's foremost aerodynamicists. Donald Douglas assigned Art Raymond to GALCIT to assist with its efforts. Aerojet Engineering Corporation, later Aerojet-General, was spun off from GALCIT in 1942. GALCIT became the Jet Propulsion Laboratory (JPL) on November 1, 1944, organizationally under NACA but managed by Cal Tech.[1] Then the first V-2 rockets launched at England, on September 8, 1944, carried portents for the future of both warfare and the aircraft industry. JPL began advanced research in rocketry that would find military applications. An early example was von Kármán's development of Jet-Assisted Takeoff (JATO) rockets.

During the war NACA, chaired by Dr. Jerome Hunsaker from 1941, was not so much concerned with advancing aeronautical knowledge as with refinements and solutions to specific problems. A major advance, however, was development of the laminar-flow airfoil, which solved the problem of drag-inducing turbulence at the wing trailing edge that had restricted aircraft performance. Rising research capabilities of the industry also contributed. The Aircraft War Production Council (AWPC) built the Southern California Cooperative Wind Tunnel at Cal Tech, used by all members.[2]

Military organizations also continued to evolve. Resolving numerous functional conflicts, the Air Technical Service Command (ATSC) was formed on September 1, 1944, by merger of the Materiel Command and the Air Service Command. Procurement and production thus were combined, and several layers of management were removed.[3] All aircraft production was managed by the ATSC Procurement Division. While research and engineering remained important functions, primary emphasis was on logistics and procurement contracts, often incorporating changes resulting from operational experience. Since designs frequently were frozen in order to expedite production, many aircraft were completed with built-in flaws or operating shortcomings, which then had to be corrected or otherwise modified before being placed in operation. All manufacturers operated major modification centers.

Several aircraft under development or test by 1940 became major production programs during the war, including the Curtiss P-40 Warhawk and the Grumman F4F Wildcat. Neither was a truly outstanding design, but they would bear the brunt of early air combat. The Douglas A-20 Havoc was the finest attack aircraft in the world and was ordered in large numbers by the British as the Boston. The AT-6 Texan, the B-17 and B-24 bombers, and the PBY have been mentioned.

By July 1941 the industry had orders for eighty thousand airplanes for the U.S. military and for Great Britain, and production was building up rapidly. Perhaps the greatest triumph beyond the sheer quantities produced, however, was the industry's ability to design and place into production and service new combat aircraft after the outbreak of war in Europe. Examples were the Boeing B-29, Douglas A-26, North American P-51, Northrop P-61, Grumman F6F Hellcat, Republic P-47, and to a major extent the Martin B-26 and North American B-25. In contrast, German industry, despite its reputation, had few designs not already in development in 1939 reach operational service in significant numbers.

A useful illustration is offered by the experience of North American. In April 1940 the British Purchasing Commission, although not greatly impressed with its performance, proposed license production of the Curtiss P-40 by North American. Dutch Kindelberger instead proposed an entirely new design to British specifications. The British, already pleased with the Harvard trainer, accepted the proposal, and Kindelberger ordered his staff into an intensive effort on the design, known as the NA-73.[4] The result became the P-51 Mustang. The prototype was flown on October 27, 1940, and was rushed into production.[5] With lend-lease, British orders were taken over by the United States, but early P-51 production still was allotted to the British. Ironically, initial AAF interest in the design was only as a fast attack aircraft, for which it served as the A-36.

In the meantime, Packard had signed a license agreement with Rolls-Royce for production of the Merlin engine, both for U.S. forces and for export, on September 3, 1940.[6] After replacement of the Allison engine with the Packard-built Merlin, conveying improved high-altitude performance, the P-51 became arguably the outstanding American fighter of the war. The change to the Merlin, at the insistence of Kindelberger, provoked opposition within General Motors, with its large stockholding in North American. Allison was a wholly owned GM subsidiary, and Packard was a competitor. The first Merlin-engined P-51 flew on November 30, 1942. The success of the P-51 also contributed to the decline in aircraft of Curtiss-Wright, which never won orders for a successor to the P-40. Efforts to advance the state of the art in fighter design and performance proceeded concurrently with the production buildup. In one such effort, the AAF awarded three contracts in 1940 for advanced pusher-engined fighter designs. The prototypes were the Consolidated (originally Vultee) XP-54, the Curtiss XP-55, and the Northrop XP-56. All were tested beginning in 1943 but encountered developmental difficulties and inadequate performance and did not receive production orders.

One of the more challenging, and subsequently controversial, programs began in the spring of 1941. At that time Britain's survival was far from assured, and strategists felt that eventual American involvement would require bombers of unprecedented size and range to take the war to the enemy. This resulted in a specification for the "10,000 miles with 10,000 pounds" bomber, an aircraft of true intercontinental range with a ten-thousand-pound bombload. The Douglas B-19 was being tested, and the Boeing B-29 was well advanced, but neither could approach the specification, and many doubted that it was attainable. John Northrop seized upon the requirement as an opportunity to demonstrate the superiority of the flying wing, and his N-1 design would evolve into the first combat flying wing as well as the largest. He first proposed the flying-wing bomber to General Arnold on July 10, 1941.[7] Development contracts were awarded to Northrop for what became the B-35 and to Consolidated for the B-36 on November 15, 1941. Later progress in both war theaters downgraded the program priority, however, and neither flew until 1946.

Major aircraft firms increasingly became assemblers, with subcontractors supplying most machined parts and major subassemblies. While enabling more efficient production, the effort required close engineering coordination between major contractors and subcontractors.[8] There was increasingly a transformation from lot to line production, in which aircraft were assembled progressively along a line, in a manner resembling automobiles.[9] Firms soon learned, however, that the aircraft industry could not directly adopt the techniques of the automobile or any other industry, since its production needs still were too different.[10] Large open-bay factory layouts were required, since aircraft extended in all directions dimensionally. Also, there were continuing problems of reconciling the need for large production volume with the need for design changes based on operational experience. The two would never be completely reconciled.

Wartime production figures remain remarkable even from the perspective of time. From January 1, 1940, until August 14, 1945, V-J Day, 303,317 military aircraft were produced, with 274,941 of that total since Pearl Harbor.[11] The peak production month was March 1944, in which 9,117 aircraft came off the assembly lines.[12] By the end of 1943, eighty-one production plants were in operation for airframes, engines, and propellers, with another five plants in Canada bringing the total to eighty-

TABLE 10. Aircraft Production and AAF Procurement during the Second World War

Production		AAF Procurement	
Manufacturer	Number	Type	Number
Beech	7,364[a]	Very heavy bomber	3,899
Boeing	16,149	Heavy bomber	31,000
Chance Vought	6,600	Medium bomber	16,070
Convair	33,000[b]	Light bomber	18,113
Curtiss-Wright	22,977	Fighter	68,712
Douglas	29,385	Reconnaissance	1,117
Grumman	17,135	Transport	22,885
Lockheed	19,273[c]	Trainer	55,712
Martin	8,983	Communications	13,591
McDonnell	66[d]	Total	231,099
North American	41,203[e]		

Sources: For production, *Jane's All the World's Aircraft,* and *Aviation Year Book,* various editions; for procurement, Smith, *Army and Economic Mobilization,* 27.

Note: Production figures are from December 7, 1941 to August 14, 1945, except as noted. Procurement figures are from January 1940 to December 1945.

a. Beech also produced 1,645 A-26 wing sets.
b. Estimated.
c. From 1939.
d. Includes drones.
e. From 1940.

six.[13] The production total of over ninety-six thousand airframes in the peak year of 1944 could have been still higher had not production been cut back late in the year.[14] By that time more aircraft were being built than could be employed. Peak manpower, attained at the end of 1943, was 2,102,000.[15] Total factory space, including engine and propeller production, was 175 million square feet.[16]

Although the Second World War production record was as significant to victory as American successes on the battlefronts and is an enduring industrial success story, the overall expansion would have been much easier if the Morrow board recommendations of 1925 with respect to the industry had been implemented. The military deserves credit for its efforts to keep the industry viable in the 1930s.

Industry Survey

Fleetwings, Inc., saw its BT-12 trainer lose out to the Vultee BT-13. Then its last design, the XBTK-1 naval-torpedo bomber, encountered delays and did not fly until 1945, ending Fleetwings' efforts in aircraft production. Fleetwings was acquired by Kaiser Cargo, Inc., a subsidiary of Kaiser Corporation, in March 1943 and continued aircraft-related subcontracting until the end of the war, when it was completely absorbed into Kaiser.[17] The firm specialized in stainless-steel and aluminum alloy parts.

The Naval Aircraft Factory was placed under the Naval Air Materiel Center on July 20, 1943.[18] It continued in aircraft production until the end of the war, however, following its N3N trainer with 136 of the PBN version of the Consolidated Vultee PBY Catalina, among other activities.[19]

McDonnell, after struggling through its first year, received military R&D and parts production contracts late in 1940, then in October 1941 received a contract to design an advanced high-altitude fighter, later designated XP-67.[20] James McDonnell had long explored advanced design concepts for fighters, and the XP-67 featured wing-fuselage blending that would appear in later supersonic aircraft. McDonnell acquired the old Monocoupe factory at Lambert Field, with thirty-six thousand square feet of floor space, also in 1941.[21] It then won a major subcontract from Douglas, the firm it would later acquire, for the C-47. McDonnell's only series aircraft production during the war was the Fairchild AT-21 Gunner crew trainer, also produced by Bellanca in addition to Fairchild.[22] A new Memphis factory was built for this task. Laurance Rockefeller eventually invested four hundred thousand dollars in the company, greatly enhancing its prospects.[23]

The XP-67 was too late for wartime production, and McDonnell remained a minor firm for the duration. Its future, however, hinged on a single event: the government's acquisition of the Whittle jet engine. Although Bell built the first American jet in 1942, development was pursued by both services and contracts awarded to several firms. McDonnell received a contract early in 1943 for development of the XFD-1 (later FH-1), the first naval jet fighter.[24]

Northrop, already thriving on subcontracting and license production, submitted its proposal for the advanced P-61 radar-equipped night fighter to the Army Air Forces on December 7, 1940, and was awarded the development contract on January 30, 1941.[25] The XP-61 first flew on May 26, 1942, and entered production immediately. Its success established Northrop as a major combat aircraft producer. Beginning only in 1940, the company completed 1,088 production aircraft by the end of the war, consisting of 674 P-61, fourteen N-3PB patrol seaplanes for Norway, and four hundred V-72.[26] Its subcontracting to Boeing, which began in 1941, would continue for decades.

In addition to production, Northrop was extremely active in experimental programs. Theodor von Kármán served as a consultant on the flying wing. Northrop was unaware of parallel flying-wing research being undertaken by the Horten brothers in Germany. The N-9M, an aerodynamic prototype for the XB-35, was first flown on December 7, 1942, and the MX-324 rocket-powered flying wing, the first American aircraft so powered, was tested on July 5, 1944. Its Aerojet rocket motor was of such low thrust, however, that the MX-324 had to be towed as a glider in order to sustain flight. Northrop's XP-56 "Black Bullet," while unstable and not a true flying wing, still possessed certain design characteristics useful in flying-wing development, knowledge later applied to the B-35.

The Chance Vought F4U Corsair first flew on May 29, 1940. When Chance Vought again became a separate division of United Aircraft in January 1943, Rex Beisel became general manager.[27] The Corsair, with its inverted gull wing and powerful R-2800 engine, became one of the outstanding combat aircraft of the war. The OS2U Kingfisher was produced through 1943, but the Corsair was Chance Vought's major war program, and for some years thereafter. Production was also undertaken by Goodyear and, more briefly, by Brewster. The Corsair was used extensively by Marine Corps units as a land-based fighter, but was initially rejected by the navy for carrier use because of its landing difficulties. After the British quickly and successfully operated their F4Us from carriers, U.S. operation followed in multiple roles.

Goodyear followed work on the Martin B-26 with subcontracting on the Grumman TBF, the Curtiss P-40, the Consolidated PBY, and others.

The firm began producing complete B-26 bombers in an expanded DPC plant in 1941, but the contract was soon canceled in favor of production of B-29 components. Goodyear expanded to four factories and 4,750,000 square feet at its wartime peak. Major airframe production was 4,006 FG-1 Corsairs under license.[28] It also developed the F2G version powered by the huge Wasp Major engine, but only ten were built.[29]

Curtiss-Wright remained the largest aircraft firm through the war period in terms of total contract value. The ranking, however, represented combined airframe, engine, and propeller business. Aircraft were produced in Buffalo, Columbus, St. Louis, and Louisville, and engines in New Jersey and Ohio. In aircraft alone, Douglas, Consolidated Vultee, and North American exceeded the output of Curtiss, which began to run down after 1943. The SB2C Helldiver and the A-25, its AAF version, were produced until the end of the war, but P-40 production was completed late in 1944.

Douglas, already in large-scale production, added the Long Beach plant in 1940. Then the Tulsa DPC plant, on which ground was broken on May 2, 1941, built Consolidated B-24s.[30] After protracted development, the XB-19 finally flew on June 27, 1941. The B-19 was almost twice the weight, at 164,000 pounds, and possessed twice the power of the Boeing B-15. But even though the Wright Twin Cyclone engines were the most powerful available, the B-19, like the B-15 also, was underpowered. Thus while not viable as a production bomber, it did provide useful research into large aircraft.

At the wartime peak Douglas operations included three California plants as well as the DPC plants in Tulsa, Oklahoma City, and Chicago. Output included thousands of military variants of the DC-3 and DC-4, Dauntless dive bombers for the Army Air Forces, navy, and marines, A-20 Havocs, and B-17 and B-24 bombers under license. The SBD Dauntless, obsolescent by Pearl Harbor and slated to be phased out in 1942, continued in production until 1944 due to the deficiencies of its Brewster and Curtiss intended successors. Production of the advanced A-26 Invader, successor to the A-20, totaled 2,501. Douglas also designed more advanced naval aircraft. Its XBT2D model, designed by Ed Heinemann, which first flew on March 18, 1945, superseded less promising designs and became the postwar AD Skyraider.

By 1944 Douglas was the fourth-largest industrial corporation in the country. But even at the wartime height Donald Douglas was pessimistic about the industry's peacetime future and was especially wary of the huge expansion in production facilities. Although the largest stockholder at the company's incorporation in 1928, Douglas subsequently sold most of his stock, which in combination with new issues left him a minor stockholder in the company he founded.[31]

Martin's major war production was of the B-26 Marauder medium bomber and the PBM Mariner flying boat. The B-26 also was noteworthy as the first aircraft for which contracts were awarded on analysis and approval of specifications, not flight testing.[32] Subsequently, off-the-shelf procurement applied to several aircraft types.[33] Martin's Baltimore complex expanded into one of the largest in the industry. While a major combat aircraft, the 5,157 B-26s produced constituted the lowest total for any American wartime bomber other than the B-29. Production of the Model 187 Baltimore for the British was completed in May 1944.

Martin maintained its interest in large flying boats throughout the war. A noteworthy development was the XPB2M patrol bomber of 1942, the largest built to that time. It was later developed as the JRM Mars transport, but did not achieve mass production, to some extent a victim of the success of the less costly PBM Mariner. A subsidiary Glenn L. Martin Nebraska Company operated a large DPC plant at Omaha for production of the B-29. While remaining primarily military, Martin designed the Model 202 airliner for postwar service.

Grumman followed its F4F Wildcat with the large XTBF-1 Avenger torpedo bomber, which first flew on August 1, 1941, and then the extremely successful F6F Hellcat, successor to the F4F, which first flew on June 26, 1942. Both enhanced Grumman's position as the leading naval-aircraft producer. Earlier, the powerful twin-engined XF5F Bobcat and its Air Corps development, the XP-50, showed great promise but were not ordered. The basic configuration was applied to the later F7F Tigercat, however.

Grumman's facilities in Bethpage expanded manyfold, largely financed directly by the navy. With priority on the F6F, Grumman ceased production of the F4F Wildcat and TBF Avenger in 1943, licensing Eastern Aircraft Division of General Motors to continue both. Grumman also licensed the J2F Duck to Columbia and produced the JRF Goose until early 1945. The advanced F7F Tigercat and F8F Bearcat fighters were too late for combat.

Lockheed, while small before 1940, became in fact the largest manufacturer at the time of Pearl Harbor, having expanded to fifty-three thousand employees and 3.5 million square feet of floor space for British as well as U.S. production.[34] While continuing the Model 18 in its many variants for the AAF, navy, and export, Lockheed focused on the P-38 and on license production of the Boeing B-17. The Model 049 airliner, with a lower priority, did not fly until January 9, 1943, and then as the military

C-69. Of greater importance was the 1943 contract to develop the XP-80 jet fighter.

By the end of 1941 Vega Airplane became a wholly owned subsidiary of Lockheed as the Vega Aircraft Corporation. Then on November 30, 1943, the subsidiary was completely absorbed and the Vega name disappeared, the plants becoming known as Lockheed A and B. The plants built the B-17 as well as PV-1 and later PV-2 patrol bombers. Unlike other major firms, Lockheed production remained entirely in Southern California and was not geographically dispersed in DPC facilities. Lockheed produced a total of 19,413 aircraft during the war, including 9,877 of the P-38.

Boeing's wartime organization included the Boeing Aircraft Company in Seattle, wholly owned by the parent Boeing Airplane Company, plus the Renton Division and the Wichita Division, reorganized from Stearman in 1941. Boeing shared in foreign orders early in the war, producing 240 Douglas DB-7B attack aircraft under license for France, but later taken over by Great Britain.[35] B-17 production finally attained a high volume in 1941, with Douglas and Lockheed (Vega) also producing large numbers under license.

The Boeing XB-29 first flew on September 21, 1942. The most advanced wartime production aircraft, the B-29 claimed most of Boeing's resources despite the volume of B-17 production. Although an outstanding design aerodynamically, the B-29 was so complex that it did not fly operational missions until July 1944, in marked contrast to simpler types that were pushed into service more rapidly. Adding to its complexity were the new and powerful Wright R-3350 engines. B-29 production originally was entirely at Wichita, where the relatively small trainer plant was supplanted by a massive new complex for the purpose. But production soon involved Renton, Martin-Omaha, and Bell-Marietta, as well as an unprecedented level of subassemblies. Wichita built large numbers of the PT-17 Kaydet in addition to the B-29.[36]

Phillip Johnson died in September 1944 at age fifty, and Claire Egtvedt offered the presidency to William M. Allen, Boeing's legal counsel and a member of the board of directors since 1930. Allen declined the offer at the time, regarding himself as unqualified.[37] Egtvedt continued to direct the company through the war. Boeing's capabilities in large aircraft would lead to major postwar success.

The most important step in Bell's expansion came on August 25, 1941, with an order for two thousand P-39s for the Army Air Forces.[38] Soon after, and also of high importance, came contracts for the XP-59 jet fighter and for operation of the Marietta, Georgia, plant. Bell produced a total of 9,584 P-39s, but most were employed in the low-altitude ground attack rather than in the intended interceptor role, and most were supplied to allied nations under lend-lease. Bell followed with the more advanced P-63 KingCobra, of which 3,303 were built, again primarily for allies, although many were used by the AAF for training and as aerial targets.

Bell expanded from just over one thousand employees at the beginning of 1940 to more than fifty thousand by 1944. The DPC Marietta plant opened on March 15, 1943, and was a notable success. The Marietta B-29s were built under a fixed-price contract rather than CPFF, and Bell produced at lower-than-anticipated cost. Larry Bell advanced the application of smooth assembly line techniques during the war to such an extent that his methods would be adopted by others. Leston Faneuf, with a varied background and later to succeed Larry Bell, joined the firm in 1943. Carl Cover, formerly with Douglas, became manager of the Georgia Division in September 1944, and James V. Carmichael, a Georgian and an attorney by training, became plant manager. Bell's postwar future, however, was to be shaped by its development of the XP-59 and the start of its helicopter division, both in 1942.

Republic became a leading fighter manufacturer, but not before resolving a troubled management situation. General Arnold and Assistant Secretary of War Robert Patterson persuaded the Republic board to install aviation veteran Ralph Damon as president in April 1941.[39] Damon was effectively drafted by the AAF for the position.[40] He then hired Alfred Marchev, a production expert, who succeeded to the presidency when Damon resigned on July 21, 1943, to return to American Airlines.

Republic produced the P-43 Lancer, a development of the P-35, for export until 1942. While the P-43 did not serve with the AAF, it did lead to development of the new and powerful P-47. The prototype XP-47B first flew on May 6, 1941, and orders for the P-47 Thunderbolt soon required a further defense plant. The Evansville, Indiana, plant began production on September 18, 1942. The P-47 production total of 15,239 was the highest of any American fighter and included those supplied to allies. Republic also began the advanced XF-12 reconnaissance aircraft and the Seabee light amphibian in 1944 for the postwar civil market.

Fairchild's principal wartime production was its low-wing PT-19 Cornell trainer. Production also was undertaken by Aeronca, Howard, and St. Louis, and by Fleet Aircraft in Canada.[41] The AT-21 Gunner twin-engined crew trainer, with construction by the Fairchild-developed Duramold process, began in 1941 and was also produced in limited numbers by McDonnell and Bellanca.[42] Fairchild continued to produce several military versions of the F-24 for U.S. and British forces. Fairchild's future appeared to lie with large transports, however. Chief designer Armand Thieblot, who had emigrated from France under sponsorship of Charles Levine and who designed the M-62, led design of the new C-82.[43]

North American, with high-priority orders for its successful P-51, B-25, and AT-6, vaulted into the top rank of wartime aircraft manufacturers. Under the dynamic leadership of Dutch Kindelberger and Lee Atwood, production rose rapidly. Kindelberger was acutely aware that manpower was a more critical production problem than either facilities or materials and devoted major attention to labor and training.[44] He also made major contributions to development of subassemblies and enhanced accessibility in maintenance. Ernest R. Breech, chairman of North American since 1933, resigned in 1942 to head Bendix Aircraft and was succeeded by Henry R. Hagan, another GM executive.

North American's defense plants were phased into production before Pearl Harbor. The Dallas plant for the AT-6 began on August 17, 1940, and the Kansas City plant for the B-25 was announced on December 7, 1940. The Dallas plant was financed by the DPC but owned by NAA, while the Kansas City plant was owned by the DPC.[45] The second Dallas plant, known as Plant B, followed and produced the B-24 under license. From 1938 through 1945 North American built 43,208 aircraft, the highest total for any U.S. manufacturer.[46]

Ryan delivered three YO-51 test models in 1940, but adaptation of light civil aircraft for observation precluded production orders. Ryan's major aircraft program during the war was the PT-22 trainer, with more than one thousand built. Ryan also became possibly the largest contract flying school in the country, as the Army Air Forces widely employed civilian schools. While remaining active in subcontracting, Ryan received a navy contract on December 21, 1943, to develop the XFR-1 compound fighter, with a nose piston engine supplemented by a jet engine exhausting through the tail.

Late in 1941, even as war approached and the military buildup and industry expansion were gaining momentum, Reuben Fleet felt increasingly pressured to sell his interest in Consolidated and to retire from the industry. Fleet, then only fifty-four, never stated publicly his reasons for so doing, but it was widely believed that he had become disenchanted with the near-confiscatory income tax policies, increasing government interference in his business, and friction with Henry Ford over Willow Run production of the B-24.[47] The government, in turn, was concerned about Fleet's efforts to maintain one-man control of a large company, pressured him to delegate authority,[48] and sought an acquirer.

Vultee Aircraft emerged as the purchaser, acquiring Fleet's 34 percent stock holding in November 1941 for a reported $10 million. The remainder of Consolidated stock was closely held. Vultee, while smaller than Consolidated, had developed a considerable capability in military aircraft. With majority ownership of Vultee by AVCO, Victor Emanuel of AVCO held operating control of the combined companies.[49] Reuben Fleet then became a consultant to Consolidated Vultee, but that proved to be a mutually unhappy experience.[50] Fleet continued to speak out on military and industry issues as a private citizen.

In January 1942 Consolidated and Vultee, under the Aviation Corporation umbrella, began to operate as a single firm managerially, and the formal merger creating the Consolidated Vultee Aircraft Corporation (ConVAir) became effective on March 17, 1943.[51] Earlier, in July 1942, Vultee purchased the Intercontinent Aircraft Corporation, organized only in 1940, while Consolidated acquired Stout Engineering Laboratories of Michigan, bringing William B. Stout to the company. Vultee president Richard Millar, along with production vice president Don Carrill and engineering vice president Richard W. Palmer, resigned soon after the formal merger. Emanuel's associate Tom Girdler of Republic Steel was effectively loaned to Consolidated Vultee from 1942 to 1944 to oversee the merger of the two firms.[52] Although age sixty-five, Girdler became chairman and Harry Woodhead, formerly with Vultee, became president. I. M. Laddon became executive vice president and general manager.

Consolidated Vultee established an outstanding war production record, delivering 28,004 complete aircraft between Pearl Harbor and V-J Day. Its B-24 production alone totaled 10,708, plus spares, and total B-24 output including Ford, Douglas, and North American, was 18,482, an American record for a basic design.[53] Further versions included the C-87 transport and the PB4Y naval patrol bomber. The PBY was produced in greater numbers, 2,140, than any other flying boat or amphibian and was famed for its rescues of downed aviators in the Pacific. BT-13 Valiant production totaled 11,357. Other production included the PB2Y Coronado and the B-32 long-range bomber, built to the same requirements as the B-29. But the B-32 was delayed in production and few were delivered. By 1944 AVCO had become the third-largest military contractor, after GM and Curtiss-Wright.[54]

Auto Industry Production

General Motors was in fact the largest American war contractor, although its activities encompassed much more than aviation. A key event was establishment of the Eastern Aircraft Division in January 1942, with five

GM automobile facilities being modified for aircraft production. Locations were Baltimore; Tarrytown, New York; and Linden, Bloomfield, and Trenton, New Jersey, with main assembly at the Trenton and Linden plants. Eastern failed in its first attempt to negotiate a deal for the Republic P-47.[55] But continued navy support helped Eastern negotiate successfully with Grumman.[56] Eastern became the principal producer of the Grumman TBF Avenger and the F4F Wildcat, freeing Grumman to concentrate on its later designs.[57] Eastern completed 5,927 F4M fighters and 7,522 TBM torpedo bombers, becoming the largest single-firm producer of naval aircraft during the war.[58] Upon completion of war production, the Eastern Division was dissolved, and GM began reconversion of the facilities to auto production.

The Fisher Body Division of General Motors also played an important role in aircraft production. The Cleveland DPC plant, originally intended to produce complete B-29s, was instead converted to production of B-29 subassemblies.[59] In an interesting but ultimately unsuccessful venture, Fisher designed a new fighter under Don Berlin, formerly with Curtiss-Wright. The Fisher P-75 Eagle was unique in that it used existing structural components, including P-40 outer wings and the A-24 Dauntless tail assembly, but was powered by a 2,600 hp Allison liquid-cooled engine. It was felt at the time that this approach would enable a new type to be put into production sooner than an entirely new design.[60] The project then was completely redesigned in 1944 as the P-75A. The P-75A met performance expectations, but the order for twenty-five hundred was canceled as the AAF decided to standardize on fewer types, ending Fisher's venture into aircraft design.[61]

As in the First World War, Henry Ford was interested in aircraft production and was confident his company's capabilities would result in major production successes. The War Department, concerned that Consolidated lacked the required capacity for the B-24, also became interested. Ford, financed by the DPC, undertook construction of the largest factory in the world under one roof: Willow Run, named after a local stream and near Ford's Dearborn headquarters. Dr. George J. Mead, the engineer who had contributed much to the success of United Aircraft, sold Henry Ford on the original idea.[62] The operation also captured the interest of the American public. Ford executive Charles E. Sorensen was placed in charge.

Willow Run, soon known as the "world's largest bomber plant," with a floor area in excess of 2.5 million square feet, was the most storied, and perhaps the most troubled, of the wartime defense plants. Its difficulties were such that wags renamed it "Willit-Run?" Its completion in May 1942 was accompanied by an announcement that B-24 production was under way, but the first aircraft was not in fact completed until September. Aircraft-engineering drawings were not easily interpreted by autoworkers, and there were major tooling differences, adding to startup problems.[63] Unlike automobile production, there was great difficulty in "freezing" a design to facilitate efficient mass production: detail changes occurred continuously. J. H. Kindelberger, no proponent of auto industry participation in airframe production, likened the automobile-aircraft transformation to that of blacksmiths to watchmakers and attacked the industry as creating a bottleneck in aircraft production.[64] The result was that B-24s produced at Willow Run were built entirely differently from those built by Consolidated Vultee. The fuselage, for example, was constructed in two lengthwise halves, outfitted with all electronics and equipment, then joined. Another problem was high worker turnover, caused in part by the plant's rural location and transportation difficulties. Problems were ironed out by the end of 1943, however, and the plant met its objectives.[65] Total B-24 production at Willow Run was 6,724.

New Aeronautical Developments

Military Air Transport

The aircraft category that directly linked the military and commercial markets was air transport. The United States, perhaps more than other combatants, required substantial air transport not only for mobility within a battle zone, but also to transport men and materiel to overseas battlefronts. While the military was interested in specialized military transports, as pioneered by the Fairchild C-31 of 1933, it depended on adaptations of civil designs for wartime requirements. The Douglas C-47 was to pass into legend both as an airliner and military transport. The Douglas C-54, from the DC-4, and the Curtiss C-46, also originally an airliner, likewise would see wide wartime service. The C-54 added transoceanic capability, invaluable at the height of the war.

The Air Transport Command of the Army Air Forces was created in July 1942, succeeding the earlier Ferry Command and taking on added functions while retaining the ferrying mission. The navy, ever intent on developing an organization equal to that of the army, earlier had created the Naval Air Transport Service, on December 12, 1941.

Specifications for specialized military transports advanced. Three new designs, the Waco C-62, developed from a glider design, the Budd RB-1/C-93 Conestoga, and the Curtiss C-76 Caravan, were authorized.[66]

All featured large cargo capacity and loading doors. None saw quantity production, although they did provide useful experience for the future.

The Budd RB-1 Conestoga involved a new design to a naval requirement by a manufacturer with no previous aircraft experience. Budd had to assemble not only a design team but a production team. An anticipated shortage of aluminum led to use of stainless steel for the design, and Budd had an impressive record with spot-welded stainless-steel fabrication.[67] But by the time Budd completed the first RB-1 in 1944, after nagging delays, the navy requirement no longer existed, and the shortage of aluminum had not occurred.[68] The army likewise lost interest in the C-93 version, cutting its order from two hundred to twenty-five.[69] The Curtiss C-76 first flew on January 5, 1943, and was originally ordered in quantity, but testing revealed assorted structural and control problems, and the entire program was canceled in August 1943.[70]

It was not until the Fairchild C-82 Packet, with its distinctive twin-boom design, that a specialized design entered quantity production. Development began in late 1941, and the XC-82 first flew on September 10, 1944, but due to various delays did not enter service until after the war.

Another military transport was derived from a bomber, as Boeing developed the Model 367 from the new B-29. Using B-29 wings and engines, the Model 367 featured a unique "double bubble" fuselage, allowing double-decker passenger carrying, or upper-deck passenger and lower-deck cargo transportation. The prototype XC-97 flew on November 15, 1944, but was too late for war service. From that point, however, the services would order, and the industry would design and produce, specialized transports.

Helicopter Progress

The development of the helicopter ranks with the jet engine among the most significant aeronautical achievements of the Second World War. Although actively experimented with in several countries for years, wartime brought both developments to the operational stage, with consequent momentum for postwar applications. While lagging in jet engine development, in the helicopter field the United States was well ahead and would remain a leader.

The helicopter sprang from the demise of the autogiro. AGA Aviation, descended from the original Pitcairn firm, had secured small orders from the British in 1941, but major orders still eluded both AGA and Kellett. Kellett received a small army order, but AGA ended development with the large, enclosed-cabin PA-36. Helicopter progress effectively ended military interest in the autogiro, and "grasshopper" aircraft were much cheaper for the army's spotter mission.

Kellett, AGA, and the new Platt-LePage firm, formed in 1940, all turned toward the development of a practical helicopter.[71] AGA was acquired by Firestone on July 30, 1942, and renamed G & A Aircraft. All patents were included in the transfer. It primarily engaged in military glider production as it continued helicopter development.[72]

Helicopter development was spurred by the congressional Dorsey-Logan Grant of 1939. The eventual three-hundred-thousand-dollar grant specified a vehicle with hovering ability, ruling out the autogiro.[73] Both Platt and LePage were engineers, and LePage had worked for both Pitcairn and Kellett. Their development, based on the German Focke design, won the award on April 15, 1940, with a commitment to deliver a practical helicopter.[74] Their XR-1 used antitorque vanes rather than the Sikorsky-patented tail rotor for lateral control. Unfortunately the first prototype XR-1 crashed, and Sikorsky received the balance of Dorsey-Logan funding after the failure of Platt-LePage to deliver its model in 1941. Successful development of the VS-300 then led to a contract for the military XR-4, which first flew on January 13, 1942, and which became the first production helicopter. The United States thus became the first country to place helicopters in service, leading to Sikorsky's position as the foremost helicopter producer in the world. The XR-4 dramatically demonstrated its capabilities with a 761-mile delivery flight from Stratford to Wright Field from May 13 to May 18, 1942. Production of the R-4 was 133, some going to Great Britain. Nash-Kelvinator later license-built 214 of the developed R-6 and YR-6 Hoverfly.[75] The larger all-metal R-5, first flying on August 18, 1943, would see widespread postwar service. Separation of Vought and Sikorsky in 1943 ended resource conflicts within United, aiding helicopter development.[76]

Although many firms worked to develop helicopters, as in the case of fixed-wing aircraft few would be long-term survivors. Further, only Sikorsky among the earliest developers would become a permanent factor. Firestone, upon its acquisition of AGA, took over the original Pitcairn Willow Grove plant, and Harold Pitcairn was effectively out of management. He then led the design of the small XR-9 helicopter for Goodyear. The developed XR-9B was successful but was not ordered by the army.[77] Kellett ended autogiro development in 1943, and a name change to Kellett Aircraft Corporation reflected its broadened activities. Kellett's first helicopter, the XR-8, an intermeshing rotor design, was successfully tested on August 7, 1944. McDonnell acquired an interest in Platt-LePage in 1942 and in 1943 formed a Helicopter Research Division in St. Louis.[78]

New firms entering the rotary-wing field included Landgraf Helicopter Company, which was incorporated on September 20, 1943, and won contracts to develop its small helicopter with twin outrigger-mounted rotors. Higgins Industries, a major shipbuilder but an unsuccessful aircraft contractor, established a division in 1943 to develop an experimental helicopter. N. O. Brantly formed a company in 1943 to develop a small coaxial-rotor design. Development would continue after the war. In 1942 the nineteen-year-old college dropout Stanley Hiller Jr. established a Helicopter Division of his father's Hiller Industries to develop his coaxial twin-rotor helicopter, which he patented. The first Hiller XH-44 Hiller-Copter was tested in August 1944, and production was licensed to Kaiser Cargo in September 1944.

The P-V Engineering Forum began in 1941 as a part-time venture by Frank Piasecki and his classmate Harold Venzie. Piasecki, a young self-taught engineer who first worked for Kellett and then Platt-LePage, brought in as chief engineer Elliot Daland, a founder of Huff Daland Airplanes.[79] As the P-V Engineering Forum, Inc., from January 1943, it engaged in aircraft subcontracting during the war but pursued helicopter development under a navy contract. Harold Venzie left in 1943 after unspecified disagreements with Piasecki, but work progressed rapidly.[80] The single-seat PV-2 on April 11, 1944, became the second helicopter, after the Sikorsky VS-300, to be publicly tested. The larger PV-3, the first tandem-rotor design, received a naval contract on February 1, 1944, flew in March 1945, and entered production postwar.

Bell, to become one of the most prolific helicopter producers, entered the field when Larry Bell hired the engineer Arthur Young in 1941. Young made major advances in rotor stability, and the small Model 30 was tested in 1943.

Jet Aircraft Progress

Although the United States lagged in the development of turbine power, or what was then known as the turbo-jet, it had not been completely inactive. Northrop conceived its "turbodyne" involving a large propeller turbine in 1939, and Lockheed proposed its L-133 jet aircraft design in 1942. Both were independent of the Whittle developments, but neither progressed to the hardware stage.[81] Northrop aspired to have its flying-wing bomber powered by the turbodyne but gained little government support. Nathan Price of Lockheed, formerly with Boeing, proposed a jet engine in 1940 but could not gain government funding for the development.[82]

The United States, in the person of General Arnold, eventually noted the potential of the Whittle engine and in 1941 made arrangements for its development and production in the United States. To insure that Pratt and Whitney and Wright, the major piston engine producers, were not distracted from their efforts, the government gave the Whittle license to General Electric.[83] Group Captain Frank Whittle traveled to Boston in secrecy to assist GE in building an American version of his design, and Union Carbine and Carbon assisted in the development of high-temperature materials.[84] Curtiss-Wright had originally negotiated with Power Jets for the license, but the U.S. government intervened to control the development. The government subsequently admitted Allison, Allis-Chalmers, and Westinghouse to the jet engine field.[85]

Bell's contract for the highly secret XP-59A jet fighter, to be powered by two GE-built Whittle engines, designated I-16, was awarded on September 5, 1941. The XP-59A Airacomet first flew at Muroc Lake on October 2, 1942.[86] While potentially a production fighter, the P-59's performance was disappointing due to its inadequate power, but the sixty-six P-59s produced provided valuable test and training experience.[87]

The government was sufficiently impressed with the jet's potential to award a contract to Lockheed for the XP-80 in mid-1943.[88] Lockheed exploited its experience with the L-133 project, and the contract also marked the birth of Lockheed's famed "Skunk Works" under the direction of Kelly Johnson.[89] The XP-80 was developed in remarkably brief time, first flying on January 8, 1944. An engine change delayed service entry until after V-E Day, although the first P-80s were arriving in Europe by war's end. McDonnell, in the meantime, was advancing with its XFD-1 carrier-based fighter, and jet aircraft contracts soon involved Douglas, North American, Republic, and Chance Vought.

Wartime Production Problems

Although the war production record of the aircraft industry remains a proud achievement, that record was not unblemished. Despite success with large-scale production, integration of large numbers of unskilled and inexperienced workers into the workforce, involvement of women workers, and major expansion of production facilities, there were significant failures. Some lay with the inexperience of senior executives in the small prewar industry with large-scale undertakings. Managing the expansion of a company from perhaps five thousand employees to one hundred thousand employees taxed the capabilities of most. Further, certain companies simply proved unable to perform to military requirements. Three examples will illustrate.

Brewster

Brewster experienced production problems of such severity that even high-priority contracts were suspended or canceled. Its problems eventually resulted in an outright government takeover. James Work resigned under pressure as president and general manager and was replaced by George Champline in 1940.[90] Dayton Brown, designer and chief engineer, departed late in 1941 for the navy's BuAer and later joined Grumman. He went on to found a successful aircraft parts firm postwar. Brewster's new Johnsville, Pennsylvania, DPC plant opened in 1941, offering hope of improved efficiency over the old Long Island City plant. But operating experience with the F2A Buffalo and SB2A Buccaneer had harmed the company's reputation, and continuing labor, production, and management problems led the navy to take over both the Long Island City and Johnsville factories on April 18, 1942.[91] Champline was forced out on January 9, 1942, and management was placed in the hands of retired captain G. C. Westervelt, who had collaborated with Bill Boeing in his first aircraft in 1916 and recently retired after seven years as head of the Naval Aircraft Factory.

Rapid expansion had strained Brewster's financial reserves, requiring advance payments from the navy, and production fell behind as well. A contract to produce fifteen hundred Vought Corsairs at Johnsville was in serious difficulty. Westervelt failed to solve the many problems in his brief tour, although the company was returned to private management. In addition to persistent production problems, Brewster was accused of war profiteering.[92] Labor relations were also strained, and the firm became the only defense contractor to suffer a strike during the war. The company's final design, the XA-32 of 1943, was not ordered.

Industrialist Henry J. Kaiser assumed control of the company in April 1943, with some success in stabilizing operations. Kaiser then stepped down as chairman on May 18, 1944, with Corsair production back on track, but by then the rate of aircraft production was being reduced, and the Corsair order carried a lower priority. The navy canceled remaining contracts on July 1, 1944, after delivery of 736.[93] The Johnsville plant remained as a modification center operated by the Naval Aircraft Modification Unit of Philadelphia, but Brewster had exited the aircraft industry.

Hughes

Hughes Aircraft Company essentially remained a hobby of Howard Hughes through 1940, but his Culver City, California, plant opened in July

1941 gave him significant production space.[94] The D-2 project, however, was finally rejected by the Army Air Forces on November 7, 1941, which Hughes regarded as a manifestation of a conspiracy against him by the government.[95] Then, ever mindful of the potential of political influence, Hughes cultivated Colonel Elliott Roosevelt to recommend for procurement the high-speed reconnaissance development of the D-2, which would become the XF-11.[96] Hughes finally received a production contract for one hundred F-11 aircraft late in 1943.

Subcontracting and component production business rose, but management remained haphazard. For long periods Hughes ignored the company, then at other times would interfere to the point of disruption. Concerns about efficiency reached a point that Hughes finally installed a professional manager, hiring Edward G. Bern in June 1943. But with entrenched resistance to change and no effective backing from Hughes, Bern resigned in frustration after only two months.[97]

Earlier, Hughes was approached by Henry J. Kaiser with a proposal to build a fleet of giant transport flying boats. Kaiser was increasingly alarmed at sinkings of his Liberty ships, then carrying the supply burden.[98] After first considering a Northrop flying-wing design, Kaiser settled on the flying boat and sold the idea to Donald Nelson at the WPB. Kaiser possessed no aircraft design capability but believed Hughes could team with him on the task. Although initially reluctant, Hughes soon agreed, but the resulting order was for only three HK-1 flying boats, not for the vast fleet Kaiser had envisaged.[99] But armed with an $18 million contract, the new Hughes-Kaiser Corporation was formed on November 16, 1942, and moved forward with design.

The AAF remained concerned about Hughes Aircraft's management and capacity to handle both the F-11 and HK-1 contracts. With delays and cost overruns the WPB canceled the HK-1 contract on March 27, 1944. Hughes, however, was immediately awarded a new contract for a prototype of the giant aircraft, redesignated the H-4 Hercules.[100]

Hughes hired Charles Perelle from Consolidated Vultee as general manager in August 1944, promising him full autonomy. Perelle, a protege of Tom Girdler, felt himself vulnerable after Girdler returned to Republic Steel, and accepted the Hughes challenge. Facing continuing resistance to new policies and an organization with no experience in mass production, Perelle nonetheless pushed development of the F-11 and H-4 projects.[101] Continuing friction with the staff and his resistance to Hughes's interference eventually led to Perelle's firing in January 1946, but in his time he had largely completed development of the F-11 and H-4, although too late for the war.[102] The production contract for the XF-11 was eventually canceled, but two prototypes were flown and tested by Hughes personally.

Curtiss-Wright

Even with wartime production urgency, Congress and the public still were highly sensitive to shoddy contract performance and possible profiteering. The Truman committee assumed a troubleshooter role, investigating contractors and programs as warranted.[103] One major investigation focused on the Curtiss-Wright Corporation and its Wright engine plant in Lockland, Ohio. The plant was criticized for deficient quality and inspections, and the corporation for its management shortcomings.[104] Guy W. Vaughan of Curtiss-Wright took a defiant attitude, setting the stage for an enduring lack of confidence between Curtiss-Wright and the government, a situation that may well have had a material effect on the company's decline in aircraft after the war.

A Summing Up

The wartime record of the aircraft industry ranks among the most remarkable achievements of the United States as a nation. From a rather small and narrow base immediately before the war, ranking only forty-first among American industries, it rose to first in less than five years. The dollar value of the industry's 1939 output, only $225 million including exports, rose to some $16 billion for 1944.[105] While there was an epic governmental effort in organization and support, especially with Defense Plant Corporation facilities, and with the participation of other large industries in aircraft production, it was still the infrastructure, the design, engineering, production, and managerial core of the prewar aircraft industry, that enabled the achievement. Men who a few short years earlier were struggling to establish or maintain their modest companies, such as Donald Douglas, Glenn Martin, Lawrence Bell, James H. Kindelberger, and Robert Gross, had become industrial titans.

The aircraft industry also became a truly national industry. At the beginning of the war the industry still was concentrated along or near both coasts. Wartime expansion, however, was strongly toward inland factories. Kansas, Texas, Missouri, and Ohio became particularly important. The massive production effort to meet war demands was so successful that production began to be curtailed well before the end of the war could be foreseen. For this reason certain marginal contractors found themselves facing cancellations even while the war raged. Wartime presented no carte blanche for military contractors.

With respect to its production success it must be noted that the American industry enjoyed an enormous advantage over those of its enemies, and in some cases over its allies, in being able to operate without the threat of air bombardment or other military damage to factories, without the widely feared shortages of critical materials such as aluminum and rubber, and without military interdiction of supply lines. In addition, labor relations were fairly stable. Organized labor viewed itself as a patriotic national resource, and the no-strike pledge was largely successful. There were several wildcat strikes, but none was of an extent to disrupt production.

In common with other military industries, the aircraft industry would face uncertainty after the war, yet there was cause for optimism, since aviation and air transportation had become firmly established, and wartime capabilities could be turned toward civil applications. Unlike the First World War experience, there was ample time to foresee the peacetime adjustment, and the government stood ready to support and assist the industry in the adjustment challenge.

Chapter 9 ✈ The Postwar Industry Adjustment

Wartime aircraft production was the greatest industrial endeavor ever undertaken, with manpower employed and the value of output far exceeding any previous industrial task.[1] Yet the accompanying massive industrial structure would not be preserved in peacetime. It was realized even at the height of the Second World War that a wrenching contraction would be inevitable, and that planning was vital.

The Industry Plans for Peace

Studies were undertaken and plans begun for adjustment as early as 1942. Professors Lynn Bollinger and Tom Lilley of the Harvard Business School published a major study in 1943, with industry financial support, entitled *Financial Position of the Aircraft Industry*. The study emphasized the rather overextended finances of the major manufacturers as a result of their feverish expansion to meet wartime requirements and projected what they could expect in the postwar period.[2] It pointed out that equity capital was very low relative to sales for the industry, especially given rapid wartime growth, and that working capital tended to be very thin.[3] There was a rather pessimistic conclusion about the postwar prospects for the industry, with the expected massive contract terminations forcing contraction, and with limited financial capacity of firms to absorb the resultant inventory losses. In addition, the prospects of raising major amounts of equity by stock sales in the postwar market were not bright.[4] These forecasts proved to be quite accurate.

Expansion of industrial capacity to meet war demands, despite many difficulties, had been largely completed by 1943. Some production actually began to taper off after mid-1944, and the total of more than ninety-six thousand aircraft in that year could have gone far higher.[5] The War Mobilization and Reconversion Act, to provide for termination of prime production contracts, was passed on October 3, 1944.[6]

The industry also planned for peace during the war, with major emphasis on managing cancellations and dislocations. On July 10, 1944, the Aeronautical Chamber of Commerce presented to the Senate Military Affairs Committee a report entitled *The Aircraft Industry Prepares for the Future,* which called for a formal postwar national air policy.[7] The government also established a Standing Subcommittee on Demobilization of the Aircraft Industry in 1944, with representatives from the Departments of War, Navy, and Commerce. On June 27, 1945, this became the Air Coordinating Committee of the Standing Subcommittee on Demobilization. The committee's efforts interestingly dealt as much with future mobilization as with demobilization. The *Report of the Air Coordinating Committee to the Standing Subcommittee on Mobilization,* issued on November 11, 1945, established an eight-point program for government action.

1. Continued research and development
2. Encouragement of the development of civil and commercial aviation
3. Provision of favorable terms for the industry to purchase surplus plants
4. Preservation of geographical dispersion
5. Planning for expansion
6. Provision of a reserve of machine tools
7. Provision of an industrial-plant reserve
8. Integration with other mobilization plans[8]

These recommendations were largely implemented.

With victory came massive contact cancellations as anticipated, and the readjustment to a peacetime economy began. The first cancellation letters were sent to the industry on August 17, 1945. Soon more than $20 billion in aircraft cancellations were issued.[9] Major firms in nonaircraft industries, such as General Motors, Ford, Goodyear, and Firestone, were only too happy to return to their traditional markets and product lines, although a then-senile Henry Ford made some rumblings of manufacturing commercial air transports. But the major aircraft firms faced a highly uncertain future. Many expected the loss of military production to be offset by a civil-aviation boom, with strong growth for private aircraft and commercial airliners. The prewar transport manufacturers Boeing, Curtiss-Wright, Douglas, and Lockheed readied adaptations of wartime transports for the airliner market, while Convair, Martin, and even Republic entered new designs. Further, North American, Douglas, Convair, Martin, Fairchild, Grumman, Republic, and Lockheed considered the private-aircraft market. Only North American and Republic moved to production, but both withdrew after losses.

Immediately following the end of the Second World War it appeared superficially that the 1919 experience was repeating itself, with the sudden lack of production work brought about by contract cancellations placing the future of many firms in doubt. The industry also anticipated severe problems not only with rapid shrinkage but with disposal of surplus aircraft on the market.[10] Firms nonetheless pursued the postwar commercial market while continuing military development. Commercial air service across the Atlantic resumed in January 1945, before the end of the war in Europe. The War Production Board order controlling output of civil aircraft was revoked on May 17, 1945, enabling firms to return to that market. But the anticipated civil boom for new airliners did not appear immediately.

Despite the difficulties in the transformation, the United States had much longer to anticipate postwar conditions and consequently planned far more effectively for the postwar world than it had with the earlier war. In addition to domestic economic adjustment, the United States had assumed the mantle of leadership of the noncommunist world and was instrumental in founding such institutions for the new world order as the United Nations, the World Bank, and the International Monetary Fund. The aircraft industry did in fact contract, but there were no dislocations of such severity as to affect the overall economy. Aviation had developed to the point that the survival of the industry was never in serious doubt. But it was only with the Korean War and recognition of a Soviet military threat that another major expansion took place.

The Aeronautical Chamber of Commerce changed its name to the Aircraft Industries Association on June 16, 1945, ending the possibility of confusion with the U.S. Chamber of Commerce. The AIA also absorbed the wartime National Aircraft War Production Council. The Manufactur-

TABLE 11. U.S. Aircraft Production, 1942–52

	Total	Military	Civil
1942	47,675	47,675	0
1943	85,433	85,433	0
1944	95,272	95,272	0
1945	48,912	46,865	2,047
1946	36,418	1,417	35,001
1947	17,739	2,122	15,617
1948	9,838	2,536	7,302
1949	6,137	2,592	3,545
1950	6,293	2,773	3,520
1951	7,923	5,446	2,477
1952	12,811	9,302	3,509

Source: *Aerospace Facts and Figures,* 1959, 6–7.

ers Aircraft Association, which had administered a heavy volume of cross-licensing agreements during the war, continued in existence, but at a sharply reduced level of activity.

In 1946 the RAND (Research ANd Development) Corporation was established by General Arnold at Santa Monica. An outgrowth of the Arnold-Douglas friendship, and initially staffed primarily with Douglas engineers, it began research on major technical and strategic questions. RAND became independent with major funding from the Ford Foundation in 1948 and gained steadily in influence. Also in 1946, General Arnold commissioned Theodor von Kármán to undertake a long-term technology plan for the future air force. With updating, the plan became the foundation of what aircraft, missiles, and support systems were developed and procured.

Merger and Locational Pressures

While there was extensive optimism for the long-term prospects of the industry, the consensus view was that there were too many manufacturers for the future market, and that rationalization was inevitable and desirable.[11] Robert E. Gross, in a statement to the President's Air Policy Commission on October 14, 1947, noted that fourteen different airframe companies produced ninety-six thousand airplanes in 1944 but delivered only 1,330 military airplanes in 1946.[12] The commission's report listed fifteen major airframe companies engaged in manufacturing in 1947.

- Bell Aircraft Corporation
- Boeing Airplane Company
- Consolidated Vultee Aircraft Corporation (includes Stinson Division)
- Curtiss-Wright Corporation
- Douglas Aircraft Company, Inc.
- Fairchild Engine and Airplane Corporation
- Grumman Aircraft Engineering Corporation
- Lockheed Aircraft Corporation
- The Glenn L. Martin Company
- McDonnell Aircraft Corporation
- North American Aviation, Inc.
- Northrop Aircraft, Inc.
- Republic Aviation Corporation
- Ryan Aeronautical Company
- United Aircraft Corporation (included for Chance Vought and Sikorsky Divisions)[13]

The situation faced by the British industry was even more critical, with no fewer than twenty-two airframe companies, and a potential postwar market only a fraction of that of the United States.[14]

With no realistic prospect of a market supporting all major manufacturers, merger discussions became increasingly active. Certain firms initiated merger proposals, while others were proposed by the military, usually discreetly, given that the industry was still legally stockholder owned. In 1946 Sherman Fairchild advocated acquiring Martin but could not persuade his board. During the same time Robert Gross was moving Lockheed toward a merger with Consolidated Vultee, which did not go forward only because of government antitrust concerns.[15] Floyd Odlum, soon after taking control of Consolidated Vultee, explored a merger with North American. There was more than one proposal to merge Curtiss-Wright. Boeing declined one proposal in November 1948 due to concerns over Curtiss-Wright management weaknesses and over airframe and engine manufacturing in the same corporation.

The most controversial merger proposal, and the subject of lingering dispute, was that to merge Northrop into Consolidated Vultee, under pressure from the air force. While neither firm ever confirmed discussions publicly, the proposal was said to revolve around the issue of Northrop and its B-35 bomber, then in competition with the Convair B-36 already in production. One proposal involved converting the wartime Fort Worth plant to produce the flying wing rather than the B-36. A meeting was held in California on July 16, 1948, involving Secretary of the Air Force Stuart Symington, Floyd Odlum, and John Northrop. Also attending were General Joseph T. McNarney, a 1917 flying-school classmate of Reuben Fleet, shortly to retire as commander of the Air Materiel Command and later to become president of Convair, and Northrop chairman Richard Millar. Symington reportedly was adamant that the number of manufacturers had to decline, and that Northrop must merge with Convair. John Northrop would not agree to being taken over by Convair, however, and his flying-wing bomber, by then the jet-powered B-49, was canceled six months later.[16] Incidentally, Louis Johnson, then secretary of defense, had been a Consolidated Vultee director before assuming that office. B-36 production was, of course, continued despite skepticism from several quarters, and Convair retained the Fort Worth factory. Allegations of political considerations in the flying-wing cancellation and further orders for the B-36 led the House Armed Services Committee to open hearings on the matter. Floyd Odlum vehemently denied any political favoritism in the B-36 program.[17] Secretary of the Air Force Symington also denied pressuring Northrop to merge with Convair.[18] Questions remained, but John

Northrop and other Northrop officers refused to comment publicly, fearful of harming future business prospects.[19]

As it developed, there were no mergers of aircraft firms during the early postwar period. Instead, the number of active manufacturers shrank slowly, as some exited the airframe industry while remaining active in related fields. Complex legal and competitive questions remained, in that the military tended to favor industry mergers to preserve a strong, if more concentrated, industrial base, but this ran counter to antitrust policy and traditional American emphasis on strong competition. Convair was acquired by the newly named General Dynamics Corporation in 1953, but further mergers would await the 1960s.

Industry contraction was indeed rapid. By December 1945 only sixteen airframe plants were in operation, down from sixty-six a year earlier.[20] Total employment had already declined to 519,900 by V-J Day and fell to only 138,700 by February 1946.[21] Financial consequences followed inevitably. General Oliver P. Echols, who became president of the AIA on January 15, 1947, noted in a major address that 1947 was the second year of general financial losses, and that total losses in the industry were approaching $100 million.[22]

TABLE 12. Industry Employment, 1939–54 (as of January 1, in thousands)

Year	Total
1939	64
1940	100
1941	253
1942	618
1943	1,609
1944	2,080
1945	1,684
1946	192
1947	195
1948	179
1949	264
1950	282
1951	343
1952	600
1953	795
1954	783

Source: Northrop, "Air Industry"; and Aerospace Industries Association, *Aerospace Facts and Figures, 1962*, 14.

The shrinking industry tended to retain those plants in which it had operated before the war. Such inland plants as those of Douglas in Chicago and Tulsa, Bell in Marietta, Martin in Omaha, Republic in Evansville, and Curtiss in Louisville and St. Louis were closed. The giant Willow Run plant was leased and later purchased by Kaiser from the Surplus Property Administration for future automobile production. But the Korean War expansion would bring the industry back to the locational pattern resembling that at the height of the Second World War, with government-owned plants reopened for production and major subcontracting.[23] Certain branch locations, such as Boeing-Wichita, Lockheed-Marietta, and Convair–Fort Worth, became almost as important as the headquarters plants.[24]

Geographical location probably was less important for aircraft than for many other industries, as production cost differentials by location were relatively minor and consequently were minor factors in purchasing decisions. Commercial aircraft especially tended to be nonstandardized products, modified to each customer's requirements. Delivery dates, performance, and operating and maintenance costs often were more critical than basic price.

Despite the postwar difficulties and uncertainties, no major aircraft firms went out of business as a direct result of the war's conclusion, as had numerous firms after the First World War. Having been forced out during the war, Brewster stockholders voted in October 1945 to liquidate, and the company was legally dissolved on February 17, 1950. By the beginning of the Cold War expansion, there was a firmly established Big Five of the airframe industry, Boeing, Convair, Douglas, Lockheed, and North American. Of the remaining manufacturers, Curtiss-Wright and Martin were the most troubled. Curtiss-Wright would soon exit the airframe industry, as would Martin in 1960, although both would prosper as corporations. By 1950, even with exits from and entries into the airframe industry, the market shares of the major firms were approximately the same as in 1940.[25]

The 1948–67 period saw the aircraft industry, after reversing the decline during the first three postwar years, exerting its greatest impact and influence on the American economy. The aircraft (later aerospace) industry became the defining industry for high technology, the major industrial employer, and leading industrial exporter. The statistical contrast of the industry in, for example, 1967, with the last full prewar year, 1938, is remarkable. The reason for the rapid rise of the industry was, of course, the Second World War, and the Cold War assured its continued prominence. There was also a direct link between military requirements and aeronautical progress, leading to rapid advances in aircraft performance and rapid diffusion of those technological advances into commercial aircraft.

Postwar Aeronautical Progress

The end of the war slowed production and new orders, but not aeronautical development. On the contrary it tended to increase the rate of change. While many aircraft designs were frozen during the war for productive efficiency, new developments were introduced quickly with wartime pressures removed. Further, many wartime developments, particularly jet-engined aircraft, reached the operational stage immediately after. Captured German technology and employment of German scientists in research enabled rapid progress in jet and rocket propulsion and swept-wing and delta-wing aerodynamics, and for the longer term, in guided- and ballistic-missile developments.[26] Both industry and government exploited these resources to the fullest. Theodor von Kármán, leading a team of engineers, had recovered German files on swept-wing research in 1945, which NACA then applied to the North American XP-86, the first Western swept-wing fighter. Incorporation of British jet engine and German wingsweep technology into U.S. designs helped NACA overcome incipient criticism that it had failed to keep apace of these developments during the war. There were even dreams of a nuclear-powered aircraft, offering almost unlimited range. Fairchild began research on the Nuclear Energy for the Propulsion of Aircraft (NEPA) program in 1946.

Air arms swiftly converted to jet power, but the day of the piston-

TABLE 13. U.S. Aircraft Exports, 1940–48

Year	Aircraft Exported		Value of All Aeronautical Exports ($)
	Number	Value ($)	
1940	3,522	196,260,556	311,871,473
1941	6,011	422,763,907	626,929,352
1942	10,448	879,994,628	1,357,345,366
1943	13,865	1,215,848,135	2,142,611,494
1944	16,544	1,589,800,893	2,825,927,362
1945	7,599	663,128,543	1,148,851,587
1946	2,302	65,257,749	115,320,235
1947	3,125	74,476,912	172,189,502
1948	2,259	66,354,000	153,629,000

Source: *Aircraft Facts and Figures*, various editions.

engined combat aircraft was far from over. The Boeing B-50, an improved, higher-powered development of the B-29, and the Convair B-36 remained in production until 1953 and 1954, respectively. The Chance Vought F4U Corsair was produced through 1952, and the Douglas AD Skyraider until 1957. The Grumman F8F Bearcat and the unique North American P-82 Twin Mustang also were standard equipment through the Korean War. Others long produced included the Lockheed P2V Neptune and Martin P5M Marlin patrol bombers.

Several military designs of the period featured compound piston and jet power, primarily for a combat speed boost or to enhance takeoff power. Jets supplemented piston engines in later versions of the B-36 and the P2V. The North American AJ Savage, primarily piston engined, contained an auxiliary jet engine in the tail, and the Martin P4M Mercator patrol bomber mounted piston and jet engines in the same nacelles. There were also new fighters appearing in 1945, the Ryan FR-1 Fireball and the Curtiss F15C, with compound power, and the Convair XP-81, with combined jet and turboprop power. Of these, only the Fireball saw production.

Several producers pursued jet engine development, and interservice rivalry extended even to jet engine programs. When jet designations were given the postwar *J* prefixes, even numbers represented navy-developed engines, while odd numbers were air force. General Electric's heavily produced J33 was developed from the original Whittle design, thereby sharing a common ancestry with the de Havilland Goblin. Allison moved into axial-flow development with the J35, and GE's later axial-flow J47 was an outstanding success. Westinghouse, a veteran supplier of naval turbines, was established by the navy in jet aircraft engines as well. Its early J34 and J46 axial-flow engines were fairly successful, but the larger J40, intended for advanced naval aircraft of the 1950s, never achieved performance expectations and was instrumental in the exit of Westinghouse from the field.

Pratt and Whitney finally entered the jet field with the centrifugal-flow J42, a license-produced version of the Rolls-Royce Nene, followed by the higher-powered J48, from the Rolls-Royce Tay. Development of the large axial-flow J57 beginning late in 1947 enabled P & W to gain jet leadership. The first production engine in the 10,000 lb. thrust class, the J57 would power such important aircraft as the B-52, F-100, F-101, and F-102, the naval F4D, A3D, and F8U, and in its civil version the first generation of jet airliners. Wright did not enter jet engine production until 1950, with the British Armstrong Siddeley Sapphire under license as the J65. By the mid-1950s the separate air force and navy engine programs were largely combined.

A weakness of early jet engines, in addition to relatively low power and high fuel consumption, was their low rate of acceleration, which posed major problems with carrier operation and with land close-support operations, where speed was less critical than rapid acceleration. Jet aircraft thus were somewhat slower in coming into shipboard operation than for land-based operation. Also for this reason, many Second World War–era combat aircraft were pressed into service in the Korean War in preference to faster jets.

For airliners, piston engines were further developed pending the development of jet engines that could permit economical commercial operation. The Wright Turbo Compound variant of the R-3350, powering the DC-7 and the Super Constellation, was the ultimate in piston power and was produced until 1958. In the Turbo Compound, exhaust gases, trapped by an arrangement of three turbines, fed the extra power back to the main gear, resulting in greater fuel economy and power for no increase in weight.

The turbine geared to drive a propeller, variously called the propjet or turboprop, was developed almost concurrently with the turbojet, but was much slower in entering operation. Although many postwar combat aircraft, including the North American B-45 and the Boeing B-52, were originally designed for turboprop power, developmental problems and advances in turbojet power effectively precluded turboprop combat aircraft. The turboprop-powered Douglas A2D Skyshark, the intended successor to the Skyraider, was canceled due to engine problems, as was the North American A2J. When perfected, the turboprop was applied to military transports and small airliners, as well as to helicopters and business aircraft, where it offered a more justifiable compromise between performance and economy.

Aircraft capabilities continued to advance. The Lockheed P2V Neptune "Truculent Turtle" gained attention with its flight of September 29–October 1, 1946, of 11,236 miles, a record straight-line distance. Later the Boeing B-50 "Lucky Lady" set a new nonstop around-the-world flight record, covering 23,452 miles with four aerial refuelings, lasting ninety-four hours and completed on March 2, 1949. The Boeing B-47 jet bomber set a West-East nonstop record on February 8, 1949, covering 2,289 miles at an average speed of 607 mph. The Lockheed P-80R set a new speed record of 623 mph on June 19, 1946; then the experimental naval Douglas D-558–1 Skystreak raised the record to 650 mph on August 25, 1947. Such records carried strategic implications given the Cold War tensions building at the time.

Both the British and the Germans had envisioned supersonic flight in their wartime research, but defeat prevented German efforts from reaching fruition, and the British failed to support their industry's efforts. Thus it would fall to the United States to be first in this field. A new aircraft definitely was needed for supersonic research since wind tunnels of the time were not capable of providing data in that realm, and even the fastest piston-engined aircraft were limited by compressibility problems. John Stack of NACA was an enthusiastic supporter of supersonic research, but NACA lacked the budget to undertake such research. Accordingly, Stack decided to collaborate with air force efforts.[27] NACA developed further advanced projects in cooperation with the military services.

The Army Air Forces turned to Bell, as it had for its first jet, to design a supersonic aircraft. It had been decided that the craft would require rocket power, despite extremely high fuel consumption. To provide sufficient duration it would have to be carried aloft and launched by a mother aircraft. Bell obtained data from the German engineer Georg Madelung soon after V-E Day that it applied directly to its XS-1 (for experimental supersonic) project.[28] Later redesignated X-1, it was first air-launched on February 15, 1946, then flew supersonically on October 14, 1947, although the flight remained secret for several months. Larry Bell, John Stack, and pilot Chuck Yeager won the 1948 Collier Trophy for the achievement. Three X-1s were built and were followed by more advanced models that extended the performance envelope. The XP-86A achieved the first jet supersonic flight, exceeding Mach 1 in a shallow dive on April 25, 1948.

The period of rapid technological advances also saw many whose efforts influenced the industry pass from the scene. Dr. Robert H. Goddard, the rocket pioneer whose work would influence the space program, died on August 10, 1945. At the dawn of the supersonic era, Orville Wright died in Dayton on January 30, 1948, at the age of seventy-six. He had last flown in 1944 in the experimental Lockheed C-69 transport that visited Dayton. He regretted to the end of his life that his invention had been developed into a weapon of war. General of the Air Force H. H. Arnold died on January 15, 1950, age sixty-three, the first of the wartime chiefs of staff to die. Samuel S. Bradley, for many years head of the Manufacturers Aircraft Association and Aeronautical Chamber of Commerce, died on April 9, 1947, age seventy-eight.

Military Transport Developments

While continuing to operate versions of commercial transports, the military emphasized specialized designs, pioneered by the Fairchild C-82.

This, in turn, led to a distinct and major new market for the industry. The need for massive capability had been dramatically demonstrated by the Berlin Airlift of 1948, and by the global reach of American military responsibilities. American responsibilities in the Far East and Europe required long-range air transport to deploy, supply, and reinforce military forces.

The Martin JRM-5 Mars, although only five were built, did set several flying-boat records for lifting personnel and cargo. Three giant land transports were developed for heavy airlift. Douglas developed the C-74 Globemaster for the air force, which first flew on September 5, 1945, while Lockheed countered with the R6O, later R6V, Constitution for the navy. Convair also developed the six-engined C-99 transport from the B-36. All were powered by the huge P & W R-4360 Wasp Major engine. None was to be produced in quantity, but the single C-99 and the two R6Vs served for several years, and the fourteen C-74s equipped a squadron.

It was not until Douglas developed its even more capacious C-124 Globemaster II, with its front clamshell loading doors and double-decker fuselage, that the giant military transport became firmly established. Such airliners as the Convair 240, Douglas DC-6, and Lockheed Super Constellation also were adapted for a variety of roles.

Boeing proceeded with its C-97 transport for the air force. The first production C-97A flew on September 15, 1949, but after fifty were delivered, production shifted to the KC-97 for in-flight refueling of Strategic Air Command bombers. C-97/KC-97 production totaled 888, and the type became the standard tanker for B-47 and B-52 jet bombers.

Northrop, needing diversification, was encouraged to develop a rugged trimotor transport for use in less-developed areas. The Pioneer failed, however, for the same reason that troubled new smaller airliners: availability of cheap surplus C-47s and other types. But an air force order for the C-125 Raider rough-field version partially rescued the program. The C-125 first flew on August 1, 1949, and thirty-three were built, including ten C-125Bs for Arctic rescue duties. Working under a fixed-price contract, however, Northrop sustained considerable financial losses on the venture.[29] While troubled, the Northrop effort did help establish an important new aircraft category: the assault transport.

The Industry and the Airlines

Airlines overexpanded in the postwar euphoria and incurred serious losses during 1946 and 1947. But the industry recovered rapidly from 1948 as air travel became more commonplace and increasingly within reach of the middle class, leading to enhanced airliner demand. Although most airlines resumed service with reconditioned and reconfigured surplus military transports, production of improved designs was under way. Many orders for new airliners were delayed or canceled in the financial difficulties, but the recovery led to a resumption of orders and to increased interest in the commercial sector. Manufacturers saw not only a large domestic but a worldwide airliner market. The commercial air transport boom was in fact worldwide, and American producers would dominate the world market as well.

Commercial service also began to be extended to smaller cities heretofore unserved. The CAB had established the Feeder Airline classification on July 11, 1944, which served as the basis for development of the local-service sector.[30] New airlines were formed, and the universal equipment was the DC-3, available cheaply and in quantity. Beech and Lockheed were interested in the potential of the feeder airline market and readied new designs. The twin-engined, fourteen-seat Lockheed Saturn and the innovative twenty-seat Beech Model 34 "Twin-Quad," powered by four engines geared to drive two propellers, were tested in 1946. Again, availability of C-47s dimmed market prospects for new feederliners, and neither was produced. Specialized designs would not find market success for another fifteen years.

A bright postwar future was forecast for commercial air freight, with large-capacity cargo aircraft enabling it to be competitive with surface transport. Freight airlines such as Slick and Flying Tiger were established, and the industry anticipated a market for specialized air freighters. A major market was slow in developing, however, and while cargo versions of airliners such as the DC-6A were produced, freight remained an adjunct to the passenger business.

Heavy regulation of commercial aviation was largely to the industry's benefit at the time, helping ensure financial stability. Regulation extended to routes, fares, and crew staffing. The Civil Aeronautics Administration was under the Department of Commerce, but the CAB remained independent. In 1946 Congress added a Federal Aid program for airport development to the CAA. International service also was regulated. The International Air Transport Association (IATA), essentially a route- and rate-fixing cartel, and the Provisional International Civil Aviation Association (PICAO) were formed after the Chicago convention for civil aviation of November-December 1944. The permanent ICAO followed in 1947, with Edward P. Warner as president.

The link between commercial airlines and national security was widely recognized, as manifested by the establishment in 1953 of the Civil

Reserve Air Fleet (CRAF). Under this arrangement, specified airliners were designated for emergency military duty and were delivered with strengthened cabin floors for heavy military cargo. Higher commercial operating costs resulting from the weight penalty were reimbursed by the Defense Department. Aircraft of the CRAF were used in several foreign crises.

American domination of the airliner market became a major frustration to the British. American market supremacy was commonly believed to extend from a wartime agreement between the U.S. and British governments in which the American industry would produce air transports, while the British would concentrate on combat aircraft. This in turn extended directly to an American postwar airliner lead. While such an agreement would have had a certain rationale, given the American need for overseas airlift, there is a serious question as to whether such an agreement was ever executed.[31] Certainly discussions took place. Assistant Secretary of State Adolf A. Berle even expressed concern that postwar employment of American transports as airliners to seize a commercial advantage would be viewed as a betrayal of the understanding "entered into in good faith" by the U.S. and British governments.[32] But Donald Douglas in particular felt that American firms need not conduct themselves as "gentlemen" in competition with their ally.[33]

Actually, the British Air Ministry, by stopping development in 1939 of the promising Short S.32 and Fairey FC.1 four-engined airliner designs, which would have been competitive with the DC-4 and the L049, probably was most directly responsible for the British postwar competitive disadvantage.[34] In any case, U.S. and British wartime aircraft production was not closely coordinated, and British perpetuation of the myth that a wartime agreement led to their competitive difficulties seems self-serving.

Major airlines eagerly reequipped. In addition to the ubiquitous DC-3, surplus DC-4s and L049 Constellations were becoming available. The first reconditioned Lockheed L049 entered service with TWA in February 1946. Douglas had produced 1,163 C-54s during the war, and many were converted to commercial use. Progress soon led to more advanced airliner versions, however, and Douglas produced only seventy-four postwar DC-4s before production ended in August 1946, switching to the improved DC-6.[35] The DC-6, originally developed as the military XC-112, first flew on February 15, 1946, and marked the beginning of the practice of "stretching" a basic airliner design. On the DC-4 wing, Douglas fitted a longer fuselage, more powerful engines, and pressurization. Lockheed quickly responded with a luxury Model 649 and longer-range Model 749

Constellation. Douglas and Lockheed competed directly in the large four-engined airliner market over the next decade.

Less successful was the Boeing Model 377 Stratocruiser, the commercial model of the C-97, which first flew on July 8, 1947. The Stratocruiser was larger and longer ranged than the DC-6 and Constellation, offering nonstop West-to-East transatlantic capability. While it won orders from several international airlines, the Stratocruiser's uneconomical P & W R-4360 Wasp Major engines made it uncompetitive, and its production life was relatively short, the last being delivered in May 1950.[36]

Convair and Martin targeted the smaller twin-engined segment, but both efforts were troubled. Convair's Model 240 eventually led to success in the lengthened, much-improved Model 340 and later 440, and Martin's Model 202 led to the improved Model 404. But both faced severe competition from surplus DC-3s. Douglas, which had not readied a new smaller airliner, instead offered the Super DC-3, with new outer wings, new vertical stabilizer, expanded passenger capacity, and more powerful engines. Airlines preferred the newer designs, however, and few new Super DC-3s were produced. But many were converted from old DC-3s, including a large fleet for the navy as the R4D-8.

Commercial versions of the three giant postwar military transports were proposed, but none was to enter service in the airline environment of 1946 and 1947. The Douglas C-74 (originally designated DC-7, a designation reapplied to the ultimate development of the four-engined DC-4 family) was not proceeded with. The Convair CV-37, based on the military XC-99, was found to be underpowered and uneconomical, and Convair ended development.[37]

Martin, after severe losses, exited the commercial market permanently, but Convair gained a commanding position in the short- to medium-range market that it sought to extend. Douglas and Lockheed searched for ways to maintain or retain market share at the large, long-range end of the market. After producing only 233 C-69 and postwar commercial Constellations, Lockheed flew its lengthened, more powerful L-1049 Super Constellation on October 13, 1950, which gained instant popularity and led a long series of progressive developments. Total production, including military variants, was 623. Douglas, then behind Lockheed, responded with the DC-7. DC-7 development was instigated by C. R. Smith of American Airlines, who almost twenty years earlier had initiated the sleeper transport that became the DC-3.[38] The DC-7 involved another fuselage stretch and the addition of Turbo Compound engines, offering true nonstop coast-to-coast service for the first time. The first DC-7 flew on

May 18, 1953, and entered service with American before the end of the year. Total DC-6/DC-7 production reached 1,041.[39]

Although many foresaw the advent of jet airliners, the operating economics of early jet engines made it difficult to justify rushing ahead with development, as airlines necessarily were more concerned with economics than was the military. The first jet airliner designs, moreover, were not American but British and Canadian. The Comet was first in production and, despite its rather short range, entered scheduled service on May 2, 1952. As is well known, this first venture in jet airliners ended tragically, but the British did establish and hold for a time a lead in turboprop airliners. The Viscount entered service on April 18, 1953, and gained substantial orders in the American market as well. As it had gained from foreign technology in so many other areas, the U.S. industry benefited from the Comet tragedy, not only in engineering knowledge but in the marketplace. Service entry of the Boeing 707 and the Douglas DC-8 in the late 1950s ensured that American domination of the global airliner market would be permanent.

Chapter 10   Industry Survey, 1945–54

The postwar aircraft Big Five dominated the industry into the 1960s, but rankings shifted over time. While all aircraft manufacturers faced adjustment, Boeing, with continuing bomber and transport production, probably was least affected of the major firms. Grumman and Chance Vought continued to specialize in naval aircraft, McDonnell, Northrop, and Republic advanced as fighter manufacturers, and Fairchild specialized in transports. Such wartime manufacturers and subcontractors as Columbia, St. Louis, Globe, and American Aviation were liquidated.[1] The Naval Aircraft Factory was reorganized under the Naval Air Materiel Center and no longer produced complete aircraft.

One new entrant mounted a strong bid to succeed in the face of doubts about market prospects. Chase Aircraft Corporation was founded in 1944 in the Bronx, primarily by former Kaiser industrial associates. The principal development and production interest initially was in all-metal assault gliders, designed under the direction of Michael Stroukoff, president and chief engineer.[2] But original investors withdrew after losses on experimental contracts, and Stroukoff relocated to Trenton, New Jersey, in 1946, where he developed the XC-122 transport from the CG-18 glider design, which first flew on November 18, 1948. Then a new, larger glider, the CG-20, led to the twin-engined XC-123, which first flew on October 14, 1949. This fully developed assault transport featured a low-slung fuselage and a large rear loading door through which vehicles could be driven directly on and off, and capability to operate from short, unprepared battle zone airstrips.[3] Chase also tested the first American jet transport. The XC-123A, the CG-20 prototype with four J47 engines mounted underwing, first flew on April 21, 1951.

The C-123 showed great promise, and the air force placed an order for three hundred. On that basis Kaiser acquired a 49 percent interest in Chase in 1951.[4] Edgar F. Kaiser then became president of Chase, followed in 1952 by Clay P. Bedford, a veteran of Kaiser shipbuilding and automobile operations. Stroukoff remained as chief engineer. But with extremely limited production capacity at Trenton, the order was to be fulfilled at both Trenton and at the Kaiser Willow Run factory. Then the C-123 order was canceled in June 1953, ostensibly over capacity concerns, but there also had been recurring conflicts with Mike "The Mad Russian" Stroukoff.[5]

While the C-123 later succeeded, Chase itself would not survive. After the Chase cancellation, the C-123 program was put out for bids and was won by Fairchild, which offered lower costs. Kaiser then acquired the remaining 51 percent of Chase stock and completely absorbed Chase in 1953. Mike Stroukoff subsequently formed his own Stroukoff Aircraft Corporation, located in the former Chase facilities. He pursued developments of the basic C-123 under an air force contract for his "Pantobase" conversion, with the loan of three early production C-123s. Testing was promising, and the air force ordered the YC-134 production version, but the program was later canceled during a period of numerous aircraft cancellations. The venture ended in 1959, leaving an embittered Mike Stroukoff.

Hughes Aircraft ended the war without ever attaining mass production. Then, amid allegations of partisan politics, Howard Hughes became a target of the Republican-controlled Special Senate Committee Investigating the National Defense Program, which had gained wartime prominence under Senator Truman of Missouri.[6] Ostensibly to investigate the F-11 and H-4 Hercules contracts, hearings were held during the summer of 1947 and in fact produced some embarrassing revelations concerning Hughes's elaborate entertainment expenses for various officials. But Hughes was a defiant witness, strongly counterattacking the committee at every turn.[7] The committee announced in August 1947 that the hearings would be suspended temporarily, but in fact they were never resumed. The committee report was critical of Hughes and his contract activities, but no one was indicted, the report had no impact, and the public regarded the episode as a fiasco.[8]

Howard Hughes, still regarded as a hero, returned to California and flew the mighty *Hercules* for the first and only time on November 2, 1947. Although the *Hercules* was seriously underpowered even with eight engines, had no prospect for further development, and had cost Hughes $7 million of his own funds, he remained determined to prove its viability.[9] In his obsession with aviation and determination to prove doubters wrong, Hughes exhibited a kinship with many industry pioneers. A more rational businessman, concerned with cash flows, creditors, and shareholders, would not have persevered. Granted, Hughes had millions to spend whereas most had not, but his actions still evoked those early aviators. Dimensionally still the largest aircraft ever built, the *Hercules* was preserved.

Afterward, the erratic Hughes gradually lost interest in his aircraft company, still a subsidiary of Hughes Tool Company (Toolco), and it became a steady drain on his finances. In a restructuring effort, Hughes in 1948 placed in charge the team of Harold L. George and Ira C. Eaker, both retired air force generals. Eaker became liaison between Culver City and the Houston headquarters, while George became general manager of aircraft operations. The scientists Simon Ramo and Dean Wooldridge also joined and led Hughes Aircraft to a near-monopoly in air force electronics.[10] By 1950, with Cold War orders increasing, the firm was profitable with air defense fighter radar and had begun development of the Falcon air-to-air missile.[11]

Hughes did not interfere with his new management, but by 1951 he had become almost completely inaccessible and showed interest only in moving his R&D operations to Las Vegas.[12] Management problems by 1952 were critical, as the scientific staff could not obtain expansion capital from Hughes. Further, Noah Dietrich, Hughes's acerbic second-in-command, increasingly intruded into the aircraft company operations.[13] On September 20, 1952, Eaker, George, Ramo, and Wooldridge met with Hughes in an attempt to iron out the numerous problems, even attempting to force Hughes to choose between them and Dietrich.[14] With no resolution, Ramo and Wooldridge resigned on August 11, 1953, to establish Ramo-Wooldridge Corporation, eventually TRW. George, with little remaining authority, also resigned. The vital scientific staff was in virtual revolt, and the company's spectacular growth from 1948 through 1953, to a force of seventeen thousand employees, was endangered.[15] Air force secretary Harold Talbott, with Assistant Secretary Roger Lewis, traveled to California and quietly gave the firm ninety days to resolve the situation or face the loss of all contracts.[16]

Hughes had earlier considered selling Hughes Aircraft, even receiving an offer from Bob Gross of Lockheed, but under the air force ultimatum decided to withdraw from direct ownership. He transferred military activities to a new Hughes Aircraft Company that was in turn 100 percent-owned by the Howard Hughes Medical Institute, established, however dubiously, as a charitable foundation. Toolco transferred all patents and trademarks and leased the Culver City facilities to the new company effective December 31, 1953.[17] Hughes was sole trustee of the institute and still president of the aircraft company. Management remained in considerable turmoil through 1954 but eventually stabilized under Pat Hyland, and Hughes Aircraft became a leading defense contractor in missiles and electronics.[18]

Consolidated Vultee, popularly known as Convair, appeared to be in a reasonably favorable postwar position. At war's end it held $38 million in cash and had $56 million in working capital.[19] The giant B-36 first flew on August 8, 1946, and was ordered into production at Fort Worth. The L-13 liaison model and airliners were produced in San Diego.

It soon became evident, however, that Convair had serious problems, and losses mounted. The CV-240 airliner, developed from the Model 110 thirty-passenger model of 1946, entered service in June 1948 but experienced production problems. B-36 production was proceeding in fits and starts, and the Stinson operation in Detroit was incurring heavy losses. The elegant B-46 four-jet bomber appeared to be a successful design, but the North American B-45 was ordered instead. Convair, in common with others, had been too sanguine about the postwar future, especially about the civil-aviation boom that never materialized.

Reuben Fleet resigned as a Convair consultant on April 12, 1946, ending involvement with the company he built. Victor Emanuel changed the name of the Aviation Corporation to Avco Manufacturing Corporation on March 25, 1947, reflecting diversification. Emanuel eventually brought some measure of stability, but by that time planned to dispose of Convair, apprehensive over the future of the industry.[20] Coincidentally Floyd B. Odlum, multimillionaire aviation investor and corporate turnaround specialist, and husband of famed aviatrix Jacqueline Cochran, saw Convair as a special situation that interested him. Odlum, through his Atlas Corporation, began buying Avco stock in the open market in 1946.[21] Odlum soon opened direct negotiations with Emanuel, and on September 4, 1947, swapped his Avco Manufacturing stock for all of Convair's aviation operations, in San Diego, Fort Worth, and Detroit. AVCO retained nonaircraft operations, including the Nashville plant.[22] Atlas Corporation took control of Convair on November 20, 1947.

Odlum immediately immersed himself in Convair's problems, replacing Harry Woodhead in the spring of 1948 with La Motte T. Cohu, from TWA and earlier with Northrop. Most important, he imposed strict cost controls and reduced the engineering manhours on the CV-240.[23] He then closed the Stinson factory and sold the rights and inventory to the lightplane manufacturer Piper for stock. Convair later sold the one hundred thousand shares of Piper at a loss but would have realized a large profit if it had held the stock longer.

Odlum overcame initial skepticism that a Wall Street financier could run an aircraft company. But even as he delivered results, Odlum considered merger, as had Emanuel earlier, regarding Convair's long-term prospects as dubious with so many firms in the industry. He explored a merger with North American before the more serious discussions with Northrop.[24] Despite several proposals, none was completed at the time, even with allegations of air force influence.[25] Convair's survival rested with the commitment by the air force to B-36 production, despite vehement opposition by the navy.[26] Benefiting from a buildup of military orders, Odlum returned Convair to profitability in 1949.[27]

The Finletter commission report of 1948 paved the way for a major expansion of air power, and the Korean War raised aircraft production to even higher levels. Convair, with strengthened management and a strong backlog of orders, profited along with most of the industry after 1950. The T-29 crew trainer, developed from the CV-240, first flew on September 22, 1949, and enjoyed a long production run. The extensively redesigned CV-340, which first flew on October 5, 1951, became a best-seller. But Odlum foresaw that prosperity would not last. Further, internal tensions were not resolved, especially between the San Diego headquarters and the giant Fort Worth plant. Odlum, moreover, remained oriented toward special situations and was not committed to permanent control of Convair. In March 1952 he initiated merger discussions between Convair and the Kaiser-Frazer automobile venture, but these did not reach fruition.[28] Afterward, he became more determined to sell Convair while the timing was most advantageous, leading to the eventual acquirer, John Jay Hopkins of Electric Boat.

Hopkins, a millionaire Wall Street lawyer and former official in the Hoover administration, joined the submarine manufacturer in 1937. Hopkins effectively became chief executive after Pearl Harbor and officially took office as president in 1947.[29] Electric Boat had purchased 10 percent of Canadair, the Montreal aircraft manufacturer, in 1944 and in 1946 gained government approval to acquire all of Canadair, then suffering losses with license production of a modified version of the Douglas DC-4.[30] Initially leasing the factory and installing new management, Electric Boat completed the purchase on January 25, 1947, for $10 million, denounced by some in Canada as less than half the firm's value.[31] But the Cold War led to prosperity under Electric Boat, with license production of the Lockheed T-33 and North American F-86 for Canadian forces as well as for export to the United States and other nations.[32]

Odlum and Hopkins first made contact in 1951 about a possible affiliation, leading to their 1953 agreement. In the meantime, on April 24, 1952, Electric Boat had become General Dynamics Corporation, reflecting its broadened activities. Canadair at that time was much larger than the original Electric Boat and was also the major contributor to profits.[33] With Convair's sales at some $400 million, General Dynamics acquired a firm almost three times its size, but on March 31, 1953, purchased 17 percent of Convair's stock, becoming the largest stockholder.[34] Odlum remained a

Convair director, but Hopkins controlled a majority of board seats and had authority to select the chief executive. The firms officially merged on April 29, 1954, with General Dynamics creating a Convair Division with plants in San Diego and Pomona, California, and Fort Worth, Texas.[35] General Dynamics then ranked among the largest defense contractors and pursued Hopkins's declared goal of becoming the General Motors of defense. Convair, by that time, had undergone its fourth ownership and control change in twelve years, from Fleet with Consolidated to AVCO with Consolidated Vultee, to Odlum's Atlas Corporation with Convair, and finally to General Dynamics as a division.

In military aircraft, Convair quickly established delta-wing leadership derived from German technology, especially the work of Dr. Alexander Lippisch, pioneer of the delta wing.[36] The XF-92A delta design, which first flew on September 18, 1948, was originally intended as an air force interceptor but was employed only for research. The concept led to a contract for the advanced, supersonic F-102 Delta Dagger in 1950. Another delta development was the XF2Y-1 Sea Dart for the navy, an experimental amphibious fighter with retractable "hydro-skis" that extended for takeoff and landing, which first flew on April 9, 1953. Doubts about the mission of such a fighter and the loss of the second prototype led to the cancellation of the program.

Flying-boat development also continued. The advanced XP5Y-1 Tradewind patrol bomber, powered by four turboprops, first flew on April 18, 1950. It was followed on February 25, 1954, by the R3Y-2 transport development, but neither was ordered into production. Production of the L-13 Sentinel also ended. Of greater future significance for Convair was its leadership in guided-missile research.

Along with others, North American was concerned about military prospects but declined to enter the commercial field, a decision it would later rue.[37] Employment dropped as low as five thousand in 1946 but soon rose as successors to its wartime mainstays, the P-51, B-25, and AT-6, entered production, strengthening NAA's aircraft leadership. It had also developed from the P-51 the twin-fuselage XP-82 Twin Mustang, which first flew on June 16, 1945. Originally intended for Pacific escort duty, instead it saw combat as a radar-equipped night fighter in Korea. The B-45 Tornado, with four J47 engines, became the first operational jet bomber. It first flew on March 17, 1947, but its performance was not outstanding, and orders amounted only to 143. Tactical bombers also did not enjoy a high priority with the air force, then emphasizing the strategic mission, but the B-45 was successful in the photoreconnaissance role.

Cold War demand led North American to acquire the old Vultee-Downey plant and later the Curtiss-Columbus plant, primarily for naval aircraft. The B-45's naval contemporary, the AJ-1 Savage, first flew on July 3, 1948. The AJ-1 did not serve in its originally intended, and controversial, strategic nuclear role from aircraft carriers, but as an in-flight refueling tanker. The advanced trainer successor to the T-6 was the T-28 Trojan, with tricycle gear and a fighter-style bubble canopy, which first flew on September 26, 1949. It also served later as a light attack aircraft.

North American built thirty FJ-1 Fury carrier-borne jet fighters. The prototype first flew on September 11, 1946, and the FJ-1 became one of the trio of first-generation naval jets, with the Vought F6U Pirate and McDonnell FH-1 Phantom. It was from the Fury that North American's most famous postwar design would spring. The air force became interested in a version of the Fury incorporating swept-wing technology. Chief engineer Ray Rice and aerodynamicist Larry Green, using captured German test data, then redesigned the Fury into the advanced XP-86 Sabre.[38] When first flown on October 1, 1947, the XP-86 bore only a vague resemblance to the FJ-1. Redesignated on June 1, 1948, as the F-86, it entered large-scale production as the standard fighter, succeeding the Lockheed F-80, and went on to win fame in Korea and to serve with numerous allied air forces. No fewer than 6,656 F-86s in all versions were produced in the United States, the greatest production total of any postwar Western military aircraft. License production in Canada, Italy, Australia, and Japan brought the total to more than nine thousand.

The success of the F-86 was transferred to the navy Fury series. The FJ-2, which first flew on February 14, 1952, was not a development of the FJ-1 but an adaptation of the F-86, and was followed by the more powerful FJ-3. All were built at the former Curtiss-Columbus plant.

North American finally became completely independent of General Motors on June 3, 1948. GM sold its holding to the public, having determined that its auto expertise could not be directly applied to aircraft, and that a mass general-aviation market would not arrive.[39] Dutch Kindelberger then became chairman of the board in addition to chief executive, and J. L. Atwood became president. The team would continue to lead the company into the 1960s.

North American's fighter reputation won it the contract for the first supersonic fighter, the F-100 Super Sabre, the first of the famed Century series. While owing something to the F-86, the F-100 was a completely new design built around the powerful afterburning P & W J57 engine, with a forty-five-degree wingsweep and extensive use of titanium in the structure. The prototype F-100 was flown on May 25, 1953, and received large-scale orders. The company remained strong in defense, but lack of diversification led to longer-term problems.

Curtiss-Wright, despite its wartime importance, faced severe postwar difficulties. It had lost out on orders for successors to its large-scale P-40, SB2C, and C-46 programs, and the massive cancellations after V-J Day left Curtiss in a weakened position. Of its remaining projects, the XP-55 Ascender and pressurized P-62, the naval XBTC and XBT2C, and an airliner development of the C-46, none would see production.[40] The naval SC-2 Seahawk also ended in 1945.

The Louisville plant closed first with the approaching end of the war, and the St. Louis plant followed on July 31, 1945. St. Louis was returned to the DPC but was eventually acquired by McDonnell.[41] Guy Vaughan was criticized as closing the St. Louis plant prematurely, but local anger eventually was tempered by McDonnell's later success.[42] The CW-20E airliner development of the C-46 held promise and received an order from Eastern Airlines on the strength of its design. Many felt that if produced it would have headed off the Convair 240 and the Martin 202 in the short-range market, but the program was canceled by Vaughan in 1946.[43] The Buffalo plant was phased out and remaining aircraft production centered at Columbus. The first Curtiss pure jet, the large XF-87 night fighter, flew on February 15, 1948, and showed enough potential that eighty-eight were ordered initially by the air force. But various delays and design problems led to its cancellation in favor of the Northrop F-89 Scorpion, marking the exit of Curtiss-Wright in aircraft.

The Curtiss Airplane Division closed in March 1951, upon completion of remaining subcontracts and after transfer of the Columbus factory to North American. Curtiss-Wright announced that it would concentrate on engine and propeller manufacturing in New Jersey. It had lagged in jet engines, but the license-produced Sapphire was ordered as the J65. Wright piston engines, chiefly the advanced Turbo Compound, continued both for commercial and military aircraft, and the company moved into the Wood-Ridge engine plant purchased from the War Assets Administration in late 1945.[44]

Why Curtiss-Wright, bearing the two greatest pioneering names in American aviation, should have made such an ignominious exit from the aircraft industry is a subject for some contemplation. The slide had begun perhaps in the late 1930s, as the firm lost design leadership. After the war, even though flush with cash, the company chose to diversify into nonaircraft ventures.[45] By one account, Guy Vaughan was intimidated by Donald Douglas, who threatened to take his engine business to Pratt & Whitney if

Vaughan entered the airliner market in competition with Douglas.[46] Many felt that Vaughan simply was not a true aircraft man, more influenced by bankers and financiers than by designers and engineers. The remoteness from operations of the New York headquarters was a possible factor. There were suggestions that the company's conflicts with the Truman committee during the war weighed against it after Truman's succession to the presidency. Finally, it was possible that Curtiss-Wright simply was outcompeted in aircraft, particularly by North American, making its exit inevitable.

Although Curtiss-Wright was still profitable in 1949, Vaughan faced increasing stockholder dissent and was replaced as chairman in April 1949 by Paul Shields. Roy T. Hurley, with an automotive background, became president. T. Roland Berner, later to head the firm, also joined the board. But Shields resigned on October 1, 1951, under cloudy circumstances, and was replaced by Hurley, who retained his position as president. Hurley completed the shift from aircraft to mechanical and electronics interests.

Clement M. Keys, founder of Curtiss-Wright and a major developer of the structure of the modern industry, died on January 12, 1952, age seventy-five.

Donald Douglas, as he had been during the war, remained pessimistic about the long-term prospects of the industry, stating in 1946, "The future is as dark as the inside of a boot."[47] Production of the DC-3 ended with 10,926 built, of which 10,123 were military versions. Douglas delivered only 127 aircraft in 1946.[48] Employment was down to 15,400 in 1948, but recovery followed quickly with the commercial DC-6 and with naval aircraft. The AD Skyraider gained strong support from the navy, and production built up rapidly from 1947. It replaced all wartime torpedo and dive bombers and became a long-term success. The jet F3D Skyknight also saw production and service. But Douglas remained disturbed over sparse air force business, having nothing to follow the A-20 and A-26. The innovative XB-42 pusher-engined light bomber and the XB-43 jet-powered development, the first American jet bomber, remained experimental. Orders for the C-74 were cut from fifty to fourteen, but Douglas then received its first major air force contract in more than three years, for the C-124 Globemaster. The C-124 first flew on November 27, 1949, and 447 were produced. Douglas also gained recognition with the jet D-558–1 Skystreak research aircraft and its rocket-powered successor, the D-558–2 Skyrocket. By 1951, with the Cold War expansion in full swing, Douglas was in production at Santa Monica, El Segundo, Long Beach, and Tulsa. Business increased further with license production of the Boeing B-47 at the reopened Tulsa plant. Combining strong military and commercial business, Douglas was again the largest firm in the industry, and the DC-6B gained an enduring reputation as the finest piston-engined airliner ever built.

Design leadership continued under Ed Heinemann with the unusual F4D Skyray, a tailless jet fighter with a low-aspect-ratio swept wing inspired by the work of Alexander Lippisch, which first flew on January 23, 1951. Douglas engineers had found the Lippisch papers at the end of the war and gained added knowledge from Lippisch personally.[49] The F4D entered production and service but was not regarded as highly successful by the navy. The large A3D Skywarrior, which first flew on October 28, 1952, and succeeded the North American AJ-1 Savage, gave the navy a true strategic capability from aircraft carriers.

The outstanding Douglas combat aircraft of the day was the XA4D-1 Skyhawk, which first flew on June 22, 1954. The Skyhawk was a particular triumph of Ed Heinemann, who responded to a Korean War requirement for an effective attack aircraft that would avoid the constantly increasing weight and complexity of others. The Skyhawk bettered all specifications and became known as "Heinemann's Hot Rod" but also became the last Douglas production combat design.

Douglas, possibly the most prestigious name in aircraft, and with strong orders, still faced a cloudy future. Douglas undoubtedly benefited from product line diversification, with no undue dependence upon any market segment, but in consequence had not developed a clear identity. Its resources were spread over airliners, military transports, naval fighters and attack aircraft, bombers, experimental aircraft, and missiles. Further, its late start in jet airliners would bring crushing financial problems.

Martin experienced problems perhaps more severe than others in determining its future. Martin even conducted design studies for personal aircraft in 1945 but decided not to enter the market. Its airliner entry, the Model 202, experienced numerous problems, and that venture more than any other factor ended the long career of Glenn Martin. Martin felt that airlines would want a simple, austere smaller airliner, and the 202 was unpressurized, but the airlines actually wanted luxury. The 202 also gained a poor reputation in service, leading to cancellations, and only forty-three Model 202s and 202As were built. Then the pressurized Model 303 was canceled in 1947.

The company lost $36 million in 1947, largely due to the 202, and losses continued into 1948, even though Martin remained profitable on military business. Martin had problems obtaining bank financing and was forced to turn to the RFC for a $28 million loan. With respect to Martin's problems precipitated by cancellations of the Model 202, Harvey J. Gunderson, a director of the RFC, stated,

> Mr. Martin, gentleman that he is, let them cancel and refunded their down payments. He took an inventory loss of from 13 million dollars to 16 million dollars. That's when they began to need help.[50]

President Glenn Martin was increasingly pressured to give up active management after the financial reverses and moved up to the position of chairman of the board in 1949. Executive Vice President H. T. Rowland eventually departed, and C. C. Pearson became president and took over active management in 1951.[51] Rowland went on to found Hayes Aircraft Corporation, a modification and conversion firm. Eventually the improved Model 404 came into service, and 103 were ordered for TWA, Eastern and the U.S. Coast Guard, with deliveries completed in 1953. Cumulative losses, however, led Martin to leave the market to others.

Military business also declined. The naval AM-1 Mauler was not fully successful in service, and orders were reduced in favor of the Douglas AD Skyraider. The P4M Mercator received only limited orders, and Martin essentially had no air force business. The XB-48 six-jet bomber, developed largely as insurance against the failure of the advanced Boeing B-47, was never a serious contender for production. The XB-51 three-jet tactical bomber, which first flew on October 26, 1949, possessed many advanced features and appeared promising but was dropped by the air force. Some held that the rejection reflected the company's disfavor with the air force: Martin was suspected of taking the navy's side in the B-36 debate. Others felt that Martin simply did not understand the changed postwar procurement environment. For diversification, Martin acquired the Rotorwings operation as a division on February 14, 1947. Agnew Larsen, who had been associated with Harold Pitcairn in autogiro development, worked for Martin in helicopters from 1948 until 1951, then repurchased Rotorwings.

With cumulative commercial and military problems, Martin was again near bankruptcy in 1951.[52] The RFC, then headed by former air force secretary Stuart Symington, initially denied a further loan on the grounds that there were too many firms in the industry. President Truman ultimately approved the loan, but the price was Martin's control of his company.[53] Creditors, including the RFC and the Mellon Bank, forced Martin to place his 26 percent stock holding into the control of trustees, and management in the hands of George M. Bunker, an engineer recruited from outside the industry.[54] Bunker had gained a strong management reputation as the "boy wonder" president of the Trailmobile Company. Pearson resigned.

Martin offset its decline in aircraft with success in guided missiles, first with a contract on April 22, 1946, for what was to become the Matador ground-launched missile, and then with the Viking research rocket for the navy. Bunker accelerated the move into missiles and electronics, and the firm's financial recovery was complete by 1954.

Martin was not yet finished in aircraft, however. The P5M Marlin patrol bomber became a long-term production program, and an air force contract for license production of the British Canberra bomber as the B-57 followed on March 6, 1951. The order marked the first U.S. purchase of a foreign military aircraft since the First World War, but the Korean War revealed a need for a fast tactical bomber or "night intruder" to replace the Second World War–vintage Douglas B-26, and the Canberra filled that need. The first Martin-built model, modified to American production techniques, flew on July 20, 1953, and entered production and service.

Glenn Martin, while never a true engineer or an innovative designer, nevertheless was instrumental in development of the aircraft industry. Financial problems and lack of adaptability forced him out of the company he founded. Afterward Martin took little interest in his position as chairman emeritus. Although possessing a considerable personal fortune, he became bitter and withdrawn, especially after the death of his elderly mother in 1953, with whom he had lived almost all his life.

Lockheed's postwar prospects were stronger than most, with production building up of the Constellation, the P-80, and the P2V Neptune. Lockheed decided, wisely in retrospect, not to produce the Saturn feederliner or personal aircraft. The Saturn would have been priced in the one hundred thousand dollar range, compared to twenty-five to thirty thousand dollars for surplus C-47s. It did offer its single-seat personal design, functionally a flying motorcycle, to the army as the AirTrooper, but it was not ordered. New Constellations were delivered to the airlines from 1947, followed by ever enlarged and improved versions. Strong orders followed for the successful P2V patrol bomber, for which Kaiser was a major subcontractor.

Air force combat business became a more pressing concern. The F-80 Shooting Star became the standard jet fighter but began to be replaced in the late 1940s after 1,732 were produced. The large, heavy XF-90 penetration fighter, which first flew on June 4, 1949, was canceled when the air force eliminated that mission. Facing formidable competition from North American, Northrop, Convair, and Republic in the fighter field, Lockheed survived through the success of the T-33, the advanced trainer version of the F-80 that went into production in 1948. No fewer than 5,691 of the T-33 series were produced over ten years, and it became the most widely used jet trainer in the world. It provided the platform for a more powerful, radar-equipped, all-weather fighter. The resulting F-94 Starfire first flew on July 1, 1949, and became a standard interceptor, with 853 produced by 1954. While a derivative design and lacking outstanding performance, the F-94 was rushed into production due to delays with the Northrop F-89. The F-94 was noteworthy as the first missile-armed fighter and gave Lockheed a basis for more advanced developments.

Lockheed began research on an advanced supersonic fighter in 1951 and was awarded a contract for the radical Mach 2 XF-104 in 1952.[55] The XF-104 first flew secretly on February 28, 1954, although with the low-powered J65 engine. When fitted with the advanced General Electric J79 and when publicly revealed, the F-104 was widely thought to represent the ultimate in manned-fighter development. While far from that, the F-104 enhanced the already strong reputation of Kelly Johnson and his small staff at the Skunk Works, as well as Lockheed's overall technological leadership.

The old DPC Marietta plant was reopened in January 1951 for Lockheed, with a contract to refurbish stored B-29s for Korean War service, but then to produce the Boeing B-47 under license. The first 150 employees were transferred from California, and the Marietta operation later was named the Lockheed-Georgia Company. James V. Carmichael, so successful with Bell's B-29 production during the war, was general manager, with Daniel J. Haughton, later to become Lockheed's chief executive, as his deputy. Lockheed also added major capacity in California with the Palmdale factory in 1951.

In common with other major firms, Lockheed entered the missile and space programs. The Lockheed Missile Systems Division (LMSD) was established on January 1, 1954, with its first project the X-7 ramjet high-altitude vehicle.[56]

Another contract enabling Lockheed to establish a leadership position was that for the next-generation assault transport, which Lockheed won in 1952 over spirited competition from five other bidders. The design built on operating experience with specialized military transports to that point and also took advantage of the Allison T56, the first large turboprop to enter service in the United States. The prototype XC-130, designed in California, first flew on August 23, 1954, and entered large-scale production at Lockheed-Georgia following completion of the B-47 contract.

Clairmont T. Egtvedt relinquished most executive functions at Boeing after finally persuading William M. Allen to accept the post of president and chief executive officer. Allen was unanimously elected at the board meeting of September 5, 1945.[57] Egtvedt remained chairman. Although Allen had no aeronautical training, he was to prove an astute leader, especially in airliners. But with B-29 orders sharply reduced with the end of the war, Allen soon faced the impact of thirty-two thousand employee layoffs in the Seattle area.

Boeing appeared to be well positioned with its C-97/Stratocruiser orders and continuing bomber contracts. But only fifty-five Stratocruisers were sold, and Boeing's earlier design studies for the Model 417, a small feederliner, and the Model 431, a thirty- to forty-passenger local-service airliner, did not proceed. The large piston-engined F8B naval fighter was not ordered, and only a test series of the twin-boom YL-15 liaison aircraft from the Wichita division was procured. Thus Boeing appeared precariously dependent upon the strategic bomber, but that field was the highest military priority of the United States at the time.

TABLE 14. Largest Aircraft Manufacturers, 1954

Fortune 500 Rank	Firm	Sales (in millions)	Employees
20	Boeing	$1,033	65,000
26	Douglas	915	71,900
34	Lockheed	733	46,000
39	United Aircraft[a]	654	45,800
41	General Dynamics[b]	649	56,000
42	North American	646	54,000
94	Republic	323	21,300
111	Martin	271	18,300
129	Grumman	235	11,800
173	Bell	186	18,900
183	Northrop	171	24,700
206	Chance Vought	150	13,600
221	Fairchild	140	13,700
252	McDonnell	123	12,300
351	Beech[c]	78	5,600
432	Temco[d]	59	8,000

Source: *The Fortune 500*, supplement to *Fortune*, July 1955. (© 1955 Time Inc. All rights reserved.)
Note: Ryan, at $45 million, did not make the *Fortune 500* list.
a. Included for Sikorsky.
b. Included for Convair.
c. Primarily general aviation.
d. Primarily a subcontractor.

B-29 production, including Martin and Bell, ended late in 1946 with 3,970 built. The B-29 was followed by 370 more powerful B-50s, the first flying on June 25, 1947. The wholly owned Boeing Aircraft Company was absorbed into the parent Boeing Airplane Company in December 1947, and Boeing continued to produce at Seattle, Renton, and Wichita. Labor disagreements resulted in a bitter strike in April 1948, which did not end until October. William Allen maintained a tough line in negotiations, and labor relations eventually stabilized. The company increasingly shifted military production to Wichita, largely at the insistence of the air force, leading to fears about its commitment to Seattle. In fact there was no compelling rationale for the Seattle location, but its relative isolation from the rest of the industry worked to Boeing's advantage, eliciting strong employee loyalty and widespread regional support.[58]

The most important program for Boeing was the six-jet, swept-wing B-47 strategic bomber, which first flew on December 17, 1947, the anniversary of Kitty Hawk. Like many early jet aircraft, the B-47 underwent extensive design changes during development, taking advantage of German jet engine and wingsweep technology. A medium bomber, the B-47 lacked the range for intercontinental missions, but with in-flight refueling became the most formidable weapon in the strategic arsenal in the early 1950s. It possessed sufficient speed to elude Soviet jet fighters of the day, unlike the slow and extremely vulnerable B-36.[59] The first major production version, the B-47B, first flew on April 26, 1951. Total production at Boeing-Wichita, Lockheed and Douglas, was 2,040, a record for a Western jet bomber.

Boeing had begun development of the B-52 intercontinental bomber in 1946, before completing the B-47. Originally intended to be powered by six turboprop engines, access to swept-wing technology and progress with the J57 engine led Boeing to reconfigure the B-52. Although appearing superficially to be a scaled-up B-47, the B-52 was in fact a completely new design. It first flew on April 15, 1952, entered production at both Seattle and Wichita, and began replacing the B-36 during 1955 and 1956.

Boeing was possibly the strongest aircraft firm during the 1950s, but it also saw a future in the missile field. Its first program, the XF-99 (later IM-99, still later CIM-10) Bomarc ramjet interceptor, was first tested in 1952.

Bell's prospects looked particularly dim in the postwar period, with no production orders. Sales were only $11.5 million in 1946. Some even felt that Bell should take advantage of its strong cash position and liquidate, given the large number of firms and the limited market. Bell, which had unionized in 1939, suffered a UAW strike late in 1945.[60] Then a dissident group of New York stockholders, including famed investor Benjamin Graham, made an effort to oust Larry Bell, but management won a proxy battle in April 1947.[61] Dividend policy was a key issue. Bell's treasurer Harvey Gaylord was instrumental in winning the battle. But in 1948 the Equity Corporation of New York acquired 51 percent of Bell by tendering for the shares of the dissident group plus others.[62] This change brought Jack Bierwirth, president of New York Trust, to the board. New York Trust had financed initial P-39 production in 1939.[63]

Bell returned the Marietta plant to the DPC on March 1, 1946, and centered its activities in Buffalo.[64] The P-59 was followed by the large, long-range XP-83 fighter, which first flew on February 25, 1945. But the production P-83 was canceled in favor of the Lockheed P-80, leaving Bell with no aircraft production prospects. But experimental aircraft soon would gain Bell major recognition. Also significant for Bell as well as for the industry was its contract on April 1, 1946, for a one-hundred-mile-range air-to-ground missile, the first guided-missile contract ever awarded.

Larry Bell, while one of the last old-line aircraft romantics, also was a strong believer in the helicopter and turned his company toward that development. The near-term problem, however, was that large-scale helicopter production would not be achieved until after 1950, with Korean War demand. Another, more bitter, strike occurred in the summer of 1949, marked by violence and requiring arbitration.[65] Operations would not stabilize until the Korean War, with experimental aircraft and rising helicopter production.

Helicopter operations moved to Fort Worth, Texas, in 1952, and Bell would achieve major success in rotary-wing development and production at that location. Bell agreed to move and to invest $5 million to convert the former Globe plant in return for a navy loan of necessary equipment.[66]

Grumman could not rest on its wartime laurels, as production ran down rapidly postwar and its future was uncertain. The last of 12,275 F6F Hellcats was delivered in November 1945. Remaining production consisted of the F7F Tigercat and F8F Bearcat advanced fighters. Grumman tested two designs for the personal-plane market but did not proceed with either, and new civil amphibians also found a small market. It began its well-known line of aluminum canoes in 1946 and later entered aluminum truck bodies, but these lines comprised only a small percentage of total business. Dependency on the U.S. Navy was undiminished.

Leroy Grumman turned the office of president over to Jake Swirbul in 1946, but remained chairman and chief executive. Swirbul had gained a strong reputation for cutting through the navy bureaucracy and getting results.[67] With its first jet fighter and its new amphibian still under development, 1947 was a lean year for Grumman, although it managed a profit.[68] Employment was down to little more than three thousand, and only the F8F Bearcat remained in full production, with a total of 1,265 completed by 1949.

The XJR2F-1 Albatross, Grumman's largest amphibian and originally developed for the navy, first flew on September 1, 1947. The newly independent air force, which had funding at a time when the navy did not, ordered the Albatross in 1949 for the air-sea rescue role as the SA-16. The Albatross later entered service with the navy and Coast Guard and was widely exported, largely replacing the wartime PBY. With redefinition of naval combat roles, Grumman converted its follow-on torpedo bomber design, the large XTB3F-1 that first flew on December 1, 1945, into the AF-1 and later AF-2 Guardian antisubmarine hunter-killer aircraft. Antisubmarine warfare enjoyed an increasing priority with the Soviet submarine buildup, and Grumman commanded a position in this field as well as in fighters. Missile development began with the experimental Rigel ramjet surface-to-surface model.

The F9F Panther, which first flew on November 21, 1947, helped secure Grumman's future. The F9F won the first major naval jet order and was widely used in the Korean War. Early jet engines were of such low power that the XF9F-1 was designed for four small Westinghouse J30s, but the first Panther built, the XF9F-2, was powered by a single imported Rolls-Royce Nene. Production Panthers were powered by the license-built P & W J42 and later by the more powerful J48. By 1949 Grumman production had again attained a high level, with amphibian, antisubmarine, and jet fighter programs. Production of the Panther totaled 1,382 before switching to the more advanced, swept-wing F9F-6 Cougar, virtually a new design. The Cougar first flew on September 20, 1951, and entered service in 1953.

Grumman succeeded the AF-2S and AF-2W with the twin-engined S2F Tracker, combining the hunter-killer functions in one aircraft. Once again Grumman's reputation enabled it to win a contract against strong competition, and the first XS2F-1 flew on December 4, 1952. Grumman's workmanlike, utilitarian designs earned the company the navy nickname of "The Iron Works." Grumman used subcontractors extensively during the Korean War, so that afterward it could take over work previously subcontracted in order to avoid layoffs.[69] Rising business strained the firm's capacity, but expansion was limited by urban development in Bethpage. Accordingly, in 1953 the navy acquired four thousand acres in Calverton, in rural eastern Long Island, for a new Grumman production facility.

An unsuccessful, but significant, program for Grumman and the navy was the XF10F-1 Jaguar, the first variable-wingsweep fighter. While building on progress with the Bell X-5 to that point, the Jaguar was an ungainly design, as variable-wingsweep technology was still in its infancy, and was further handicapped by its underpowered Westinghouse J40 engine. The XF10F-1 first flew on May 19, 1952, but the program was canceled in May 1953. The variable wingsweep experience would later serve Grumman well in the F-14 program, however.

Chance Vought remained a division of United Aircraft along with Sikorsky, Hamilton Standard, and Pratt and Whitney. While continuing production of the F4U Corsair, Chance Vought developed the XF5U "Flying Pancake," and the F6U Pirate, one of the trio of first-generation naval jets. The major event for Chance Vought in the early postwar period was its relocation to Dallas in 1948, taking over the wartime North American "B" plant. The largest industrial move in history to that time, it involved fifteen hundred key employees, two thousand machines, and fifty million pounds of equipment.[70] The flow of work was minimally disrupted. The move was initiated and financed by the navy, concerned that its two main aircraft suppliers were located on the East Coast, still considered more vulnerable than the interior.[71]

Chance Vought separated from United after rising concerns that airframe competitors, in dealing with Pratt and Whitney and disclosing competitive information regarding their engine requirements, could find such information used to advantage by Chance Vought.[72] The spinoff thus was held necessary to avoid any conflicts of interest. Rex Beisel, strongly opposed to the separation, left in 1952.[73] Although only fifty-six, Beisel retired and was succeeded by his deputy, Fred O. Detweiler. Chance Vought first became a wholly owned subsidiary of United in January 1954, then an independent corporation on July 1, 1954.

The radical twin-jet, swept-wing, tailless F7U Cutlass naval fighter, which first flew in 1948, was Chance Vought's major aircraft development. The design drew heavily from German technology but required considerable redesign before service entry. The resulting F7U-3, first flying on December 20, 1951, became Chance Vought's only aircraft program.

Fairchild had built a strong wartime record, and C-82 production proceeded postwar. J. Carlton Ward, uninterested in the personal-aircraft market, licensed the veteran F-24 to the new Texas Engineering and Manufacturing Company (Temco), which produced three hundred when the program was completed in April 1947.[74] Fairchild Engine and Airplane was one of only four firms to report an operating profit in 1946, the others being North American, Martin, and Grumman. While not managing the company, Sherman Fairchild held strong views as to its postwar direction, bringing him into open conflict with the autocratic Ward. Fairchild had chafed for some time at the style of Ward, feeling ignored and patronized, and also was interested in a merger or sale. There was a public breach by mid-1946, but Ward controlled the board, which among other matters dismissed Fairchild's proposal to buy Martin.

Fairchild resigned as chairman and director on October 22, 1946, stating that he had little opportunity to employ his experience for the betterment of the company and for the establishment of policies.[75] Ward, while expressing regret at Fairchild's resignation, emphasized the need for more formal management given increased industry complexity.[76] With Fairchild's departure the position of chairman was eliminated, although restored under Ward in 1948.

Fairchild remained the company's largest stockholder and was determined to gather enough support to regain control. By the spring of 1949 he felt strong enough to move. He organized pickets for the 1949 annual meeting, urging stockholders not to attend, and as a result the meeting was twice postponed, lacking a quorum due to withheld or revoked proxies.[77] When the meeting was finally held, Fairchild prevailed in the voting and elected a new board of directors, including himself and Grover Loening. One issue was Ward's generous pension benefits. Ward resigned on July 6, and Richard Boutelle, a senior executive recently fired by Ward for favoring the Fairchild slate, was elected president and chief executive. James A. Allis of the Chase National Bank became chairman. Sherman Fairchild took no formal position but pursued other business interests.[78] Headquarters also moved from New York to Hagerstown in July.

The Ranger division was changed to Fairchild engine division in 1950. The C-119, the improved, more powerful successor to the C-82 Packet, flew late in 1947 and was ordered in quantity. Production in several versions amounted to 1,112 when completed in 1955. But the XC-120 development with detachable cargo pack under the fuselage, while gaining wide attention at the time, was not proceeded with, nor was the extensively redesigned C-119H Skyvan. In the meantime, Fairchild gained the Chase C-123 contract, and the first Fairchild C-123B flew on September 1, 1954.[79]

Fairchild had felt for some time that it was in official disfavor. From its XNQ-1 naval basic trainer it had developed the air force T-31, which won over the Beech T-34 and Temco T-35. But the initial order for one hundred was canceled in September 1949, and its eventual replacement by the Beech T-34 Mentor led to considerable controversy. In addition, the air force had initially awarded a C-119 production contract to Kaiser-

Frazer in December 1950, to be fulfilled at the firm's Willow Run factory and sharing space with Kaiser-Frazer automobile production. But the contract was canceled as production was under way, after an finding, hotly disputed by Kaiser, of high costs at that plant. Fairchild thus retained the C-119, but it was destined to be the last Fairchild design to see mass production. Some viewed the C-123 program as a consolation prize for Fairchild, since it had been effectively frozen out of the bidding for the C-119's successor, the C-130.

While still ranked as a major aircraft firm, Ryan did not achieve major success with production of its own designs but remained a successful subcontractor. The unstable FR-1 Fireball was withdrawn from service in July 1947, after only 103 were produced, and Ryan's major manufacturing activity was the Navion business and military liaison aircraft, acquired from North American. With Korean War priorities, Ryan sold rights to the Navion. While out of aircraft production, Ryan gained important experimental aircraft contracts and was to forge a bright future in the emerging missile and unmanned-aircraft fields.

The McDonnell FH-1 (originally XFD-1) achieved the distinction of first jet aircraft to take off and land on a carrier, achieved on July 21, 1946.[80] Although heading a young company, James S. McDonnell gained a strong reputation as a meticulous manager with a firm control of costs.[81] McDonnell produced sixty FH-1 Phantoms for the navy, the last being delivered on May 27, 1948. The program that firmly established the company, however, was the F2H Banshee, developed from the Phantom. The early F2H-1 and F2H-2 were followed by the redesigned F2H-3, and total production was 895. In the late 1940s the firm brought in the designer Don R. Berlin, formerly with Curtiss-Wright and Fisher, who contributed to further success in jet design.

The XF3H-1 Demon, the next-generation naval fighter, first flew on August 7, 1951, and after its J40 engine was replaced by the Allison J71, entered service. The XF-85 Goblin, a compact fighter designed to be carried by the B-36 and air launched in its defense, proved to be unneeded with jet bomber progress. The XF-88 Voodoo, designed to the same requirement as the Lockheed F-90, first flew on October 20, 1948. Its design potential induced the air force to support a supersonic F-101 development for the long-range fighter-bomber, interceptor, and reconnaissance roles. The F-101A first flew on September 29, 1954, and entered production. Its success also reduced McDonnell's dependence on the navy.

In 1953 McDonnell was among the losers in the navy competition for a supersonic fighter, won by Chance Vought. McDonnell continued its design, however, and in 1954 the navy awarded a contract for a prototype. The result was the XF4H-1 all-weather fighter, and its success would make McDonnell the strongest military firm in the industry.

Northrop still regarded the flying-wing concept as the key to its longer-term future. Following the experimental MX-324 rocket aircraft, Northrop won a contract to develop the jet-powered XP-79B flying-wing interceptor, which first flew on September 12, 1945. An unusual design, in which the pilot lay prone, it was intended to ram enemy bombers, its heavy magnesium wings being able to withstand the strike. Unfortunately the XP-79B was destroyed on its first flight, and development was not pursued.

Northrop's major program was the F-89 Scorpion, the first specialized all-weather intercepter to enter service. The XF-89 prototype flew on August 16, 1948, and a total of 1,052 F-89s was produced in several versions, securing Northrop a position as a major combat aircraft supplier for the Cold War era. It was also the last aircraft that John Northrop had a direct hand in designing.

Northrop's XB-35 bomber first flew on June 25, 1946, but was troublesome in tests, the major problem being the gearing for the contrarotating propellers for the pusher engines, for which there was no precedent. The concept still appeared promising, leading to a jet conversion as the YB-49, which first flew on November 1, 1947. The airframe required little modification, and jet power bypassed the gearing problem of the piston engines and improved speed markedly. But high fuel consumption reduced its range to that of a medium bomber, placing the B-49 in competition with the new B-47 rather than with the B-36. Despite major concerns over its stability, the design still had strong advocates, and the air force ordered thirty RB-49 reconnaissance versions. But a fatal crash of a YB-49 on June 5, 1948, damaged its prospects. Among those killed was Captain Glen Edwards, for whom Edwards Air Force Base, center of flight testing, was named. The flying wing was canceled early in 1949, with the air force relying on the B-36 and the B-47 for strategic missions.

The end of the flying-wing program was one of the more enduring controversies of postwar aviation, with many feeling that political rather than technical factors were paramount. The air force ordered all airframes destroyed, and there were even suspicions of sabotage in the YB-49 accident. One outcome was a thirty-year hiatus on flying-wing development, but more immediately the cancellation left Northrop an extremely troubled company. The F-89 was the only production program, and development of the Snark ballistic missile, a forerunner of the later cruise missile, was troubled. The company had lost heavily on the Pioneer transport and could not afford to develop civil versions of the flying wing without a base of military orders. Attempts at nonaircraft diversification were unprofitable, but the company acquired Radioplane Corporation as a subsidiary, also returning to Northrop its president, Whitley Collins.

John Northrop, always a strong believer in technical training for the aircraft industry, founded the Northrop Training School for that purpose during the war. It became the Northrop Aeronautical Institute in 1945, eventually becoming independent of Northrop Aircraft. Ultimately it became the degree-granting Northrop University in 1975.

Management changes occurred frequently, beginning with La Motte Cohu's return to TWA in 1947. Richard Millar served as chairman from 1947 to 1949, then returned to investment banking while remaining a board member. He was succeeded by General Oliver P. Echols, who had joined Northrop after his resignation as president of the AIA in September 1948, and who also became general manager. Echols brought in Edgar Schmued from North American as assistant chief engineer in 1952, over Jack Northrop's objections, ostensibly to resolve problems with the F-89 program. Many senior managers left after the B-49 cancellation, and more North American engineers and managers moved to Northrop, leading to loss of the familial atmosphere built by Northrop and Cohu.[82]

John Northrop resigned as president and chief engineer on November 20, 1952, selling his stock and completely removing himself from the company. Northrop, although a private man who said nothing that could damage the company, was widely felt to have become disenchanted by the flying-wing controversy as well as by loss of executive authority. Toward the end of his life Northrop admitted that his company was not politically astute and that he believed the flying wing had been canceled for political reasons. He felt he had committed perjury in 1949 by denying those political motives during congressional hearings.[83] His confession also brought into question the truthfulness of Odlum and Symington in their testimony at that time.

Echols succeeded Northrop as chief executive but was to serve only briefly as he died suddenly on May 15, 1954. Whitley C. Collins became president and chief executive. The firm's production future beyond the F-89 looked uncertain.

Production of the Republic P-47 ended in November 1945. While developing a jet fighter, Republic made two attempts at diversification: the Seabee amphibian, for the private aviation market, and the Rainbow long-range airliner. The Seabee sold well initially, but production ended after large losses. The advanced F-12 long-range reconnaissance aircraft first flew on February 7, 1946. With four P & W Wasp Major engines, it was the fastest multiengined piston-powered aircraft ever flown. Although almost the size of a B-29, its clean lines made it appear smaller. The order was canceled with the end of the war, however, and Republic transformed the design into the forty-seat Rainbow. The Rainbow appeared promising and was ordered by American and Pan Am, but costs and airline financial difficulties led to its cancellation.

Had the Seabee and Rainbow succeeded, Republic would have joined the first rank of aircraft manufacturers.[84] Instead, their costs, along with the purchase of Aircooled Motors, which produced engines for the Seabee, left the company with precarious finances. Paul Moore remained the largest stockholder as Alfred Marchev was moved to chairman of the board at the end of 1946, and Mundy L. Peale became president and chief executive. Marchev's support of the Seabee was regarded as a major contributing factor in his eventual firing on May 1, 1947.[85] He died on December 6, 1947. Peale, who ran the Evansville plant during the war, was oriented exclusively toward the military sector, but Alexander Kartveli was regarded as the firm's major asset.

Fortunately, Republic was in a strong position with its P-84 Thunderjet, which first flew on February 28, 1946. Republic originally had considered a straightforward jet conversion of the P-47 but instead designed a completely new aircraft. Being relatively small and financially weak, however, Republic needed advance payments from the air force in 1947 to operate and to enable F-84 production to build up. As straight-winged F-84 production progressed, Kartveli, impressed by the North American F-86, felt that Republic should have a swept-wing fighter as well. The prototype F-84F Thunderstreak, essentially a converted F-84, first flew on June 3, 1950, but required considerable development before entering production. Initially designated F-96, the design later returned to the basic F-84 designation, the company feeling that Congress would be more likely to fund development of an existing design rather than a new program in the financial climate of the day.[86] But serious F-84F delays and attendant criticism of Republic led the company to develop the improved F-84G straight-winged model, the first fighter to possess in-flight refueling capability. Production of F-84 Thunderjets in all versions totaled 4,453 when completed in 1953, and the type was widely exported.

The production F-84F and the photoreconnaissance RF-84F Thunderflash, by that time effectively new designs, finally entered service in 1954. The designation inconsistency was equivalent to that in naval aviation between the F9F Panther and Cougar and between the FJ-1 and FJ-2 Fury.

Republic's experimental XF-91 short-range interceptor, to meet an unfulfilled air force mission, featured unusual inverse-taper swept wings and combined afterburning jet and rocket power, giving a rapid climb rate on interceptor missions. Ultimately unsuccessful, the XF-91 nonetheless led to overall industry progress in both afterburner development and in manned rocket power, as well as assisting Republic in the extremely advanced XF-103 and XF-105 fighters. By 1954 Republic was well under way with both, and its production future appeared favorable for the next decade.

Helicopter development progressed rapidly, with gains in performance, reliability, and lifting capacity. The helicopter truly came of age during the Korean War, adding to the rescue/evacuation role such missions as cargo transport, liaison/observation, assault transport, and antisubmarine warfare. The civil market also held promise.

The Sikorsky S-51, developed from the wartime R-5 model, was produced both for military and commercial use. The S-51 was licensed to Westland in Great Britain and was also widely exported. The S-55, which first flew on November 7, 1949, was a major advance in that, in addition to being larger and more powerful than previous models, its engine was mounted in the nose in the manner of fixed-winged aircraft, allowing much greater cabin room for passengers and freight.[87] The S-55 saw wide military service and was the first helicopter to receive commercial certification for passenger carrying. Production was also licensed to Westland. Sikorsky entered the turbine field with the XH-35 of 1954. Igor Sikorsky continued as engineering manager of the division bearing his name and took pride in that his helicopters had saved many lives in combat situations.

The P-V Engineering Forum, with promising helicopter experiments during the war, reorganized as the Piasecki Helicopter Corporation in 1946, supported financially by Laurance Rockefeller and Felix DuPont Jr.[88] Frank Piasecki focused on the twin-rotor configuration, and the firm achieved production success with the "Flying Banana" during the 1947–54 period. The HRP-1 and more powerful HRP-2 for the navy, from the early PV-3, and the developed H-21 Work Horse for the air force were produced in quantity. They were joined by the smaller PV-14, originally designated XHJP-1 by the navy, in February 1946. As the navy HUP-1 Retriever and the H-25 Army Mule, it was produced until 1954.

Piasecki became a developmental leader, and facilities expanded steadily. In 1949 Piasecki undertook the ambitious YH-16, the world's largest helicopter, intended to transport forty troops. The YH-16 proved to be unstable, however, and development was dropped by the air force after the loss of the second prototype. Failure of the H-16 and crashes of the H-21 led to a troubled period for Piasecki.[89]

Kellett, although a leader in the field and with an order backlog, suffered bankruptcy in 1946. The familiar problems of high costs and inadequate working capital contributed to the failure.[90] The company had developed the ten-passenger XR-10, which first flew on June 14, 1947, the largest helicopter built to that time. Kellett then sold its huge XH-17 design to Hughes in August 1947. The company later reorganized and continued on a reduced scale with the intermeshing rotor concept under contract.[91] W. Wallace Kellett died on July 23, 1951, but the company emerged from bankruptcy in 1952 and continued small-scale development.

Glidden S. Doman, an engineer who formerly worked for Sikorsky, organized Doman-Frazier Helicopters in Connecticut on August 31, 1945. He held some helicopter patents and began development on his own. As Doman Helicopters, Inc., the firm tested its LZ-5 model on April 27, 1953, and limited production followed. G & A Aircraft, descended from the original Pitcairn, could not find a postwar market, and its owner Firestone dropped the helicopter program in 1947.[92] The Pitcairn airport at Willow Grove became a naval air station. Platt-LePage, its development contracts canceled in 1945, sold remaining rights to McDonnell in that year and liquidated on January 13, 1947.[93] Landgraf Helicopters also liquidated in 1947.

Bell intensified its helicopter development. The small Model 47, derived from the Model 30 and also designed by Arthur Young, first flew on December 8, 1945. The Model 47 gained a commercial certificate from the CAA on March 8, 1946, the first helicopter in the world to do so.[94] It began both military and civil service in 1948 and was produced in record numbers and widely exported.

The Model 48, a scaled-up Model 47 with a military designation of YH-12, was ordered in small numbers in 1946. Bell followed with the large tandem-rotor Model 61, the first antisubmarine helicopter, for the navy as the HSL-1. Fifty were produced from 1953. Bell followed Sikorsky internationally by licensing the improved Model 47G, appearing in 1953, to the Italian firm Agusta. It later was produced in Japan and Great Britain. The Model 48, which Bell had abandoned, was also licensed to Agusta.

Charles H. Kaman, formerly head aerodynamicist for Hamilton Standard, founded his company on December 12, 1945, financed by venture capital, and began development with a small workforce. Applying the intermeshing rotor developed with the assistance of the German engineer Anton Flettner, his demonstrator Model K-125 flew on January 15, 1947.[95] Kaman sought military contracts but suffered from inadequate financing and certification problems.[96] The firm began development of a cropduster model but finally received a naval contract in May 1947 for rotor development. This gave Kaman sufficient backing to issue a public stock offering in November 1948, and to complete certification of his initial design by May 1949.[97] A major disappointment followed with the loss of a navy contract to Sikorsky. The navy, however, later awarded Kaman a development contract for the model HOK-1, and finally, in September 1950, Kaman won a production contract for HTK-1 training helicopters.[98]

Kaman went through a period of managerial strain, with Charles Kaman assuming more personal control. The company achieved the distinction on December 11, 1951, of testing the first turbine-powered helicopter in the world, the Model K-225. With production increasing, Kaman secured financing from the navy to build a new 155,000-square-foot plant in Bloomfield, Connecticut, in 1953.[99] The firm continued to build the HOK-1 and then the H-43A, its last piston-engined helicopter, for the navy, marines, and air force, following with the turbine-powered HH-43B Husky for the air force.[100]

When affiliated with Kaiser, Stanley Hiller developed the coaxial X-235 for the navy, but the project eventually was canceled. Hiller then developed the single-rotor UH-5 of 1947 into the successful Model 360 for the civil market. A major advance was his Rotormatic design, in that small paddle-shaped airfoils between the rotors enhanced stability. After a name change to United Helicopters, Inc., Hiller exported the Model 360, redesignated UH-12, and produced navy HTE-2 and air force H-23B trainers.[101] The military models entered service in 1952, and Hiller became a significant factor in small helicopters, producing the basic design into the 1960s. The small Hornet, with rotor tip ramjets, was tested in October 1954.[102] Hiller also developed the Model J5 with a jet torque compensating device replacing a more conventional tail rotor, and later the innovative "flying platform" ducted fan for the army.

Howard Hughes began his involvement with helicopters by purchasing the giant Kellett XH-17 Sky Crane. Featuring a main rotor with a span of 130 feet, the XH-17 was powered by two jet engines modified as gas generators supplying gas to the tips of the rotor blades. The design was static tested from 1950 and first flew on October 23, 1952, but ultimately proved unsuccessful.[103] Hughes Aircraft was then focused in missiles and electronics, but with the interest of Hughes a Toolco Aircraft Division was formed and began development of light military helicopters in 1955.[104]

McDonnell tested the twin-rotor XHJD-1 Whirlaway in 1946, then the largest helicopter flown and owing much to Platt-LePage design concepts. McDonnell also tested the diminutive XH-20 "Little Henry," with rotor tip ramjets, but both remained experimental.

By 1948 Harold Pitcairn determined that Bell and others had infringed on his rotary-wing patents and were profiting from government contracts on that basis. He sued for compensation. The government ruled that it held its contractors harmless and would indemnify them for any losses. Pitcairn therefore would have to recover from the government.[105] Pitcairn began legal actions through his Autogiro Company in 1951, and proceedings lasted for years.

Chapter 11 ✈ The Cold War Industry and the Transition to Aerospace

The Cold War soon led to a military expansion and major increases in defense appropriations, and the industry began to operate in what appeared to be a permanent Cold War environment. It became a major employer and economic force as well as a technological leader. The rate of technological progress became the most rapid in aviation history. Two government initiatives especially affected the structure and operations of the American industry during this period.

The Finletter Commission

The President's Air Coordinating Committee, popularly known as the Truman committee, was established in 1946. Membership consisted of ten individuals from government agencies concerned with aviation. Its focus was primarily on civil concerns such as exports, air traffic control, and airport development. There was also a Joint Congressional Air Policy Board. The aircraft industry, however, was initially uninterested in government postwar planning, as it looked to a booming civil market for its future. But by 1947 the industry desperately needed government contracts.[1] It was the Truman administration that fully integrated the aircraft industry into national defense planning.[2]

Concerns over the long-term survival of major aircraft manufacturers and the lack of a coherent aviation policy were so great that President Truman in 1947 established the President's Air Policy Commission to make recommendations on the industry for the postwar world. Although originally concerned only with the condition of the industry, the commission eventually gave attention to five major areas, air power and national security, the aircraft-manufacturing industry, aeronautical research and development, civil aviation, including personal aircraft, and government organization.

The commission was officially created on July 18, 1947, and became known as the Finletter commission after its chairman. At that time the commission classified fifteen manufacturers as major, ranging alphabetically from Bell to United Aircraft (see page 152). In addition, nine companies were classified as major light-aircraft manufacturers, from Aeronca to Temco.[3] The commission held extensive hearings involving most industry leaders, and its report, *Survival in the Air Age,* was submitted on January 1, 1948.[4] The report began by noting that there were several inherent characteristics of the industry that would prevent it from operating in what might be described as a "normal" environment.

1. A product that is, almost indivisibly, a weapon of war and a carrier of commerce
2. A market with but one major customer, government, which purchases 80 to 90 percent of its entire output
3. A violently fluctuating demand, due to the uncertainty of requirements of its major customer
4. A lack of the production continuity that is vitally important in sustaining a trained workforce and in keeping production costs to a minimum
5. A rapidly changing technology that causes a high rate of design obsolescence and abnormally high engineering costs
6. An extremely long design-manufacturing cycle
7. An organization in excess of present requirements[5]

It is but a short step to the realization that the 1948 findings describe almost precisely the industry's situation in the 1990s.

The report of the Finletter commission, coming shortly before the occurrence of major Cold War crises, the Berlin blockade and airlift, the fall of the Iron Curtain over Eastern Europe, and the Communist triumph in China, provided a major impetus to a buildup of air strength. Then the disclosure of the first Soviet atomic bomb test in 1949 presented a new and ominous threat. While perhaps alarmist in tone, the Finletter report stimulated support for air power beyond its concerns for the industry. At a time when plans were in place to reduce air strength from fifty-five groups to forty, the Finletter commission recommended a buildup to seventy, with concurrent increases in reserves and the Air National Guard.[6] The buildup to seventy groups, also endorsed by the Joint Congressional Air Policy Board, was to be attained by January 1, 1950. The congressional board was terminated in mid-1948 after issuing its report. The Finletter commission also recommended the retention of the President's Air Coordinating Committee and the repeal of the Vinson-Trammell act and noted the severe airline losses in 1947 as being worthy of government attention.[7] The concern for the airlines was not only for their importance as a public service but also for their wartime potential. The report emphasized that the majority of air force and navy combat aircraft were in urgent need of replacement. These recommendations, largely endorsed by the administration, led to widespread optimism for the future of the industry.

The air buildup still fell short of the 1950 goal, but the Korean War broke out on June 25, 1950. The threat of a Soviet-equipped opponent gave the military buildup a high priority. The concurrent industry expansion resembled in certain ways that of the Second World War. The Defense Production Act was passed in 1950, providing for so-called V-Loans to finance production. A Defense Production Administration was formed, with a subordinate Aircraft Production Council. The APC operated essentially as a staff function dealing with resource allocation for the industry.

Production facilities expanded. Defense plants in Marietta, Georgia, and Tulsa, Oklahoma, reopened for the Boeing B-47, probably the highest priority aircraft program of the day. Selection of Lockheed and Douglas as second-source contractors also reflected the desire of the Pentagon to limit the growth of Boeing relative to others.[8] Boeing insisted, however, that its erstwhile competitors be deprived of details of how its production was organized.

Other moves included the 1950 North American takeover of the former Curtiss plant in Columbus and the 1952 Bell takeover of the former Globe plant in Fort Worth. In addition, 599 Republic F-84F fighter-bombers were built by the General Motors Buick-Oldsmobile-Pontiac plant in Kansas City during 1951–55. GM actually completed its fixed-price contract early but found itself the target of a General Accounting Office (GAO) finding of excess profits.[9] The real problem may have been excessive costs by Republic. Regardless, investigations of firms for excess returns would recur.

The Independent Air Force

Plans for the postwar military structure were discussed from 1944. In early 1945 legislation was proposed for the establishment of a unified defense structure, involving an air arm equal in status to the army and navy. Along with the establishment of major combat commands, the Air Materiel Command was established on March 9, 1946, succeeding the wartime Air Technical Service Command. The AMC became the primary point of interaction between the Army Air Forces and the aircraft industry. Shortly thereafter the AAF was given further autonomy under the War Department in May 1946 and was effectively operating independently by the time the Department of Defense was officially established.

There was spirited opposition to certain aspects of the plans within the navy, which regarded an independent air force as a threat to its status, although it would retain control of naval aviation. There were broad concerns that the future organization would institutionalize four separate air arms, those of the air force, navy, Marine Corps, and army, which in fact occurred.[10] On July 26, 1947, the National Security Act became law. The act implemented four major national security objectives: a comprehensive national-security program, three equal military departments, coordination and unification of defense activities, and strategic direction and operation of forces.[11]

The act also created the National Security Council and the Central Intelligence Agency. Concurrently, President Truman signed Executive Order 9877, which established more specific functions of the services.[12] The first secretary of defense, James Forrestal, who as secretary of the navy had bitterly opposed certain aspects of the new structure, now was charged with making that same structure work. On September 18, 1947, the United States Air Force came into being, and Stuart Symington was sworn in as first secretary of the air force. The newly independent air force enjoyed strong political support and received the major share of military R&D funding, which among other things led to rapid progress in the nascent missile and space programs.

Also important to the industry, the Armed Services Procurement Act of 1947 addressed the preservation of a defense mobilization base. While emphasizing competitive bidding, it permitted numerous exceptions upon determination of a national emergency. The air force invoked the national-emergency criterion routinely until 1956.[13]

The air force, in possession of the strategic mission, had largely preserved the bomber force, but Korea resulted in expansion of all aspects of military aviation. Heavy transports and helicopters became increasingly prominent. Thus the industry experienced increased business across a wide field. There was also increasing competition in the rapidly developing avionics field by nonaircraft firms, and the industry was increasingly pressured toward greater geographical dispersal, as in the Second World War.[14]

Naval and Marine Corps aviation also expanded, both performing a variety of carrier-based and land-based combat missions. The wartime torpedo and scout (dive) bomber designations were abolished on March 11, 1946, in favor of a straightforward attack designation. Coastal patrol and antisubmarine warfare using shore-based aircraft became a specified function of the navy. The navy gained a share of the strategic nuclear mission with carrier-based strike aircraft. Some progress was made in eliminating duplication, however, with the establishment of the Military Air Transport Service (MATS) on June 1, 1948. MATS was under operating control of the air force, with the naval air transport function merged. But the navy retained authority in the development and operation of flying boats.

Army aviation also expanded. Organization became more formal following the independence of the air force, but aviation units still were integral to field units. An army–air force agreement in 1949 restricted fixed-wing aircraft to an empty weight limit of twenty-five hundred pounds and helicopters to four thousand pounds. The air force controlled close tactical or battlefield support of the army, as well as transportation of army troops and materiel, including parachute dropping. The army was limited to utility, casualty evacuation, and most important, liaison, succeeding the earlier spotter function, but still made sporadic attempts to expand its air missions. The air force managed army procurement, but the army established fixed-wing and helicopter training schools.

Army aviation rapidly modernized by replacing its wartime "grasshopper" spotter aircraft with such models as the Aeronca L-16, the North American/Ryan L-17 Navion, and the Piper L-18 and L-21, all adaptations of civil models. The Bell H-13, the military version of the Model 47, firmly established the army in helicopters. Then the competition for a purpose-designed liaison aircraft, won by the Cessna L-19 in June 1950, marked a major step forward for army aviation as well as giving the general aviation firm Cessna a major military market stake.

After further discussions on the role of army aviation, a definitive agreement between air force Secretary Finletter and Army Secretary Pace on November 4, 1952, established a fixed-wing weight limit of five thousand pounds empty, but weight restrictions on helicopters were eliminated.[15] The army gained expanded air transport and medical evacuation

TABLE 15. Evolution of the U.S. Air Force

Date	Designation
August 1, 1907	Aeronautical Division, United States Signal Corps
July 18, 1914	Aviation Section, United States Signal Corps
April 6, 1917	Aeronautical Division, United States Signal Corps[a]
May 21, 1918	Division of Military Aeronautics, United States Army
June 4, 1920	Army Air Service
July 2, 1926	Army Air Corps
June 20, 1941	Army Air Forces
September 18, 1947	United States Air Force

Source: Air Force Magazine, May 1975, 128.

a. During the First World War, the air arm of the American Expeditionary Forces (AEF) was designated the Air Service, but this designation did not apply to the entire Aeronautical Division of the Signal Corps.

roles but was effectively limited to fixed-wing aircraft no larger than the de Havilland (Canada) Beaver. In consequence the army increasingly emphasized helicopters, where rapid development lessened interest in fixed-wing aircraft.

Aeronautical Progress, 1949–59
The navy, somewhat in competition with the air force, pursued high-speed flight research in cooperation with NACA. The Douglas jet- and rocket-powered D-558-2 Skyrocket, which first flew on February 4, 1948, set many records. The jet engine was later removed, with the Skyrocket air-launched in the manner of the X-1. It was first tested as a pure rocket on August 15, 1951, and became the first manned aircraft to reach twice the speed of sound, attaining Mach 2.01 on November 20, 1953. The developed and more powerful Bell X-1A reached Mach 2.4 on December 12, 1953 and went on to set a new manned altitude record of ninety thousand feet. On October 3, 1953, the Douglas F4D Skyray raised the official level speed record to 753 mph, followed on October 29, 1953, by the North American F-100 record of 755 mph, officially a tie under FAI rules. Dutch Kindelberger and Ed Heinemann shared the Collier Trophy for the records. In another flight regime, two Sikorsky H-19s with long-range fuel tanks completed the first helicopter transatlantic crossing on July 31, 1952.

John Stack of NACA led the development of a supersonic wind tunnel, speeding the advent of operational supersonic aircraft. The first combat aircraft with true supersonic capability, the North American F-100, moved rapidly into production and service.[16] It was soon followed by further air force and navy supersonic designs. Although publicized by the air force as an "air superiority" fighter, the F-100 was instead a capable fighter-bomber in operation. Bell's next experimental aircraft was the variable wingsweep X-5, which first flew on June 20, 1951. Variable wingsweep, enabling wings to pivot from straight to highly swept back, offered both short takeoffs and landings and high-speed flight performance. The X-5 owed much to the Messerschmitt P.1101, discovered by Robert J. Woods of Bell and brought to the United States after the war.[17] While not so designed by the Germans, the P.1101 configuration was adaptable for variable sweep. Testing established, however, that practical variable wingsweep required much further development. Douglas did not fare well with the radical X-3 Stiletto, which first flew on October 20, 1952, featuring a razor nose and small, thin, straight wings similar to those later employed on the Lockheed F-104. Underpowered, the X-3 could barely attain supersonic speed, even though designed to explore the Mach 2 region. Northrop's less-ambitious X-4 Bantam, a tailless aircraft for transsonic research, flew on December 15, 1948, and conducted a successful test program.

The Convair F-102 delta-winged interceptor was the first aircraft designed from the outset as an integrated weapon system, combining electronics, radar, and weaponry. It was also designed for supersonic speeds in level flight but could not attain that performance. With a conventional streamlined fuselage, the F-102 experienced a rapid rise in drag as it approached Mach 1, and the design clearly was in crisis. It fell to Richard T. Whitcomb, a young NACA engineer, to find the solution in 1953, with his development of the "area rule." Whitcomb had been aware of transsonic drag buildup even before the F-102 experience, and with extensive wind tunnel and flight testing found that for minimum drag at transsonic or supersonic speeds, the cross-sectional or frontal area of a body should expand and diminish in an even manner, or be consistent throughout the length of the aircraft. Typically, however, frontal area was maximized at the center of the aircraft, as fuselage and wings were joined at the point of maximum thickness, causing a sharp increase in drag. To minimize the aircraft's total frontal cross-section, and hence drag, Whitman designed the fuselage with a concave taper at the wing roots, where the frontal area was greatest. Thus the fuselage cross-section was reduced where the wings joined the fuselage, offsetting their increased cross-section.[18]

The concave taper, known as the "wasp waist" or "Coke bottle," enabled the aircraft to slip easily through the sound barrier. The F-102 was quickly and successfully redesigned, and area ruling became a standard feature of supersonic aircraft. The first aircraft designed from the outset with area ruling was the Grumman F11F-1 (originally F9F-9) Tiger, which first flew on July 30, 1954. NACA's aeronautical contributions in the 1950s in area rule and variable wingsweep design were significant, but it was evident that the R&D capacity of the industry far exceeded that of the government.

Much aeronautical development during the 1950s was concerned with the expansion of the flight envelope. The Bell X-2, swept-winged successor to the X-1, with a structure of Monel-metal and stainless steel, reached Mach 3 (1,980 mph) on November 18, 1955. The X-2 reached 2,148 mph on September 27, 1956, but the flight, and the program, ended with the craft's destruction and the death of pilot Captain Milburn G. Apt. Actually, delays in the X-2 program saw its potential research contributions superseded by operational supersonic aircraft. Progress by the early 1950s had largely solved the problems of high-speed flight.

The F-100 established a supersonic official speed record of 822 mph on August 2, 1955. Another major record was the nonstop flight of three B-52Bs of Strategic Air Command around the world in forty-five hours and nineteen minutes, completed on January 18, 1957. Marine Corps major John H. Glenn Jr., later an astronaut, broke the transcontinental speed record for the 2,445-mile New York–Los Angeles route in his Chance Vought RF-8A Crusader on July 16, 1957, taking three hours, twenty-two minutes, for an average speed of 734 mph. Then the McDonnell F-101 set a round trip transcontinental speed record in November 1957, flying West to East at 781 mph and East to West at 671 mph, averaging 721 mph for the 4,981.8 miles. An F-101 followed on December 12, 1957, with a new official speed record of 1,207 mph.

Rapid aeronautical progress was accompanied by rising concerns over the sharply escalating cost and complexity of aircraft. Transsonic and supersonic combat aircraft were impressive, but single examples began to cost millions. Service and maintenance problems also rose, forcing more sophisticated training requirements.[19] The performance advantage enjoyed by the lighter and simpler MiG-15 over the heavier, more complex F-86 in early encounters in Korea heightened the debate. The debate would intensify in the future, as cost and complexity of weaponry became a prime target of military reformers. Even the Douglas A4D Skyhawk, intended to counter the trend, grew steadily heavier and more complex with development.

The decade of the 1950s probably was, in aeronautical development, the most exciting and fruitful in the history of American aviation and represented the apogee of the aircraft industry. The number and variety of new models developed and the pace of progress was equaled in no other decade. By 1960 the transition to jet combat aircraft was almost complete, and fighters of not only supersonic but of Mach 2 capability were increasingly commonplace. Throughout the decade the industry's orientation and principal activity still was in aircraft development and production, although the future increasingly pointed to missiles, electronics, space vehicles, and other high-technology areas. The Cold War kept factories humming with military production, and the airliner, helicopter, and general-aviation sectors also were booming. Exports were strong in all major categories.

The industry lost five major figures during a relatively brief period in 1955–56, Glenn L. Martin, Frederick B. Rentschler, Lawrence D. Bell,

William B. Stout, and William E. Boeing. Rentschler was still in active management of United Aircraft at the time of his death, and Bell had only recently stepped down due to health reasons.

The Industry Environment, 1954–60

With its rapid expansion during the Korean War the aircraft industry led all industries in employment, R&D investment, and sales. It was comparable in economic importance to such industries as automobiles, steel, electronics, and chemicals. On June 24, 1954, the Aircraft Industries Association announced that the industry had become the nation's largest industrial employer, with eight hundred thousand persons employed, the first time it had attained that distinction in peacetime.[20] Yet the industry remained less stable than others and faced an increasingly uncertain future. Only ten manufacturers survived into the jet age with production of their own designs: Boeing, Convair, Douglas, Lockheed, McDonnell, Northrop, North American, Republic, and the principally naval firms Chance Vought and Grumman. Success in the early jet age was no guarantee of stability or survival, however, and many were later forced to exit or merge due to a lack of follow-on contracts.

In the mid-1950s, at the height of the Cold War, no fewer than six firms, North American, Northrop, Republic, Convair, Lockheed, and McDonnell, were producing fighters for the air force. Similarly, five firms, Douglas, Chance Vought, McDonnell, North American and Grumman, were producing naval fighters and attack aircraft. There were also five major engine producers, Pratt and Whitney, General Electric, Allison, Westinghouse, and Wright. Were so many models and manufacturers needed at the time? In retrospect, the answer clearly is no. Did this represent an effort to maintain several contractors in business as a national-defense resource, to maintain production skills and capacity in time of war? Undoubtedly. Charles E. Wilson, secretary of defense in the Eisenhower administration, was a believer in the concept of a defense mobilization base, although the paramount strategic retaliatory mission emphasized forces-in-being. Wilson ordered the armed forces to plan procurement around a sound mobilization base, including geographical dispersal.[21]

Preserving a mobilization base, however, did not preserve all aircraft manufacturers. Some thirty years later, by the mid-1980s, with escalating costs and lengthening aircraft service lives, rationalization had reduced the number of active fighter suppliers to the air force to only two, McDonnell Douglas and General Dynamics. By that time also, only two naval combat aircraft suppliers, McDonnell Douglas and Grumman, and only two major engine manufacturers, P & W and GE, remained. Were there risks in reliance on so few contractors? Unquestionably. The conundrum of maintaining a strong but cost-effective military establishment, given competing claims on the national income, while preserving an irreplaceable national resource, as the aircraft/aerospace industry is, continued.

The industry had reached maturity, but as with most mature industries, adaptation was necessary for long-term survival. Managerial effort and investment became oriented toward competitiveness in the markets of the future. While the aircraft industry of the late 1950s was not threatened either by imports or loss of technological leadership, other factors clouded its prospects. Perhaps foremost, many senior managers of leading firms were slow to accept inevitable changes, including the financial challenges of design and production leadership or of simply surviving in the airframe industry. Mergers, acquisitions, and restructuring reminiscent of the wave of such developments in the 1928–34 period were increasingly seen as imperative, yet several firms still were dominated by one personality, in certain cases the company founder, making transformations more difficult. Most were reluctant to merge or restructure even when those measures clearly were called for, primarily due to a stubborn determination to remain independent or to maintain established product lines.

In retrospect the Convair–General Dynamics combination was healthy for the industry, whatever the management and control difficulties following. Smaller firms such as Republic, Fairchild, Martin, Northrop, and Chance Vought probably should have sought merger partners from the 1950s, and those eventually merging should have done so sooner. Of the industry's pioneering giants, Glenn Martin was forced to step aside by creditors, and later Donald Douglas was forced aside as well, in both cases painful but unavoidable measures. Few of the early industry leaders were able to retire or bow out gracefully. Seversky, Fleet, Fairchild, Piasecki, and Ryan all lost control of their firms at some point and in varying degrees. Vaughan of Curtiss-Wright was forced into retirement, and Northrop retired prematurely in disenchantment. Only the Boeing, Grumman, and McDonnell companies were able to retain their basic strategy and structure and enjoy management stability into the aerospace era, while moving strongly toward diversification. Boeing successfully transformed from a primarily military to a commercial firm, and Grumman was especially noteworthy in that Roy Grumman both initiated diversification and shared control with other company executives, insuring an orderly management transition.

Yet with the technological challenges faced, enormous R&D expenses, and the long lead times involved in bringing new programs to production and service, it became more difficult for even the largest firms to strengthen capital structures and to gain or retain access to external capital. In addition, the aerospace industry possessed the highest ratio of technical employment to total employment and total sales of any industry.[22] Both the government and the airlines were increasingly reluctant to finance development efforts in the manner of an earlier era. Adding to the challenges facing the industry, the massive military expansion sparked by the Korean War and the Soviet threat began to taper off from 1955. Rapidly increasing cost and complexity of aircraft and the promise of missiles as principal weapons for the future led to greater production and financial pressures on the industry. The Eisenhower administration, which felt that the Truman administration had starved defense capabilities, entered office in 1953 determined to increase military strength with its "New Look" program, which emphasized nuclear weapons. But Eisenhower soon found it necessary to operate under a fixed budgetary ceiling for defense. Among other changes resulting, no longer would two or more broadly comparable aircraft be produced for the same mission or missions, no longer would older aircraft necessarily be replaced on a one-for-one basis, and existing production was continued at slower rates. The industry thus was pressured both to contract in capacity and to diversify.

Several firms simply found themselves with no new aircraft production to follow those programs being completed. This describes first Ryan and Martin in the 1950s, then Republic in 1964, North American in the late 1960s, and later Fairchild and Vought in the 1980s. The "weapon system" concept begun in 1953, in reaction to the muddled B-36 development program, presented a further challenge to contractors, making them responsible for the entire system rather than only for the airframe component.

From that time the term *systems* became increasingly used in both the military and the industry to refer to a grouping of components to constitute an integrated whole. Thus in addition to weapon systems, aeronautical systems, missile systems, space systems, and electronic systems appeared. Defense contractors increasingly referred to their subordinate organizations as systems divisions.

A wave of contract cancellations fell in the 1957–59 period, coinciding with completion of the 137-wing program of the air force. The cutbacks also reflected the growing realization that continued proliferation of aircraft and other weapon programs simply was not enhancing overall defense capabilities. Total defense spending as a percentage of gross national product declined steadily from fiscal year 1955 to fiscal year

1965.[23] Yet while new aircraft programs declined, those undertaken were of such magnitude that a single program could have a major impact on an entire state or region. The first multibillion dollar aircraft and missile development programs appeared. Consequently, social considerations attending a reduction or cancellation often became more significant than military or strategic needs, a phenomenon that would only increase in intensity.

Even as the industry recognized the necessity of extensive restructuring, no firm voluntarily abandoned the aircraft field. Those that succeeded did so as diversified high-technology firms while retaining aircraft development and production as the largest distinct activity. During 1957 the government phased out its practice of furnishing plant, equipment, and working capital to manufacturers, slowed down progress payments, and became more selective in providing financial assistance.[24] Government-owned plants still were operated, in part because the industry was fearful of being burdened with excess capacity. Thus the government still effectively subsidized the industry in three major areas: research-and-development funding; government-furnished plant and equipment; and cost-reimbursement procurement, especially with Cost Plus Fixed Fee contracts.

Given the risks and costs of defense contracts, the industry was unable to operate without such government support, but changes in the development, production, and procurement environment since the Korean War period increased pressures felt by the industry. Future contract reforms served to further increase risks to contractors.

From the late 1950s it was foreseen that then-current aircraft programs, both military and commercial, would have ever-increasing production and service lives, not needing replacement every few years, as

TABLE 16. Industry Employment, 1955–61 (as of January 1, in thousands)

Year	Total
1955	761
1956	837
1957	895
1958	784
1959	755
1960	674
1961	669

Source: Northrop, "Air Industry"; and Aerospace Industries Association, Aerospace Facts and Figures, 1962, 14.

had been the case. With aircraft generations lengthening, new program opportunities would become less frequent. But increasing exports tended to lengthen production lives, and aircraft also began to be exported earlier, especially to major military allies and major international airlines. NATO and other military alliances enhanced military export markets, especially for the rearmament of Germany and Japan. The Lockheed C-130 became a prominent example, remaining in production after more than forty years, increasingly for export. The Douglas A-4 was produced for twenty-four years, although at low volume in its final years, and the McDonnell F-4 Phantom II was produced for twenty-one years. Both were widely exported. Both the Boeing 707 and 727 were in production over twenty years.

Missile development progressed rapidly, although initially with greater emphasis on pure research than operational deployment. While the first guided-missile development contracts had been awarded to the industry in 1945 and 1946, it was not until the Korean War period that military missiles began to reach operational status. Early operational missiles included the Sperry Sparrow and Hughes Falcon air-to-air and Douglas Nike ground-to-air interceptor models. Missiles bid to succeed expensive manned aircraft.

Development of space exploration rockets, largely from German technology, involved all military services. The Martin Viking research rocket, which owed much to the V-2, attained an altitude of 135 miles in August 1951. Strategic missiles had been studied since the end of the war, and 1954 saw the development of the first miniaturized inertial guidance systems, which would lead to the intercontinental ballistic missile (ICBM).[25] Contracts for the Convair Atlas and Martin Titan were awarded also in 1954, the feeling at the time being that two were needed to ensure success. The space and military missile programs became closely related in that first-generation ICBMs served as launch vehicles for satellites and manned orbiting capsules. The "space race" with the Soviet Union also led to an increased priority.

Significantly, the aircraft industry did not take the initiative in this area. Both the space program and early missile development originated with the military, with the industry tending to view the developments as impractical or as a threat to aircraft programs.[26] Douglas, Martin, Bell, and Ryan were early leaders in missiles, but it was not until after 1954 that most firms began to regard the field as critical to their futures. By the late 1950s the missile and space programs had grown to the point that much of the increased spending in those areas came from funding previously applied to manned aircraft procurement. After a flood of fighter production, only the Republic F-105 remained in the air force procurement budget for fiscal year 1960.[27]

The aircraft industry also realized belated that, even with aircraft production declining, it might not necessarily fall heir to the missile and space programs. Technological industries such as automobiles and electronics could establish strong positions in those fields as logically as aircraft firms. Guidance and propulsion systems were more critical in missiles than aerodynamics, so that aircraft firms held no special advantage. Further, the fact that the government awarded missile contracts to nonaircraft firms clearly indicated a desire for greater competition, and that the aircraft industry held no place of preference in the new fields. Nonaircraft firms that became missile contractors included Bendix, Chrysler, Goodyear, Honeywell, Hughes, Martin Marietta (created in 1961), the Naval Weapons Center (NWC), which developed the widely employed Sidewinder air-to-air missile, Philco-Ford, Raytheon, Sperry, and Western Electric. Nonaircraft firms involved in the space program included Aerojet-General, Thiokol, and Ramo-Wooldridge Space Technology Laboratories (STL). Missile and space programs thus represented both a threat to the continued viability of the aircraft industry as well as a major future opportunity.

One researcher likened the evolution of the aircraft/missile/space industry during the late 1950s and early 1960s to the "creative destruction" process described by the noted economist Schumpeter. In a competitive marketplace with rapidly evolving technology, some firms would inevitably fail or be destroyed, while others would transform and flourish, and new competitors would appear to serve the transformed markets.[28] Thus missiles represented the new technology, threatening to displace aircraft, the old technology. In addition, new competitors held a potential advantage in not carrying the disadvantages of aircraft experience.[29]

The combination of declining military procurement, contract cancellations, and heavy development expenses for the new generation of jet airliners gave the aircraft industry unwanted prominence among the financially troubled. For three consecutive years aircraft firms, Douglas in 1959, Lockheed in 1960, and General Dynamics in 1961, led the list of money losers among the *Fortune* 500 industrials.[30] Douglas led the list again in 1966, its last year of independent existence. The traditional aircraft industry thus faced unprecedented challenges, and survival led to three broad strategic responses.

- Diversification within the aerospace industry
- Diversification into other technological fields
- Diversification into nonaerospace commercial and industrial product lines

It was clear that traditional aircraft manufacturers would become aerospace firms, but organization and product activity could vary widely. McDonnell pursued the first alternative by acquiring Douglas, as did Fairchild by acquiring Hiller and Republic. North American, Ryan, and Chance Vought were acquired by diversified but largely nonaerospace companies. General Dynamics and Lockheed diversified into nonaerospace but still defense-related activities, as well as attempting to compete in commercial aircraft. Grumman and Boeing attempted to redirect their capabilities into such commercial activities as urban mass-transit vehicles, but without great success. A recurring problem with aircraft firms was that they simply lacked experience in other markets or endeavors, and management could seldom exploit the technological capabilities of their firms in other areas. While it remained difficult for the aerospace industry to diversify successfully, the primary survival criterion remained adaptability, involving very selective diversification into specialized areas. Nonaerospace diversification would be by acquisition.[31]

While there is no specific date or event to mark the transformation of the aircraft industry to the aerospace industry, 1958 and 1959 probably represent the principal years. The term *aerospace* also proved to be difficult to define precisely, but generally encompassed development and production of manned and unmanned aircraft, missiles, propulsion systems, space vehicles, and associated electronics.[32] Accordingly, nonaircraft firms such as General Electric were included. Actually, the air force was as responsible as the industry for the term, since it was pursuing a military role in space. Fortunately for the industry, declining aircraft procurement was almost directly offset by rising missile and space program procurement. While combat aircraft would remain the bread and butter of the industry for years to come, missiles already comprised one-fourth of the industry's sales in 1958.[33] A long-term problem, however, was that heavy procurement of such strategic missiles as the Polaris and Minuteman would not continue indefinitely due to a low rate of obsolescence.

Organizational structures reflected the increasing missile and space emphasis, accompanied by corporate name changes. The Glenn L. Martin Company became the Martin Company early in 1957, and Northrop Aircraft, Inc., became Northrop Corporation in 1959. Similar changes by others would follow. The aerospace industry thus consisted of the older, primarily airframe companies, along with major sectors of the electronics, communications, chemical, and automotive industries. The transformation of aircraft firms was as extensive and significant as, for example, that in the office equipment industry, where such firms as IBM and NCR transformed into primarily computer manufacturers.

Government and industry associations also reflected the changing environment. NACA was succeeded on October 1, 1958, by the National Aeronautics and Space Administration (NASA), largely as a response to Soviet space achievements. The new acronym quickly became a part of the language. All NACA activities and facilities, including Langley, Ames, Wallops Island, Lewis, the Flight Research Center at Edwards Air Force Base, the Jet Propulsion Laboratory, and the Marshall Space Flight Center, were folded into NASA. The major focus became space research, but aeronautics continued to be important. The increased space budget also gave the industry a fourth major government customer after the three military departments. Procurement activities were generally similar, however, so that the industry still could view government as a single large market. NASA also had the salutary effect of removing the space program from potential interservice rivalry, although jurisdictional disputes still appeared, especially with the air force over space activities.

The Federal Aviation Act of August 23, 1958, created the Federal Aviation Authority. The act repealed the Air Commerce Act of 1926, the Civil Aeronautics Act of 1938, and the Airways Modernization Act of 1957. Removed from the Department of Commerce, the FAA began operating as an independent agency on December 31, 1958. Continuing its activities in air safety, licensing, and registration, the FAA further developed the long series of Federal Air Regulations (FARs) with comprehensive coverage of aviation.

In 1959 the Aircraft Industries Association became the Aerospace Industries Association, recognizing the changed structure and nature of the industry. The Air Coordinating Committee, which in later years had focused on international civil-aviation activities, was terminated in October 1960.

Helicopter and Convertiplane Advances

Concurrently with the decline in military aircraft production and the growth of missile and space programs, there were highly significant developments in two other key sectors: helicopters or vertical-takeoff aircraft; and airliners.

Turbine power for helicopters became increasingly commonplace after the mid-1950s, and helicopter capabilities steadily advanced. Turboshafts enabled the power-to-weight ratio to increase, overcoming that limitation of piston engines. A single main rotor with antitorque tail rotor and tandem rotors became the dominant configurations, while the intermeshing, coaxial, and twin outrigger configurations were largely abandoned for the time.

Even after its success in the Korean War, military helicopter procurement dropped sharply in 1954, and a mass commercial market had yet to appear.[34] All helicopter producers suffered. But competitions for advanced military designs were under way, and the design spinoff for numerous commercial and business applications also lay ahead.

Bell, with large-scale production of the Model 47G continuing for both military and civil markets, won in 1955 the important army contract for an advanced utility helicopter with its Model 204 turbine-powered design. As the XH-40, it first flew on October 20, 1956, and entered production for the U.S. Army in 1958. Later redesignated UH-1, it became the famed "Huey" of Vietnam and the most widely used helicopter in the world. By that time totally committed to helicopters, Bell in 1958 considered acquiring Vertol, with a complementary range of helicopters, before its acquisition by Boeing. The UH-1 was succeeded by the Model 204B, or UH-1B, in 1961, and this variant was license-produced by Agusta in Italy and Fuji in Japan. The more advanced Model 205, or UH-1D, first flew on August 16, 1961, and was license-built in Germany as well as by Agusta.

Bell also developed the improved four-seat Model 47J Ranger for both military and civil markets, beginning production in 1956. Bell Helicopter Corporation, a wholly owned subsidiary, was formed in Texas on January 1, 1957, to control helicopter operations, and Bell employment and production were increasingly centered there. Bell succeeded the Model 47J with the streamlined Model 206 JetRanger light turbine helicopter for the commercial market. First flying on February 10, 1966, the Model 206 won a major market both in civil and military roles.

The Sikorsky S-58, successor to the S-55, first flew on March 8, 1954, and was widely adopted by the military services and by many other nations, as well as finding a civilian market. It became Sikorsky's largest helicopter program and joined the S-51 and S-55 in being produced under license by Westland in Great Britain. The S-56, a contemporary of the S-58, held the distinction for a time as the largest and most powerful production helicopter in the world, with twin two-thousand-horsepower P & W R-2600 piston engines mounted in external pods. It carried twenty-six troops in its Marine Corps HR2S-1 assault version. Igor Sikorsky retired as engineering manager in July 1957 at age sixty-eight but remained active in research until the end of his life.

In 1958 Sikorsky developed the S-62, the first amphibious helicopter, with a flying boat hull for water landings and employing many components of the earlier S-55. The S-62 also was licensed to Mitsubishi. Then the S-61, designated HSS-2 (later SH-3) by the navy, first flew on March 24, 1959. A more advanced amphibious design for antisubmarine and other missions, the S-61 also was licensed to Westland, which pursued a long line of development. The commercial S-61 became widely used by the early helicopter airlines. Sikorsky also made major advances with the heavy lift helicopter. The experimental S-60 Flying Crane was developed from the S-56. Then the larger S-64 Skycrane, first flown on May 9, 1962, and powered by twin JFTD-12 turboshafts, could lift heavy cargo in a sling or in a detachable container. Five were later employed by the army in Vietnam as the CH-54, and the developed CH-54 was powered by T73 turboshafts. The large S-65 transport, derived from the S-64 and powered by twin T64 turboshafts, first flew on October 14, 1964, as the CH-53A for the Marine Corps.

Sikorsky still experienced difficulty in winning major contracts from the largest helicopter user, the U.S. Army. The utility helicopter program was awarded to Bell, the transport helicopter program to Vertol, and the Light Observation Helicopter (LOH) program to Hughes. Sikorsky remained the largest helicopter producer in the 1960s but no longer enjoyed the dominant position it held earlier. With the S-61, S-62, and S-65, however, it led in large helicopters. Production of the veteran S-58 was suspended in December 1965 but was later resumed for export.

Piasecki, although having experienced phenomenal growth in the early 1950s, faced major retrenchment after 1954. Laurance Rockefeller held policy disagreements with Frank Piasecki and had become disenchanted with his management skills as well. The Rockefeller interests hired Don Berlin as president and chief executive, then in May 1955 deposed Piasecki from the company he founded.[35] Embittered by his ouster, he then formed a new Piasecki Aircraft Corporation while retaining his large stockholding in the original Piasecki Helicopter Corporation.[36] In March 1956 the original company was renamed Vertol Corporation. Frank Piasecki then launched a bid to regain control of Vertol, but the subsequent elections produced a board majority supporting Rockefeller and Berlin.[37] Piasecki returned to his independent developments, acquiring the old Bellanca facilities in New Castle, Delaware, for expansion. He engaged in component production and subcontracting in addition to helicopter development. From company resources he developed the experimental Model 16H Pathfinder compound helicopter, with a ducted pusher propeller in addition to the main rotor, which first flew on February 21, 1962.

Berlin led Vertol into the commercial helicopter field with the Model 44, developed from the H-21. The major success, however, was development of the advanced Model 107, beginning in 1956, which took full advantage of turboshaft progress. The Model 107 first flew on April 22, 1958, and was ordered by the navy as the CH-46A Sea Knight. The similarly configured but substantially larger and more powerful CH-47 Chinook followed in 1961, and became the army's standard heavy assault helicopter.

By 1959 Vertol had recovered to the point of being the largest specialized, independent helicopter company in the industry and thus became an attractive acquisition candidate. On March 31, 1960, Vertol became a wholly owned subsidiary of the Boeing Company.[38] The acquisition also included the subsidiaries Allied Research Associates and Canadian Vertol.[39] Inevitable tensions rose between Vertol veterans and Boeing management but eventually abated. Don Berlin ran the Vertol Division under Boeing until his retirement in 1963 and was succeeded by Boeing executives.

The Hughes Tool Company, Aircraft Division, tested its first design, the two-seat Model 269A, in October 1956. The Model 269A, later redesignated Model 200 for the civil market, won military orders as the TH-55A trainer, 792 being delivered. Hughes then won a major victory with the army LOH contract on May 26, 1965, for its OH-6, thereby becoming a major manufacturer in the growing field. The OH-6 won over the Bell OH-4 and Hiller OH-5. The OH-6, which first flew on February 27, 1963, was developed from the civil Model 300 and in turn from the original Model 269A and was to replace the Cessna L-19, Bell H-13, and Hiller H-23. The civil version of the OH-6 became the Model 500.

Howard Hughes, foreseeing a vast worldwide market for a new light helicopter, was so determined to win the LOH order that his bid, at a unit price of less than twenty thousand dollars, factored in a loss of ten thousand dollars per unit based on an initial production run of one thousand, or $10 million. He felt that the loss could be recouped, and profits earned, on follow-on orders and civil sales.[40] But this scheme, violating Defense policy on "buy-in" bidding, was to cause him serious financial difficulty.

Hiller continued to produce the basic H-23 light helicopter for the armed forces while pursuing civil markets and further designs. In the summer of 1960 Stanley Hiller negotiated a merger of Hiller into the Electric Autolite Company, with Hiller continuing to operate as a separate subsidiary. Doman suspended helicopter production in 1961. The D-10B, developed from the LZ-5, was ready for production in 1965 but did not find a market. Doman's plans for license manufacture of the D-10B in Italy likewise did not proceed, but development later resumed in the United States.

Kaman progressed as a major manufacturer with its intermeshing-rotor military designs. Ralph Damon, president of TWA and a director of Goodyear, met with Charles Kaman in 1956 concerning Goodyear's interest in acquiring Kaman. Kaman declined the proposal, feeling that his stockholders would be better served by the firm's independence.[41] Production of the H-43A, Kaman's last piston-engined helicopter, ended in July 1959, and Kaman then focused on turbine-powered single-rotor designs. Its successful entry in the LAMPS (Light Airborne Multi-Purpose System) competition, the advanced HU2K-1 (later SH-2) Seasprite, first flew on July 2, 1959, and entered quantity production for the navy. The type was particularly successful in the long-range sea rescue role and would enjoy a long, if sporadic, development and production life.

By the 1960s, with the army the dominant customer and with increasingly centralized procurement in the McNamara defense era, Kaman felt itself becoming frozen out of new helicopter business. McNamara's systems analysis attempted to prove quantitatively that one system was superior to another in cost-effectiveness, and Kaman suffered under this system by being ranked as too expensive. Such producers as Bell, Vertol, and Hughes were favored for army orders, and Charles Kaman suspected that President Johnson favored Bell due to its Texas factory.[42] Kaman then found itself the object of another unwelcome takeover attempt in the early 1960s, this time from the newly formed Ling-Temco-Vought Corporation. The attempt was initiated by Paul Thayer, acting for James J. Ling. Charles Kaman especially disliked Ling's arrogance and determined to remain independent.[43]

Beginning in the early 1950s the armed forces sponsored industry research and development in one of the oldest dreams of powered flight, an aircraft that could combine the best features of fixed-winged aircraft and helicopters. Helicopters had the ability to take off and land vertically and to hover, but were difficult to pilot and were severely limited in speed and lifting capacity. Conventional aircraft were easier to control and did not present such constraints on speed and capacity but were restricted to long runways, expensive to build and militarily vulnerable. Both military and civil operators dreamed of craft that could escape the shortcomings or capitalize on the virtues of each. The autogiro was an earlier approach that had failed to win a market. Now, technological advances promised to bring the concept to reality. All three services were involved.

There were numerous approaches to the vertical-takeoff-and-landing (VTOL) aircraft, or convertiplane. Some combined a large rotor for takeoff and landing with short wings with conventional engines for forward flight. Others used separate, vertically mounted jet engines for takeoff and landing, and others for forward propulsion. Still others combined propeller blades and rotors, tilting from vertical to horizontal, or fixed propeller/rotor blades mounted on tilting wings, or were tailsitters with fixed lift and propulsion systems. One successful approach involved a single jet engine with thrust vectoring, with exhaust nozzles rotating from vertical to horizontal.

McDonnell tested its hybrid Model XV-1 in 1954. A relatively simple design with short fixed wings, it was powered by a single pusher piston engine for forward flight and geared to a conventional rotor for takeoff and landing. The XV-1, probably the first successful convertiplane, took advantage of wartime German research in the field.

There was strong interest in VTOL combat aircraft of high performance, yet without restriction to long runways or aircraft carriers. The air force sponsored the Ryan X-13 Vertijet, while the navy sponsored the Lockheed XFV-1 and Convair XFY-1 "Pogo" turboprop designs, all of which were literally tailsitters. First flying in March and August 1954, respectively, both "Pogo" models succeeded in taking off vertically and making the transition to horizontal flight, but were underpowered and unstable.[44] The Ryan X-13 made its first conventional flight on December 10, 1955, and its first horizontal-vertical transition and back on November 28, 1956. It achieved its design goal of vertical takeoff to horizontal flight to vertical landing on April 11, 1957, but also remained strictly experimental.

The first fixed-wing tilt-rotor design, the Bell XV-3, first flew on August 23, 1955. While an experimental proof-of-concept aircraft only, powered by a small piston engine, numerous successful flights proved its practicality. The military later returned to support of the tilt-rotor concept. Bell also developed the X-14, powered by two small jet engines, which employed the jet-deflected-thrust concept. The X-14 achieved hovering flight on February 19, 1957. By 1960 Hiller had developed the X-18 tilt-rotor convertiplane transport, using the airframe of the old Chase XC-122. Considerable expense was saved by utilizing the Allison T40 engines and Curtiss-Wright propellers of the navy's earlier "Pogo" program.

Ryan also developed the VZ-3RY convertiplane employing a single centrally mounted shaft turbine engine driving two large wing-mounted propellers, with exhaust deflection vanes enabling the craft to hover as well as achieve horizontal flight. Early tests were successful, but the prototype was destroyed in a crash in February 1960. Fairchild also tested the deflected-slipstream concept in 1960 with its VZ-5FA.

U.S. Army–sponsored convertiplane development began with the Lockheed XV-4A Hummingbird, with small fixed wings and twin jets employing nozzles and ducts for thrust deflection. The XV-4A was first flown as a conventional aircraft on July 7, 1962. The Ryan XV-5A, another VTOL research aircraft for the army, employed fans mounted in the broad wings for lift. The XV-5A first flew on May 25, 1964.

Curtiss-Wright developed several convertiplanes beginning in 1960, marking its return to the aircraft field, although only on an experimental basis. The Curtiss-Wright X-19A featured four lifting propellers, mounted in tandem on each side of the fuselage on stubby wings, which then tilted forward for forward flight. The Bell X-22A, which first flew on March 17, 1966, featured four large ducted propellers, which tilted for takeoff and hovering, with the ducts providing lift for horizontal flight. The LTV-Hiller-Ryan XC-142A, a triservice transport owing much to the X-18, made its first conventional flight on September 29, 1964, followed by a successful vertical-horizontal transition on January 11, 1965. While a major advance toward operational application, the craft experienced stability and vibration problems and ultimately became a budget victim.

The industry pursued the convertiplane concept aggressively, fulfilling many R&D contracts and achieving promising results during the 1954–64 period, but none resulted in an operational application or production. The concept would wait years to be realized.

The American Jet Airliner

Jet airliners, with their unprecedented technological and financial risks, posed a major challenge for both the industry and the airlines, but the day of the piston-engined airliner clearly was ending. The stretched Douglas DC-7C "Seven Seas" and the extended-wing Lockheed L-1649A Starliner, both introduced in 1956, were the first true intercontinental airliners and the ultimate piston-engined airliners. But by that time American jet airliners were under development, and orders for both were relatively low.

Further, airliner development had long since attained a broad technical plateau, with only incremental improvements appearing. The most advanced piston-engined airliners essentially were descendants of designs of twenty years before. It was felt that a major design advance was overdue. In addition, by 1953, before the disastrous crashes, there was considerable optimism about the Comet.[45]

Airliner manufacturers, while recognizing the risks, definitely were encouraged to move into the jet generation by two further factors:

- The United States was a growth market for air travel over the long term, despite cyclical fluctuations.
- Exports also were a growth market, given the worldwide expansion in airline travel, the international competitiveness of airlines, and the U.S. lead in airliner development and production.

Major foreign carriers began to demand the latest airliners as soon as U.S. carriers, adding large export orders to initial home production. Yet the risks remained formidable. In addition to the technical risks, demand for new airliners was always highly cyclical. Airlines tended to engage in buying sprees during periods of growth and prosperity and then cancel, reduce, or postpone reequipment in the periodic downturns. Major airlines rushed to order the first-generation jets, fearful of being left behind competitively. While providing a strong stimulus to the industry, there were doubts that the volume of orders could be sustained. Still, recurring periods of overcapacity led to disposal of relatively new airliners onto the used-airliner market, which in turn cut into new orders. Since airlines were inherently high-fixed-cost operations and tended to be highly leveraged financially, business downturns quickly precipitated financial crises. Such crises carried an immediate negative impact to the builders.

Also weighing against the long-term growth scenario was another inherently dangerous situation, that of the long lead time between commitment to a new design and the market conditions prevailing by the time of service entry. Quite apart from any threat posed by competing models, economic or market conditions could deteriorate from those projected by the manufacturer and the airlines when the new model was launched, leaving the manufacturer with huge development costs and no immediate prospect of recovery. Survival in the commercial market almost demanded major diversification.

Whatever the risks, no established manufacturer was willing to abdicate the field. All were acutely aware that failure to enter the jet market would mean their permanent withdrawal from that important sector. The early lead established by the British also forced the industry to take notice. Accordingly, there was a flurry of activity in the 1952–55 period. An informal jet airliner Big Four, Arthur E. Raymond of Douglas, Hall L. Hibbard of Lockheed, Wellwood E. Beall of Boeing, and Thomas G. Lanphier of Convair, emerged.⁴⁶ A civil development of the J57 engine provided the propulsion basis for the first large jetliners, and most believed that jets could achieve higher productivity than piston-engined models due to their speed and capacity. This became an urgent necessity since the initial Boeing 707 and Douglas DC-8 would cost in excess of $5 million each. By comparison, the DC-7C cost approximately $2 million.

Boeing's entry, leading other American competitors, was one of the greatest risks ever taken by an aircraft company and remains a classic case of a major gamble that succeeded. Boeing had studied advanced airliners since the end of the war, both with turboprop and turbojet developments of the Model 377 Stratocruiser and new, smaller, swept-wing designs under the Model 473 series. The parallel studies were consolidated in 1951 as the Model 367–80.⁴⁷ The designation, extending from the military Model 367, was somewhat misleading, probably intended to deceive potential competitors. William Allen authorized development on April 22, 1952, soon after the giant B-52 had first flown. Had the $16 million venture failed it probably would have bankrupted the company.⁴⁸ Instead, the eventual Model 707 made Boeing the leading airliner builder, and the only consistently profitable one, in the world.

Boeing's success was by no means a foregone conclusion. While possessing great expertise in large aircraft, and valuable jet development and production experience with the B-47 and B-52, Boeing's airliner record was lackluster. Its Model 40 single-engined and Model 80 trimotor models were beaten by others in the small market of the 1920s. The Model 247 was bested by the DC-2 and DC-3, and the 307 Stratoliner and 314 Clipper, while successful designs, had their prospects curtailed by the Second World War. The postwar Stratocruiser then lost out to the DC-7 and Super Constellation, carrying a large financial loss as well.

The prototype Model 367–80, after extensive ground testing and with great fanfare, made its first flight on July 15, 1954. Among the spectators was the seventy-two-year-old Bill Boeing, long retired.⁴⁹ His wife Bertha launched the prototype with a bottle of champagne across the nose in the manner of ship launchings. In building the prototype, Allen had gambled that the air force, although it had not issued a specification, would need a fleet of jet tankers for its jet bombers to replace the slow KC-97.⁵⁰ Jet tankers would enable refueling operations to take place at higher speeds and altitudes, with greater mission effectiveness. The air force in fact ordered an initial production batch of the KC-135 variant in June 1955, and the program was under way.

Pan American, also in June 1955, became the launch customer for the commercial Model 707, followed quickly by United and Eastern. In another coup, Allen negotiated an agreement with the air force that permitted the tanker and airliner models to be produced concurrently on the same assembly line, resulting in greater production efficiency.[51] The Model 367–80 had in fact been constructed in Boeing's government-owned facilities. The first KC-135 flew on August 31, 1956, some two years before the 707 began commercial transatlantic service on October 26, 1958. The 707 was not, however, the first turbine-powered airliner in transatlantic service. The British turboprop Britannia 310 began December 19, 1957, followed by the Comet 4 on October 4, 1958.

The 707 enjoyed great success both militarily and commercially. While it owed much to the B-47 and B-52, the 707 still differed significantly in its low wing, found preferable for airliners, and tricycle landing gear as opposed to the bicycle gear of the bombers. Orders rose rapidly, and further developments, the 720 for the medium-range market, and the larger intercontinental 707–320 and 707–420, appeared. The initial models, with their first-generation fuel-gulping engines, possessed only marginal transatlantic capability, and Juan Trippe of Pan Am pressed William Allen to develop the turbofan-powered 707–320. Trippe, a master at playing manufacturers against each other, hinted that he would order more DC-8s otherwise.[52] Success of the 707–320 gave Boeing command of the jet market. The 707 and the DC-8 were, incidentally, the first airliners to carry more than one hundred passengers normally.

Douglas was to experience major difficulties in the jet airliner field, although attaining large-scale production and presenting formidable competition to Boeing. Douglas began studies in 1951 in reaction to the Comet, even in advance of development of the DC-7. The Douglas Special Projects Office was set up in June 1952 but took three years to decide on DC-8 production.[53] Douglas felt somewhat complacent, believing that the airlines would not order a Boeing product in quantity. Douglas also could not determine the optimal size for the new airliner or the power mode, at times considering turboprop developments of the DC-6B and DC-7.[54] C. R. Smith of American Airlines pushed for turboprops, while Pan Am wanted a larger intercontinental jet. Douglas finally announced its DC-8 in June 1955, on the tail of Boeing's announcements of the first air force KC-135 order and first airline 707 orders.

The DC-8 presented a greater corporate risk for Douglas than had the 707 for Boeing, since Douglas had no base of military orders for its design.[55] Already seriously behind, Douglas was under pressure to offer an airliner not simply the equal of the 707 but clearly better. The DC-8 prototype did not fly until May 30, 1958, incidentally some twenty-five years after the first flight of the DC-1, and won its adherents but could not be described as clearly superior to the 707. In retrospect Douglas probably maintained the DC-6B and DC-7 in production too long and should have reallocated resources toward jet development sooner. That the DC-8 was almost a year later than the 707 in entering scheduled service proved to be highly critical, and development costs had strained the company's finances. Then Boeing beat Douglas to the market with its long-range 707–320, and Douglas lacked the resources to respond quickly.[56]

In another market segment, Douglas negotiated U.S. rights to market the French Caravelle, a smaller twin-jet airliner, feeling that it might potentially save the expense of developing a new small airliner. Major sales were not achieved, however, and no license production was undertaken. Douglas later decided to develop its own twin-engined airliner, further straining its finances.

Lockheed, seeking to replace its Super Constellation series in production, differed from Boeing and Douglas in its approach. Rather than competing directly with another large, four-jet airliner, Lockheed decided on the medium-range segment. Further, Lockheed was impressed with the success of the British Viscount and cautious about pure jet power given the record of the Comet.[57] Consequently, Lockheed announced its Model 188 Electra in 1955, named for first Lockheed airliner of 1934. The Electra was powered by four Allison 501 turboprops, civil versions of the T56 of the successful C-130 military transport, giving Lockheed a direct military spinoff benefit. Orders were encouraging, and the first Electra flew on December 6, 1957, two weeks before the first commercial Boeing 707. In service, however, five crashes during 1959–60, traced to various causes, damaged the Electra's reputation and effectively ended future sales prospects.[58] Only 172 were produced, forcing Lockheed to take major financial write-offs on the program. Over the long term, however, the program proved a financial success in that it provided the basis for the P3V Orion patrol bomber, produced in large numbers.

Of all entrants into the burgeoning jet airliner field, Convair became the most troubled. Its venture ultimately forced a $425 million write-off by 1963, the biggest loss suffered by an American corporation to that time, even greater than that by Ford with the Edsel.[59] Convair wished to remain in the airliner market both for diversification benefits as well as to replace its 340/440 series, then winding down. By 1955 Convair felt that Boeing and Douglas had captured the long-range market with the 707 and DC-8, but that it had a potential advantage in the short- to medium-range segment.[60] Despite the validity of this analysis, the first Convair customer was the mercurial Howard Hughes, for his TWA. After the Boeing 707 was announced, Hughes expressed interest in a new jet tailored to his medium-range needs.[61] Hughes had also ordered the 707 but was faced with later delivery than his competitors and so was determined to beat them in the medium-range segment. Allowing Hughes to dictate specifications was to cause Convair much grief and ultimately end its role as an airliner builder.

Organizational problems resulted because Convair then was some 75 percent of General Dynamics and was too autonomous from the GD board and executives in New York. The jet airliner decision, with its enormous risks for General Dynamics, was made in San Diego. The original design, first named the Skylark 600, then Golden Falcon before becoming the Model 880, was supposedly medium range for the needs of Howard Hughes, but as developed was powered by four engines and approached the 707 and DC-8 in size and capability. Yet it was too late in timing to overtake them in the market. In hindsight Convair definitely should have focused on the shorter-range market, along the lines of the successful French Caravelle. Such a venture could have captured the smaller jet market and might well have headed off Boeing's later, and spectacularly successful, Model 727.[62]

When Convair first considered the market in 1955, General McNarney, president of the Convair Division, strongly supported the effort, despite the urgent warning of Thomas G. Lanphier, Marine Corps air ace and chief of planning, of the twin dangers of overoptimism and of moving away from Convair's dominance in the short-range market.[63] The Convair 880/990 episode became a classic case of exactly that. Specific problems were lack of managerial control, cost underestimation, market overestimation, and a product simply mismatched with market needs in size. Yet Convair originally estimated breakeven at just sixty-eight aircraft, and the total sales forecast was for several hundred, providing an estimated program profit of $250 million.[64]

The Convair 880 was officially launched in March 1956. Hughes, through Toolco, signed a TWA order for thirty Model 880s in June 1956, and his contract also carried exclusive operating rights for one year, further handicapping Convair in the marketplace.[65] In addition to its late start and the Hughes one-year exclusive, slowing other orders, Convair had only a small deposit from Hughes.[66] McNarney remained optimistic, however, that the 880 would be a strong seller and profit generator.[67] Seemingly Floyd Odlum, who had dealt with Hughes in the 1930s in the movie industry and was still a director, would have advised caution, but Convair was so eager to gain orders that it accepted Hughes's terms. Then soon thereafter, Boeing dealt the 880 program an effective death blow by launching its economical shorter-range Model 720, which United ordered over the 880.

Convair, becoming desperate for orders, rushed ahead with its larger and more advanced Model 990 (initially designated Model 600, as had been the 880), winning an order from American. But C. R. Smith, a hard-nosed negotiator, required Convair to accept piston-engined airliners in trade, adding yet another complication. The major snag, however, was that the 990 fell slightly short of the contract speed guarantee, which meant that American could not advertise that it operated the world's fastest airliner. After Smith originally canceled the entire order, Convair renegotiated, this time for only fifteen 990s, on a basis that effectively guaranteed losses on the sale.[68] A further problem was that differences between the 990 and 880 were so great that the 990 was practically a new design, with much greater development costs than projected.

Compounding Convair's difficulties, Howard Hughes stalled on 880 deliveries, citing several reasons other than the real one, that he simply lacked available cash after paying Boeing for his long-range 707s.[69] On October 5, 1959, Convair was forced to begin pulling partially completed 880s off the line, disrupting production.[70] Hughes also canceled an order for thirteen 990s. With costs mounting and sales prospects dimming, General Dynamics took the first financial write-offs late in 1960, but the eventual total rose to a magnitude that threatened its viability as a major firm. The failure also seriously damaged General Electric in the jet engine field, as losses on the CJ805 civil version of the J79 for the 880 and 990 left GE far behind Pratt and Whitney. Output of the 880/990 series totaled only 102, the last being delivered in 1964. Both proved to be good airliners in service, however, if somewhat uneconomical in fuel usage.[71]

The financial and control problems evidenced throughout the episode were exacerbated by major managerial upheavals. John Jay Hopkins, running General Dynamics with an iron hand, had brought in Frank Pace, formerly secretary of the army, as executive vice president in 1953 but never viewed him as a successor. Nonetheless, the board replaced Hopkins, then seriously ill, with Pace as president and CEO in April 1957.[72] Hopkins died only days later. Pace lacked extensive corporate experience and also found it difficult to control GD's far-flung operations, most of which were acquisitions. One result was that the GD board of directors had expanded to an unwieldy thirty-two members.[73] In addition, General McNarney, who retired as president of Convair in 1958, was succeeded by John V. Naish, an industry veteran. Naish then pushed ahead with the larger 990, compounding the original error.[74]

Fairchild, with military production running down, decided to compete in the growing local-service or feeder segment. Rather than incurring the costs and risks of a new design, Fairchild negotiated a license with Fokker of the Netherlands to produce and market its Friendship for the North American market. The Friendship, designated F-27 by Fairchild, was powered by twin Rolls-Royce Dart turboprops, the same engines used on the Viscount, and the Fairchild prototype first flew on April 15, 1958. The F-27 actually became the first American-built turbine-powered airliner in service, beating the 707 and the Electra.

The F-27 seemed a natural for its intended role of DC-3 replacement, with a capacity of forty to forty-eight passengers, but orders were slow. Local service airline economics at the time was such that greater capacity and passenger appeal simply could not overcome the capital expense of a fleet of new airliners and their greater absolute operating costs. Further, the F-27 was initially priced optimistically at only $540,000, a figure that rapidly escalated to $780,000. Used DC-3, Martin 404, and Convair 340/440 airliners continued to be the major competition. Further, there were numerous production problems, not least of which involved conversion from the metric system. Fairchild also did not use the newly developed Redux bonding construction method as did Fokker. Fairchild lost a total of $29 million, largely due to the F-27, from 1958 to 1960.[75] Production had to be suspended in 1960, and only 139 F-27s had been ordered by 1965. The market eventually strengthened with government guarantees to local-service carriers, and the lengthened, more powerful FH-227, which first flew on July 1, 1966, extended the production life of the series.

Jet airliner development was perilous, and the Convair case in particular illustrated a major problem that has long been recognized: it is exceedingly difficult to make money producing commercial airliners. Financial analysis showed that aerospace companies with the lowest returns on assets and net worth were those in airliner development, such as Boeing, Douglas, General Dynamics, and Lockheed.[76] Costs threatened General Dynamics, forced Douglas into a merger, and also brought Lockheed to the brink. Later, Boeing almost failed due to the 747 program.

Airliner market growth continued, but the appearance of new external complications, principally concerning energy costs and the environment, carried a major impact for both manufacturers and airlines. From the 1960s jet noise levels over populated areas became a major issue. Certain cities threatened to ban operation of noisy jets entirely, and manufacturers strived to develop quieter and more economical jet engines. Fuel costs were to rise sharply later.

Chapter 12 Aerospace and Defense Converge

The situation of the industry at the end the 1950s was similar to that at the end of the Second World War: there simply were too many companies for available or anticipated business. Despite a general recognition that consolidation should take place, there had been no mergers other than the General Dynamics acquisition of Convair. Concerns over inadequate or diminished competition resulting from mergers remained. Lack of progress was due in part to the view of concentrated industries taken by the Antitrust Division of the Department of Justice. The 1946 Lockheed–Consolidated Vultee merger was prevented on antitrust grounds. Only when a clearly failing firm was being acquired was the merger allowed. With a similar, if not more critical, situation faced by the British industry, the government in contrast effectively forced a series of mergers among major aircraft companies beginning in the late 1950s, eventually resulting in a single giant British Aerospace Corporation.

Industry Survey, 1954–67
Ryan by the 1950s had effectively exited aircraft production, although active in many other areas. Ryan developed the Firebee target drone and the Firebird, the first true air-to-air guided missile. The X-13 Vertijet (once incorrectly referred to as the XF-109), the VZ-3RY, and the XV-5A Vertifan convertiplanes all advanced the field. Ryan, while a smaller firm, represented one of the more successful diversification efforts in the industry, with positions in missiles, drones, navigation systems, and NASA radar, as well as experimental aircraft.

In 1955 the Emtor Holding Company, a California investment firm, acquired a 20 percent interest in Ryan. Ryan originally went public in the late 1930s with the assistance of Brashears, the firm that had also assisted Lockheed. Claude Ryan held only 12 percent of the stock by 1955, and Emtor gained effective control.[1] Emtor principal Robert Johnson joined Ryan's board and became president in 1961, with Claude Ryan continuing as chairman. Ryan acquired a 50 percent interest in Continental, the aircraft engine producer, in 1965, which led the way to its ultimate corporate destiny, with Teledyne.[2]

By 1954 Bell again faced diminished prospects. Of its notable series of experimental aircraft only the X-2 remained. Management also began a transition as Larry Bell, embattled with labor and stockholder disputes and a renowned workaholic, resigned as general manager on October 2, 1954, on health grounds, although remaining president. Bell finally resigned as president on September 18, 1956, and died on October 20, 1956, at age sixty-two.

Bell remained financially sound with a variety of R&D contracts, but such contracts tended to be low-profit areas.[3] It had delivered the first air-to-ground guided missile, the GAM-63 Rascal, to the air force on October 30, 1957, but the production order was canceled. Bell simply was unsuccessful in bidding on contracts for fixed-wing aircraft. Its X-16 high-altitude reconnaissance aircraft lost to the Lockheed U-2. Bell continued to develop advanced concepts, many under the direction of General Walter Dornberger, the German missile expert who joined Bell as a vice president. Dornberger led the Dyna-Soar manned orbital glider concept, extending from German wartime research, but Bell lost the contract to Boeing. Bell's orientation toward advanced structures rather than aerodynamics was felt to be a factor in the loss of the X-15 contract to North American. Bell thus also faced a decline in experimental aircraft, as the X-2 program ended. Many researchers and engineers departed, and new chief executive Leston Faneuf, without an aeronautical background, was regarded by many as of suspect competence.[4] Only some twenty-five hundred Bell employees remained in Buffalo–Niagara Falls by 1959.

Major reorganization followed. Bell Helicopter Corporation controlled rotary-wing development, while other operations were consolidated in Buffalo. Then the Textron Corporation of Rhode Island acquired all defense and aviation business of Bell on July 5, 1960. Nonaerospace activities were organized into an independent Bell Intercontinental Corporation. The renamed Bell Aerospace Corporation became a wholly owned subsidiary of Textron and controlled three operating companies, Bell Aerosystems, for aeronautical R&D and VTOL experiments, Bell Helicopter, and Hydraulic Research and Manufacturing Company in California. Harvey Gaylord became president of Bell Aerospace after the acquisition, but Leston Faneuf remained a director. Textron's decentralized structure allowed Bell Aerospace to retain its autonomy.

Martin made a vigorous effort to remain in military aircraft but ultimately was forced to exit. Martin's final attempt in aircraft of its own design was the advanced four-jet, swept-wing P6M Seamaster flying boat, to fulfill the stated missions of mine laying and reconnaissance, but which the navy actually hoped would enhance its participation in the strategic mission. The P6M resulted from a specification issued on July 31, 1951, and first flew on July 14, 1955. After a series of crashes, and with growing doubts about its missions, the P6M was canceled but remains noteworthy as perhaps the ultimate development of the large flying boat. Martin won a contract over Douglas for the supersonic XB-68 tactical bomber to succeed the B-57 and Douglas B-66, but that program also was canceled, in April 1957. The B-57 and the P5M were Martin's final aircraft efforts.

With the delivery of the last P5M-2 Marlin on December 20, 1960, the firm formally announced that it was leaving the aircraft-manufacturing field after almost fifty years. George M. Bunker led Martin into the missile field because he felt the firm had no alternative.[5] He had established a Missile Division in Denver, Colorado, largely at the insistence of Air Force Secretary Talbott on geographical dispersion of defense activities. Bunker then negotiated a merger with the American-Marietta Corporation of Chicago, a major producer of chemical products and construction materials, which became effective on October 10, 1961. Martin Marietta Corporation went on to success in missiles and electronics. The Titan ICBM was a well-managed and successful program and also served as a launch vehicle for manned and unmanned orbital and space flights.

An important strength of Boeing was that its large contracts for the B-52, the Bomarc, and the KC-135 tanker effectively subsidized development and production expenses of the 707 airliner. While its IM-99/CIM-10A Bomarc was withdrawn from service in 1962, it led Boeing to major involvement in the missile field. Aware of the need for further commercial diversification, Boeing established the Industrial Products Division in 1955, then established the Transport Division for airliners in 1956. The company was fortunate in that the eventual end of its military aircraft lines, with completion of B-52 production in 1962, coincided with its success in the commercial-airliner field.

By the early 1960s Boeing by most measures was the largest single aircraft or aerospace firm in the world, with more than one hundred thousand employees and major bomber, airliner, helicopter, missile, and space programs. Boeing was not only the largest industrial employer in Seattle, but also in the entire Pacific Northwest. The Boeing Airplane Company was renamed the Boeing Company on May 3, 1961. Under the corporate umbrella were six operating units: Aerospace Group, Commercial Airplane Division, Turbine Division, Vertol Division, Wichita Division, and Boeing Associated Products. In the span of a few years Boeing had transformed from a primarily military to a primarily commercial company.

Missile and space programs still rose in importance, and Boeing won the contract for the Minuteman solid-fuel ICBM program in 1958, to succeed the first-generation liquid-fuel ICBMs. The first Minuteman wing was formed in 1963, and Boeing followed with the improved Minuteman

II and more advanced Minuteman III. But Boeing suffered the cancellation on December 10, 1963, of the air force/NASA development contract for the X-20A Dyna-Soar space glider, which lacked a direct military application. With the loss of five thousand jobs the cancellation was a heavy blow, but Boeing achieved later space success with a contract for the Saturn first-stage booster on December 14, 1961, and for the robot Lunar Orbiter two years later. A large Space Center was built by 1965 to house development.

Boeing lost the supersonic strategic-bomber program to North American, but the later loss of the TFX contract to General Dynamics marked the exit of Boeing from military aircraft, apart from military versions of its airliners. Some suggested that Boeing's decline in military aircraft was directly related to its commercial success, as the military tended to steer contracts to those firms regarded as more needy, as with the B-70 to North American. In early 1965, however, Boeing formed a new Military Airplane Division, specifically in pursuit of the air force Heavy Logistics System (HLS) transport contract, incorporating the Wichita Division into the new unit. Other divisions were largely unchanged. The last of 732 KC-135 tankers was completed on January 12, 1965, and production of all military versions of the 707 totaled 820.

Boeing grew internationally by acquiring an interest in the German aerospace firm Bolkow in December 1964. Claire Egtvedt retired as chairman of the board in 1966 after almost fifty years at Boeing. William Allen remained CEO but did not immediately assume the vacant chairmanship.

Boeing proceeded to develop a family of jet airliners, with Minuteman profits effectively subsidizing new airliner development. The 707–320B and 707–420 were the first airliners to be powered by new turbofan engines, alleviating the high fuel consumption that had so troubled early jet airliner operation. The new tri-jet 727 medium-range airliner was launched on November 30, 1960, with orders from United and Eastern. In some respects the 727 was an even greater risk than the 707, facing major doubts that it could be offered at a price that airlines were willing to pay, given the availability of smaller airliners at much lower cost. The 727 made its first flight on February 9, 1963, and entered scheduled service with unusual rapidity, on February 1, 1964. It captured the rapidly growing medium-size market, where it lacked a direct competitor other than the lightly regarded British Trident, and soon became profitable. Perhaps more important, however, was that the 727 sealed Boeing's position as the dominant commercial-aircraft producer in the world. It then launched the twin-jet 737 in February 1965, for the short-range market. The 737 was also a major risk, two years behind its competitors and with Boeing's resources heavily committed to other programs.

The Convair Division of General Dynamics continued as one of the Big Five aircraft companies, with sales reaching $1 billion by 1956, but in many respects remained the most troubled. General Dynamics, in addition to its airliner disaster, became a rather conspicuous victim of the changing industry environment. Defense contractors increasingly were viewed less as national resources than as something more sinister and exploitative. The Arsenal of Democracy of the Second World War had evolved into the "military-industrial complex" of the 1960s. While all defense firms hired retired senior military officers, General Dynamics was criticized as more heavily involved in the practice than most.[6]

While committed to missiles and space, Convair continued aircraft production. Output of the CV-240/340/440 airliner series, including military variants, totaled 1,075 before ending in 1958. The YF-102A first flew with the new area-ruled fuselage on December 20, 1954, and entered large-scale production. It was followed by the more advanced F-106 (originally F-102B), powered by the P & W J75, which first flew on December 26, 1956, and was produced until 1961. Convair also pursued a nuclear-powered aircraft, the NX-2, for several years, benefiting from its corporate linkage with Electric Boat, the nuclear-submarine builder. The program finally was canceled in early 1961 as unfeasible.

The most challenging military aircraft program for the company came with the contract in 1952 to develop a supersonic bomber to succeed the Boeing B-47. This "dream bomber" would exploit Convair's delta-wing expertise and be powered by four GE J79 engines. All weaponry was carried in a jettisonable streamlined pod beneath the fuselage, and its structure employed honeycomb materials. Most significantly, the contract was the first awarded under the comprehensive "weapon system" policy in which Convair was made responsible for performance of all systems, including electronics, weaponry, and subcontracted components. The XB-58 first flew on November 11, 1956, and entered production at Fort Worth. Orders were restricted to only 116, however, due to strategic reassessments as well as some questions about the aircraft's performance. A Strategic Air Command request for a further 228 aircraft was rejected.

By 1960 General Dynamics aerospace operations consisted of three divisions, GD-Pomona, GD–Fort Worth, and Convair in San Diego, all extending from the old Consolidated Vultee. In May 1961 General Dynamics reorganized into an Eastern Group at New York and a Western Group at San Diego. The Western Group housed the aerospace activities, plus the Atomic Division. The Convair nameplate was discontinued in 1962, and all future aircraft were under General Dynamics. A 1963 reorganization then eliminated the two geographical regions and established twelve operating divisions, including those in aerospace. The protracted process of bringing the giant Convair operation under control finally was completed. But with the end of the fighter and airliner programs, employment at San Diego dropped to thirty-two hundred by 1964, and complete aircraft production ceased after some thirty years. Future production would be centered at Fort Worth.

The troubled management history of the firm continued. By the end of 1959, the Chicago industrialist Colonel Henry Crown became General Dynamics' largest stockholder, as the dimensions of the 880/990 debacle were becoming evident.[7] Crown then merged his Material Services Corporation into General Dynamics in exchange for stock. After the airliner debacle, John Naish was ousted in May 1961, and Frank Pace, with reduced authority, retired under pressure in 1962. Crown replaced him with Roger Lewis, then with Pan Am, where he had been regarded as heir apparent to Juan Trippe. Lewis, insisting on complete authority as chairman and president, encountered conflicts with Crown regarding his autonomy almost from the outset. Lewis carried out a management housecleaning, but Crown opposed his plan to buy the Quincy Shipyard operation from Bethlehem Steel. Lewis searched for ways to neutralize Crown or otherwise orchestrate his exit.[8]

The opportunity came through the nature of Crown's holdings. At the time Crown gained control, his 18 percent ownership consisted principally of a special preferred stock paying a fixed dividend, but convertible to voting common at specified prices and dates. The General Dynamics stock price by 1965 was well below that required for conversion, but Lewis confronted Crown with an offer to buy out his holding, in effect calling the stock. Crown, outvoted on the board, was forced to agree. The buyout was completed on March 14, 1966, costing the company some $120 million but leaving Lewis in complete control.[9] Crown departed, but remained determined to recapture control of the company.

Lockheed was among the first major aircraft companies to adopt a strategy of diversification from military aircraft, a strategy that assumed increased urgency with the series of cutbacks, stretchouts, and cancellations falling in 1957. Lockheed suffered $150 million in canceled orders, and employment fell by twelve thousand.[10] The renamed Missile and Space Systems Division, which relocated to Sunnyvale, California, in

September 1956, increased in importance. The Polaris submarine-launched ballistic missile (SLBM) program was a widely praised success, leading to more advanced SLBM systems. Lockheed acquired a shipbuilding company in 1959 for countercyclical diversification, and the Electra program was an attempt to remain in the commercial sector. After Robert Gross failed to acquire Hughes, he acquired Stavid Electronics of New Jersey as Lockheed Electronics. Lockheed Air Service, Inc., involved with maintenance and conversion, also was important. Lockheed's financial loss in 1960 was largely attributable to the Electra and military cutbacks.

Lockheed's veteran management team remained in place throughout the 1950s. Courtlandt Gross became president of Lockheed in 1956, and Daniel J. Haughton moved to Burbank as executive vice president. Chairman and CEO Robert Gross led the diversification strategy but died unexpectedly on September 1, 1961, at age sixty-four. He was succeeded by Courtlandt Gross, who while also a management generalist, did less to diversify further. Haughton succeeded Courtlandt Gross as president and soon dominated the company. The leadership transition continued as veterans Hall Hibbard and Cyril Chappelet retired in 1960 and 1963, respectively. Lloyd Stearman, first president of the reorganized Lockheed, rejoined the firm in 1955 as an engineer and worked on several advanced projects until his retirement in 1968.

A significant achievement of the Skunk Works, named officially the Advanced Development Projects Office, was the U-2, an apparently simple design featuring a large-area straight wing that first flew in 1955. Although announced as intended for meteorological and other innocuous missions, and given a misleading utility designation, eventually it was revealed that its true mission was high-altitude reconnaissance over the Soviet Union under control of the CIA. The success of the Skunk Works with the U-2 led to the extremely advanced CL-400 hydrogen-fueled reconnaissance project of 1956, intended to operate at altitudes above one hundred thousand feet. The CL-400 development was canceled as unattainable but ultimately led to a contract for the advanced A-11 Mach 3 project.

Lockheed began helicopter and convertiplane development with the small experimental CL-475 in November 1957. Then the joint army-navy XH-51 rigid-rotor high-speed research helicopter first flew on November 2, 1962. Lockheed entered the Advanced Aerial Fire Support System (AAFSS) competition in 1964 for an eventual successor to the army's first-generation armed helicopters. Lockheed won the contract March 23,

1966, with its AH-56A Cheyenne, featuring compound power with a pusher propeller and capable of 220 mph. Lockheed appeared to have a future in helicopters.

In the troubles it was later to experience Lockheed was to an extent a victim of its own success. It enjoyed a strong reputation with such designs as the F-104, the U-2, A-11/SR-71/YF-12A, C-130, P-3, the JetStar jet utility transport, and the C-141. The C-130 was widely exported and highly versatile, including commercial roles. Its success led to later military transport designs, to the Electra, and to the P-3 Orion to succeed the P2V Neptune. But the F-104, smaller, lighter, and in many ways less complex than others of the Century series but still capable of Mach 2, was not ordered in large numbers. The air force preferred the heavier and more complex F-105, F-4 Phantom II, and TFX.

The F-104 found wider success with allied air forces. Although the Japanese initially planned to adopt the Grumman F11F-1F as its primary fighter, Lockheed succeeded in converting them to the F-104G multimission version. Then Lockheed won the hotly contested NATO order for a standard fighter, involving license production in Germany, Italy, Belgium, and the Netherlands. The F-104G, which first flew on October 5, 1960, became a major allied fighter over the next two decades, being license-produced in Japan and Canada as well. Although not known at the time, Lockheed's marketing of the F-104 was later revealed to have included questionable if not illegal inducements, severely damaging the firm's reputation and prospects.

The A-11, another highly secret joint CIA/air force development of the Skunk Works, won over a Convair entry in August 1959. As the A-12, it first flew on April 26, 1962, and the operational SR-71 reconnaissance aircraft first flew on December 22, 1964. It set flight records that remain unbroken and was virtually unique among advanced aircraft in being fully developed from the start, not requiring later improved versions. The design also pioneered stealth, or low radar cross-section, technology. The C-141, the first jet military transport in service, was particularly noteworthy in that the prototype first flew on December 17, 1963, only thirty-three months after contract award, and the entire program of 284 aircraft was completed in 1967, unusual rapidity for such a large and complex aircraft program.

North American undertook a major reorganization in 1955, establishing four new divisions, Missile Development, Autonetics, Atomics International, and Rocketdyne, thereby positioning itself in missiles and space, atomic energy, electronics, and rocket engines. The El Segundo headquarters opened in 1957. Missile Development became the Space and

Information Systems Division in 1960, by which time aircraft had become a minority of total business. The Autonetics Division won major avionics contracts, and Rocketdyne led in large launch engines. But North American still was the most government dependent of all aerospace firms, with government sales accounting for 97 percent of the total in 1961.[11]

Despite its strong record in military aircraft, North American suffered a triple blow during the late 1950s, with the cancellations of the F-107 in favor of the Republic F-105, the Mach 3 F-108 Rapier interceptor, and the elimination of B-70 Valkyrie strategic-bomber production and restriction of the prototypes to research. Cancellation of the Navajo air-breathing ICBM in 1957 was another setback.[12] The delivery of the 2,270th and last F-100 in 1959, although not widely recognized as such at the time, marked the beginning of the decline of North American in aircraft.

North American won the WS-110A contract on December 23, 1957, for a supersonic bomber to replace the B-52. Some felt that North American was favored because of its recent setbacks. The original specification, for sustained Mach 3 performance, seventy-thousand-foot operating ceiling, and intercontinental range, undoubtedly was the most challenging

TABLE 17. Largest Aerospace Manufacturers, 1967

Industry Rank	Firm	Fortune 500 Rank	Sales (in millions)	Employees
1	McDonnell Douglas	16	$2,933	140,000
2	Boeing	19	2,879	142,700
3	NAR[a]	28	2,438	115,300
4	Lockheed	30	2,335	92,200
5	General Dynamics[a]	32	2,253	103,200
6	United Aircraft[a]	33	2,212	78,700
7	Ling-Temco-Vought[a]	38	1,833	60,300
8	Textron[a]	49	1,446	62,000
9	Grumman	87	969	35,800
10	Martin Marietta[a]	127	695	29,500
11	Northrop	181	469	23,600
12	Ryan[a]	219	393	20,400
13	Fairchild Hiller	294	289	12,400
14	Cessna[b]	342	213	11,300
15	Beech[b]	400	174	9,900

Source: "500 Largest Industrial Corporations, 1967." *The Fortune Directory,* June 15, 1968, 188–204. (© 1968 Time Inc. All rights reserved.)

a. Highly diversified. Aircraft production constituted a minority of total sales. Martin Marietta and Ryan no longer produced manned aircraft.

b. Primarily general-aviation manufacturers.

ever faced by the industry. Many regarded it as simply unattainable, but rapid aerodynamic and engine progress gave grounds for optimism. Weighing more than five hundred thousand pounds, powered by six boron-fueled GE J93 turbojets, and employing structural composite materials, the XB-70 was the most advanced aircraft ever conceived. The program became plagued by budgetary cutbacks and development delays, and plans for the chemical fuel were dropped in 1959. Then President Eisenhower, increasingly skeptical about the future of the manned bomber in the missile era, restricted the program.

With its implications for employment in Southern California, the B-70 became a minor campaign issue in the 1960 elections and appeared briefly to enjoy brighter prospects. But the projected first flight in 1962 was delayed. When the first prototype finally flew on September 21, 1964, the B-70 already was a victim of the conviction of Secretary of Defense McNamara that strategic bombers were obsolete. The air force attempted to continue the program as the RS-70, for strategic reconnaissance, but a disastrous midair collision with an F-104 on June 8, 1966, destroying the second prototype, ended air force flight research. The first prototype was transferred to NASA in March 1967, where it was employed in the supersonic airliner program.

Naval production also ran down. The final navy and Marine Corps Fury, the FJ-4, appeared in 1954, and 1,115 swept-wing Furies were produced by 1958. The remaining naval-production programs at Columbus were the T2J (later T-2) Buckeye jet trainer, which first flew on January 31, 1958, and the supersonic A3J (later A-5) Vigilante, which first flew on August 31, 1958. The T-2 enjoyed a long production life, but the A-5, as the RA-5C, was employed as a high-altitude reconnaissance aircraft rather than as a strategic bomber and was produced in relatively small numbers. The A-5's advanced supersonic configuration, however, was reflected in later Soviet and American fighters such as the MiG-25 and F-15. North American also produced the OV-10 Bronco battlefield reconnaissance aircraft for the Marine Corps and air force and for export to Thailand, Germany, and Venezuela. The successful T-39 Sabreliner air force utility aircraft and crew trainer led to a business jet version. Collectively, however, these programs could not offset what North American had lost.

North American had designed a small jet airliner, the NAC-100 Centuryliner, and later entered the supersonic-airliner competition, but neither proceeded. With no significant commercial lines, diversification was crucial, although merger talks with the Foster Wheeler industrial firm were broken off in 1959. After Dutch Kindelberger's retirement as chief executive in 1960, J. L. Atwood decided that the space program was more promising than commercial aviation and moved strongly in that direction.[13]

North American underbid Martin Marietta for the prime contract for the Apollo moon-landing vehicle, and also developed the Saturn V launch vehicle, enabling rapid growth during the decade. Kindleberger died in 1962 at age sixty-seven, ranked among the most successful aircraft industrialists. Atwood was far more cautious and analytical than the aggressive Kindleberger, but continued emphasizing space activities as he considered the firm's future.

McDonnell's military production remained strong throughout the 1950s, completing 519 F3H Demons for the navy and 807 F-101 Voodoos for the air force. The prototype XF4H first flew on May 27, 1958 and, after winning a competition against an advanced development of the Chance Vought F8U Crusader, entered large-scale production.

McDonnell still suffered setbacks, however. Its Model 119 entry in the air force light utility transport/trainer competition lost to the North American Sabreliner. The Model 220 executive development also failed to win orders, leaving McDonnell with a major loss. The F3H Demon was withdrawn from first-line service prematurely, although its problems were more attributable to specification changes during development than to any design flaws. Then, after some twenty years in the helicopter and convertiplane fields, McDonnell ended its efforts in the mid-1960s. McDonnell acquired the license and marketing rights for the promising Breguet 941S STOL turboprop transport in June 1962, which it designated the Model 188. Despite intensive market efforts, no military or commercial orders resulted.

While growing with the Phantom II program, James S. McDonnell astutely foresaw the eventual concentration of the aerospace industry along the lines of the automobile industry. Further, he saw the need for diversification away from defense, in which corporate survival depended almost entirely upon winning the next contract.[14] He assumed the title of chairman of the board in 1962, and Davis S. Lewis became president. Lewis, who began his career in 1939 with Martin, had joined McDonnell in 1946 and risen rapidly.

Exploring the commercial market, McDonnell began development of the T85 short-range airliner in 1962. Soon it became apparent that this design almost duplicated the Douglas DC-9 then under way, leading James S. McDonnell to approach Donald Douglas with a merger proposal, initially involving a fifty-fifty teaming on the DC-9. McDonnell and Douglas had previously teamed on the TFX competition, and McDonnell already owned a significant block of Douglas stock. Douglas, believing that his company could survive independently, rejected the proposal at the time.[15] Douglas continued with the DC-9, while McDonnell, with no commercial record, failed to interest the airlines and dropped the T85 development.

The major turning point for McDonnell was the decision in 1961 by Secretary McNamara that the air force adopt the F4H Phantom II as its standard tactical fighter. The air force version, initially designated F-110 but redesignated F-4 when military designations were standardized, received massive orders.[16] It also provided McDonnell with major export prospects for the first time. Given the declines of Convair, Lockheed, Northrop, Republic and North American in the fighter field, and with the future of the strategic bomber in doubt, McDonnell became the major combat aircraft supplier. Its growth was even more remarkable in view of the overall decline in aircraft procurement at the time.[17] The first F-4C for the air force flew on May 27, 1963, and the fighter remained in production for the next sixteen years. With diversification and involvement in the space program, the term *aircraft* was dropped from the company name in 1966. But despite its rise in stature in the industry, McDonnell remained essentially a single-product firm in the F-4, a continuing concern to James S. McDonnell.

In the late 1950s aircraft carriers could be seen with four types of Douglas aircraft embarked, AD Skyraiders, A3D Skywarriors, A4D Skyhawks, and F4D Skyrays. Only Grumman and McDonnell offered strong naval competition, and in a period when no company had a stronger standing with the airlines than did Douglas. Such apparent strength, however, only disguised the fact that Douglas had entered a long-term decline. Overall business fell from 1957 to 1963, and the firm experienced serious losses in 1959 and 1960, largely due to the DC-8. Only the A4D Skyhawk continued in quantity production, and the DC-8, with its enormous development costs and heavy expansion expenses at Long Beach, faced a formidable competitor in the Boeing 707. No air force contracts followed those for the C-133 and B-66, and with completion of the B-66 program the Tulsa plant was again closed. The C-133, intended to replace the veteran C-124, first flew on April 23, 1956, but orders amounted to only forty-nine aircraft, and production ended in 1961. The giant C-132 logistics transport had been canceled in 1957, and the F6D Missileer, a large, long-range fleet defense interceptor, also was canceled.

All was not tranquil within the management of the company during this period as well. Donald Douglas officially retired in October 1957 upon reaching age sixty-five but assumed the title of chairman of the board and retained veto power over major decisions. He installed as president his forty-year-old son Donald Douglas Jr., who had served for seven years as vice president for military sales. The move was widely resented within the company and adversely affected morale. During a time of enormous cost pressures affecting the entire industry, the son also proved to be less cost-conscious than the father.[18] After the financial losses the senior Douglas again became more active in management, but authority was unclear. The younger Douglas was abrasive with veteran executives, prompting retirements of Arthur Raymond and Frederick Conant, among others.

Ed Heinemann also departed in 1960, after being reassigned twice.[19] He then became vice president for engineering at General Dynamics in 1962, where he completed his career. But Wellwood E. Beall, departing Boeing after years of conflict with William Allen, was hired by Douglas late in 1964 as executive vice president for operations and focused on airliner development to his retirement in 1973.

By 1961 the Douglas organization consisted of the Aircraft Division at Long Beach, the Missile and Space Systems Division, the Tulsa Division, a modification center, and the Charlotte division, producing Nike and Honest John missiles. As military aircraft declined the El Segundo plant was closed in 1962. Douglas enjoyed a strong stake in the missile and space fields, with the Nike interceptor and Thor strategic missiles, but lost the Minuteman contract to Boeing in 1958 despite the latter's relative lack of experience in missiles. Douglas then won the contract to develop the technologically daunting Skybolt air-launched ballistic missile in 1959. But after numerous development problems and delays, the program was canceled in December 1962, costing Douglas among others thousands of jobs.

Despite the series of setbacks, by 1965 Douglas again appeared to be enjoying prosperity and bright prospects. Sales exceeded $1 billion, profits were strong in all but commercial aircraft, and the stock reached an all-time high. DC-8 sales were climbing, as were those for the DC-9 short-range twin-engined airliner, launched on April 9, 1963, with strong orders. The DC-9 first flew on February 25, 1965. Douglas introduced the stretched and more powerful DC-8 Super Sixty series, which Wellwood Beall had led, and which with a capacity of up to 250 passengers were the largest airliners in the world until the introduction of the Boeing 747. A-4 Skyhawk and Nike missile production continued, and Douglas won the large contract for the advanced and highly secret Manned Orbiting Laboratory (MOL), on August 25, 1965. Douglas also had major involvement with the Saturn launch vehicle for the Apollo, and developed the Harpoon attack missile for the navy. Donald Douglas Jr. then reorganized the company into two divisions, the Aircraft Group at Long Beach and the Space Systems Group at Huntington Beach.

Unfortunately, Douglas management had failed to fully anticipate the looming financial problems connected with the rising volume of business, and by the end of 1965 the firm was again suffering serious losses. Ironically, Douglas had not overcome a long-standing deficiency in commercial-market assessment and had seriously underestimated demand for the DC-9. Its delay in moving into military electronics, which did not occur until 1961, was another problem. Capital needs for production were simply overwhelming, and indebtedness increased steadily. The company brought in the investment-banking firm of Lazard Freres to assess the situation and received an alarming report. Not only were effective internal controls not in evidence, but labor and critical materials were in short supply due to the Vietnam buildup. Deliveries of airliners began falling behind, costing progress payments from customers.[20]

The company managed to obtain $75 million in debenture financing in June 1966, quickly following with another $100 million from a consortium of eight banks, of which $75 million was guaranteed by a government V-loan. But after initially forecasting profitability for 1966, Douglas reported an operating loss for the second quarter due to spiraling cost increases. Further, despite the massive paperwork required to support financing requests as well as for contracting, it became apparent that Douglas had little in the way of effective financial forecasting and cost determination. Douglas simply had not been financially oriented, and bankers began to distrust its figures. Capital needs eventually amounted to over $400 million, but the loss of confidence by the market made it impossible to raise additional equity capital. By mid-1966 the company was hemorrhaging cash, forcing Douglas to reconsider merger. Jackson R. McGowen, vice president for commercial aircraft, tried valiantly to save Douglas from being forced into merger, including raising further deposits from airline customers, but it was not enough.[21]

In a critical meeting on October 7, 1966, financiers determined that raising further capital either publicly or privately was unfeasible and demanded that the company seek a buyer.[22] Donald Douglas reluctantly did so, attracting bids from General Dynamics, North American, Martin Marietta, McDonnell, and Signal Oil and Gas. Lockheed did not enter. Although not primarily in aerospace, Signal owned Garrett, a major Douglas supplier, and was headed by Harry Wetzel, son of a veteran Douglas executive. The elder Douglas, however, was more agreeable to being acquired by North American, headed by his old friend J. L. Atwood, and with headquarters only a few miles away. North American appeared a logical strategic match but was very conservative financially and would not commit to providing the cash Douglas desperately needed. Sherman Fairchild made a late bid, which was rejected, and Signal was eliminated due to a perceived lack of management depth.[23]

James S. McDonnell, who had sold his earlier holdings of Douglas at a strong profit and had later assembled another position at then-lower prices, emerged as the strongest contender. By December 1966 McDonnell held more than eight hundred thousand of the 5.2 million shares of Douglas.[24] Regarded as the most efficient firm in the industry and with one of the strongest management teams, McDonnell officially announced the merger deal on January 13, 1967. McDonnell was willing to pay a fair price to Douglas shareholders and also proceeded not knowing if the merger would pass review by the Justice Department: Attorney General Ramsey Clark was regarded as suspicious of Big Business.[25] Justice approved, however, and the merger became official on April 28, 1967.

Douglas had almost twice the number of employees as McDonnell, eighty thousand to forty-five thousand, but the merger was effectively a McDonnell takeover. Donald Douglas was given the title of chairman emeritus of McDonnell Douglas Corporation. Donald Douglas Jr. transferred to St. Louis as vice president–administration but exercised little influence. While widely criticized for his management, Donald Douglas Jr. probably was most affected by unfortunate timing, moving into the presidency as the company's fortunes were declining.

McDonnell Douglas was headed by James S. McDonnell as chairman and David S. Lewis as president. A separate Douglas Aircraft Company and McDonnell Aircraft Company were formed. The Douglas Santa Monica headquarters was phased out in favor of Long Beach. Lewis also served as chairman and CEO of the Douglas Company, while Jackson McGowen, effectively running Douglas at the time of the merger, became president and chief operating officer. Sanford N. McDonnell, James S. McDonnell's nephew and heir apparent, who had managed the F-4 program since 1961, became president of McDonnell. The merged company became what Douglas had once been alone, a firm with strong military, commercial, and missile and space lines. While the merger would not be untroubled, McDonnell had attained the diversification he had long sought.

Northrop had little aircraft business in prospect after the F-89 program, as well as no significant commercial business. The Snark jet strategic missile encountered problems with accuracy and reliability, and production was limited to only one squadron. Success in aircraft came with the T-38 Talon supersonic trainer, which first flew on April 10, 1959, and while of dubious necessity was heavily ordered by the air force. From the

T-38 sprang the F-5 single-seat lightweight fighter, which was widely exported to allied nations under the Foreign Military Sales (FMS) program, and also built under license in Canada.

Thomas V. Jones, who began with Douglas during the war and had joined Northrop in 1955 as director of engineering, became president at age thirty-nine on the death of Whitley Collins. In Jones the company finally gained strong leadership, overcoming the managerial instability of the past decade. Jones emphasized T-38 and F-5 development as the niche Northrop needed to survive in aircraft, since it could not hope to compete directly with the aerospace giants. He then further redirected the company toward high technology.[26] Removal of the term *aircraft* from the corporate name in 1959 underscored that change. Northrop Corporation then consisted of four major divisions, Norair for aircraft, Radioplane for unmanned aircraft, Nortronics, and Northrop International.

Northrop's reputation also grew in space guidance systems and as a major aircraft subcontractor. Contracts for the air force/NASA M2-F2/F3 and HL-10 lifting-body space glider vehicles were important to the development of the eventual Space Shuttle orbiter. But in redirecting the company, Jones also determined that Northrop's future lay with political influence in Washington. That determination, involving among other things export efforts for the F-5 and later aircraft, would lead it into scandal.

Republic completed production of the F-84F and RF-84F in August 1957, with 2,711 and 715 built, respectively. About half were supplied to NATO allies under the Mutual Defense Assistance Program (MDAP). The supersonic F-105 Thunderchief first flew on October 22, 1955. The F-105 was designed with an internal bomb bay, the first ever on a fighter aircraft, and was capable of nuclear-strike missions. Although prototypes were powered by the J57, the production model, which first flew on May 26, 1956, was powered by the more powerful J75. The F-105 presented design and operating challenges and initially developed a poor reputation for serviceability, but it later established an excellent record in Vietnam. Production continued until 1964, but it was to be Republic's last independent design.

Republic had bid on the 1950 supersonic-interceptor program for the air force but lost to the Convair F-102. Republic's XF-103 design, which it continued to develop after the contract loss, led to a new specification for an even more advanced interceptor intended to reach Mach 3. The radical XF-103 featured a buried cockpit, small delta wings, and extensive use of titanium. It was to be powered by an afterburning Wright J67 turbo-ramjet engine, a license-produced development of the Bristol Siddeley Olympus, projected to develop fifteen thousand pounds of thrust as a turbojet and thirty thousand pounds as a ramjet. Later the design was changed to accommodate the experimental XRJ55 engine developed from the J67, as the J67 was not developing adequate thrust. The program was canceled in August 1957, however, even as the first prototype was being built. Development delays had cost Republic the chance to lead the world to Mach 3.[27]

Republic made several attempts at diversification. It established a Helicopter Division in December 1957, building the French Alouette helicopter under license for the North American market, but did not achieve major success. In December 1957 Republic announced a new Rainbow turboprop airliner project, reviving the name of its earlier airliner venture. It was sized slightly below the Lockheed Electra, for the short- to medium-range market. Airliner interest was lukewarm, however, and the project did not proceed. Republic also acquired a minority interest in the Dutch Fokker firm on March 21, 1960. A joint Fokker-Republic project for the Alliance variable-geometry attack aircraft was marketed in 1962 but attracted no support. But none of these measures, even if successful, could have assured Republic's survival as an independent firm. The end of F-105 production, loss of the TFX contract, and a roller-coaster sales record during the 1946–64 period made Republic's future doubtful.[28]

Grumman, in its thirty-year history to 1960, had built 40 percent of the fighter aircraft in the history of naval aviation.[29] But its previously secure position became seriously threatened. Before 1954 Grumman could not foresee a situation in which it would not be selected by the navy for a majority of its fighter aircraft, but the contract to McDonnell for the F4H Phantom II led to searching internal reassessments. As the researcher Randolph Kucera reported,

> Basically, Grumman misread Navy intentions regarding the firm because of significant and rapid changes in: (1) personnel at the top levels of Navy aviation; (2) Navy conceptualization of the "Industrial Mobilization Base" and its relationship to aircraft procurement practices; and (3) the technological stature of McDonnell Aircraft. An additional cause was inadequate design and engineering *team* capabilities within the Grumman organization, which did not facilitate effective corporate responses to these changes.[30]

For the first time in its history Grumman found itself without a next-generation fighter for the navy.[31] Grumman's and the navy's first supersonic fighter, the F11F-1 Tiger, was a successful design and also marked a departure from the traditional "Iron Works" philosophy, being of a light, rather elegant appearance. But the F11F was produced only in small quantities and served only some five years in first-line service, from 1957 to 1962. Although suffering from inadequate power with the Wright J65, the F11F-1F Super Tiger powered by the GE J79, lighter than the J65 but 50 percent more powerful, promised to fulfill the design potential. The Super Tiger was seriously marketed to both the Japan and Germany, but both eventually ordered the Lockheed F-104G, possibly to their eventual regret. The developed version also was not ordered by the navy, then committed to the excellent Chance Vought F8U Crusader, with the McDonnell F4H Phantom II to follow.

Grumman, acutely aware of its military dependence, continued with its aluminum boat, truck body, and postal delivery vehicle lines, but these accounted for only a small percentage of the company's total volume. Total sales declined, although the company remained profitable. Major diversification became imperative, and in 1957 Grumman moved into business and agricultural aircraft as well as other aspects of transportation.[32] By the late 1950s Grumman's activities were grouped into four major lines, military aircraft, civil aircraft, boats and hydrofoils, and space and missiles. Employment stood at more than fourteen thousand in 1960.

Even without a new fighter, Grumman still had important aircraft programs. Production of the F9F Cougar totaled 1,985 when completed at the end of 1959. The S2F Tracker continued, and new designs included the WF-2 Tracer radar early warning variant, the OV-1 Mohawk fast observation aircraft for the army, and the W2F (later E-2) Hawkeye carrier-based early-warning radar aircraft, which made its first flight on October 21, 1960. The C-2A Greyhound was a carrier onboard delivery (COD) variant. Probably most important was the A2F (later A-6) Intruder, which first flew on April 19, 1960, and would be produced for over thirty years. Grumman then cast its fighter future with the dual-service TFX program, teamed with General Dynamics.

Grumman had leased airfield facilities at Stuart, Florida, during the Korean War and in 1967 transferred Mohawk production to a new factory at that site. Another factory was opened in Savannah, Georgia, for production of its successful Gulfstream executive aircraft. Grumman's transformation into an aerospace firm was strengthened with its receipt of a contract for four Orbiting Astronomical Laboratories (OAL) in October 1960. Then came the major contract for the Lunar Module for the moon-landing program, leading to continuing involvement with the space program.

With the exception of the untimely death of Jake Swirbul in 1960,

Grumman maintained a stable management team throughout the Cold War era. Clinton Towl succeeded Swirbul as president. Leroy Grumman stepped down as chief executive in 1962, at age sixty-seven, and retired as chairman of the board in 1966, succeeded by Towl. Bill Schwendler served as chairman of the executive committee, and Llewellyn J. "Lew" Evans, who had joined Grumman in 1951 as legal counsel and moved rapidly into management, became president. With the Vietnam buildup, employment exceeded thirty-three thousand in 1966.

The last of 303 Fairchild C-123B Providers was completed in 1958. The proposed M-185 utility jet lost out to the Lockheed JetStar for air force orders, and the company lacked the resources to develop a civil version, given the priority of the F-27. Airliner financial losses were a major setback, although the company became profitable in other areas. In a move largely related to F-27 losses, Richard Boutelle was fired as president in December 1958. But a new chief executive with airline experience, J. H. Carmichael, lasted only until December 1960, unable to turn the program around. Sherman Fairchild returned as chairman and chief executive. But with his numerous business interests, Fairchild did not intend to manage the company indefinitely, but rather to set broad policies. In its transition to aerospace, the company reorganized into functional divisions, the Aircraft and Missiles Division of Hagerstown, the Armalite Division, the Astrionics Division for drones, and the Stratos Division, in air conditioning and pressurization. With no development contracts, the engine division closed in 1959.

Sherman Fairchild brought in Edward G. Uhl, formerly a senior executive of Ryan, as president and chief executive in 1961. Uhl led the diversification effort and also increased activities in repair and conversions, including a major program to convert AC-119G gunships for Vietnam service.[33] Fairchild also held a major subcontract for the B-52 program. The firm was renamed the Fairchild Stratos Corporation early in 1961 and encompassed three divisions, Aircraft and Missiles, Astrionics, and Stratos.

Fairchild acquired Hiller Helicopters in May 1964 from ELTRA, the renamed Electric Autolite, for cash. Hiller had lost money for three consecutive years, but Fairchild was attracted both by diversification and by the prospect of winning the lucrative LOH production contract. The company was renamed Fairchild Hiller Corporation in September 1964, and Stanley Hiller became an executive vice president.[34] Although the LOH contract went to Hughes, Fairchild produced the FH-1100 civil development and continued the older H-23.

Sherman Fairchild still aspired to build an East Coast aircraft giant to compete with those of the West Coast, leading to the acquisition of Republic Aviation, which faced dim prospects beyond the F-105 program.[35] Fairchild began acquiring Republic stock in 1964, then announced successful purchase negotiations on July 8, 1965. After completion of the merger in September, Republic became the Republic Aviation Division of Fairchild Hiller. Although he had been considering a merger since 1960, Mundy L. Peale became upset by Fairchild Hiller's plans to downsize Republic, and resigned.[36] Fairchild's aircraft aspirations also led to the later bid for Douglas.

While an excellent design, production of the F-105 was cut short by Secretary of Defense McNamara in favor of the McDonnell F-4. Republic had a reputation as a high-cost producer, and Fairchild did not gain an active production program, but it did acquire a capable staff of engineers and researchers. Many Republic personnel transferred to Hagerstown, while others went to Grumman, but the Republic Division continued as a factor in combat aircraft.

After 1965 Fairchild Hiller became a complete aerospace firm with creation of a Space and Electronic Systems Division. Aircraft production was limited at the time to the F-27 and developed variants, the FH-1100 civil helicopter, and license production of the Swiss Pilatus Porter in small numbers. Production of the H-23 was phased out in 1967.

TABLE 18. U.S. Aircraft Production, 1953–68

Year	Total	Military	Civil
1953	13,112	8,978	4,134
1954	11,478	8,089	3,389
1955	11,484	6,664	4,820
1956	12,408	5,203	7,205
1957	11,943	5,198	6,745
1958	10,938	4,078	6,860
1959	11,076	2,834	8,242
1960	10,237	2,056	8,181
1961	9,054	1,582	7,472
1962	9,308	1,975	7,333
1963	10,125	1,970	8,155
1964	12,492	2,439	10,053
1965	15,349	2,806	12,543
1966	19,886	3,609	16,277
1967	18,660	4,000	14,660
1968	18,976	4,000	14,976

Source: *Aerospace Facts and Figures,* 1969, 27.

Temco, the former Texas Engineering and Manufacturing Company, founded in 1946 and originally in general aviation and military subcontracting, attempted to enter the jet military trainer market with its Model 51, designated TT-1 Pinto. The first turbojet primary trainer, the TT-1 first flew on March 26, 1956. An evaluation batch was ordered by the navy, but the type did not enter service. Subcontracting continued as the major part of Temco's business.

Temco's neighbor, Chance Vought Aircraft, saw its troublesome F7U Cutlass withdrawn from service in 1957 but achieved a major success with the F8U Crusader, which won a navy contract in 1953 against strong competition from seven firms. The prototype first flew on March 30, 1955, and rapidly succeeded the Cutlass in production. The first operational aircraft to employ a variable-incidence wing, valuable for shipboard operation, the F8U (later F-8) was produced until 1965, including export to the French navy.

Reflecting increasing diversification, the corporate name became Chance Vought Corporation on December 31, 1960. Vought's independence was soon to end, however. The Texas industrialist James J. Ling, through Ling-Altec Electronics, acquired Temco in 1960 in a friendly takeover, in the process vaulting the Ling name into the *Fortune* 500. Ling-Temco Electronics then began acquiring Vought stock in the open market, having received information that certain major shareholders were open to a sale.[37] Vought's military aircraft and the Corvus missile also looked promising. Although forced to borrow heavily to do so, since Chance Vought was much larger than Ling-Temco, Ling made a tender offer in January 1961. Fred O. Detweiler strongly opposed the Ling bid, but company officers did not control the stock. A civil antitrust suit by Vought against Ling failed, and the merger took place in March 1961, whereupon Detweiler resigned.[38] The Antitrust Division challenged the takeover, also unsuccessfully. The Ling-Temco-Vought Corporation then was formed effective August 31, 1961. Ling, however, was forced to step aside as CEO by creditors and was succeeded by Robert McCulloch, with Ling since the Temco acquisition.

A new Chance Vought Corporation was formed as the aerospace unit of the parent, but the name was eliminated in a reorganization on October 20, 1963. The parent then became the LTV Corporation in January 1965, with three operating divisions: LTV Aerosystems, primarily the old Chance Vought, LTV Electrosystems, and LTV Ling-Altec. Continuity in aircraft was assured on February 11, 1964, when LTV (Vought) Aerosystems won, against strong competition, the contract to develop a successor to the A-4. The navy required adaptation of an existing design to save time and money, and Vought developed the A-7 Corsair II from the F-8, unprecedented in that a supersonic design was adapted into a subsonic design. The A-7 made its first flight on September 27, 1965, and was later adopted by the air force for the attack mission.

Aeronautical Progress

Interest in convertiplanes gradually declined, due in some degree to helicopter and fixed-wing STOL advances, but also to recognition of the enormous expense of developing such aircraft to operational status. Testing revealed especially delicate problems of stability and control in the vertical to horizontal transition. Yet convertiplane development and operation undoubtedly could have progressed faster and further than they did. Most experimentation through the 1950s simply was wasted by lack of follow-through, even though turboprop progress in particular made V/STOL more practical than before. Air force priorities of the day, however, were toward ever-increasing speed and range rather than slow-speed and short-field performance.[39]

The British P.1127 VTOL attack fighter received considerable U.S. government support during the 1960s, particularly for the Bristol Siddeley BE.53 rotating-nozzle engine that became the production Rolls-Royce Pegasus and eventually resulted in the Harrier operational fighter of 1969.[40] A Marine Corps version was license-produced by McDonnell Douglas. International collaborations would continue to characterize VTOL efforts.

The frontiers of flight continued to be pushed back. A Lockheed F-104C established a world altitude record for manned aircraft of 103,389 feet, on December 14, 1959, using the zoom-climb technique. The straight-line nonstop distance record was broken by a turbofan-powered Boeing B-52H, which flew the 12,519 miles from Okinawa to Spain on January 11, 1962. In the mid-1950s a joint Air Force/Navy/NACA program for a rocket-powered hypersonic aircraft resulted in the North American X-15. The X-15, carried aloft under the wing of a B-52, made its first powered flight on September 17, 1959, and soon was exceeding Mach 3 routinely. The fully developed Reaction Motors XLR99 engine enabled the X-15 to attain its design potential. On June 17, 1962, it attained Mach 5.92, and on August 22, 1963, reached 354,000 feet, or 67 miles. The McDonnell F4H Phantom II raised the FAI speed record to 1,606 mph on November 22, 1961. Then the Lockheed YF-12A interceptor version of the A-11 set a new record of 2,062 mph on May 1, 1965, and with a different crew set a horizontal altitude record of 80,258 feet the same day.

While it had required forty-four years of progress in powered flight to achieve Mach 1, Mach 2 was achieved only six years later, Mach 3 only two years after that, and then Mach 6 in 1962. Hypersonic flight was proved to be feasible. In the meantime, combat aircraft had progressed in speed capability from about 650 mph for the F-86 to beyond Mach 2 (1,320 mph) for the F-104, F-105, and F-106. But the widely anticipated progression to Mach 3 for the next generation of combat aircraft was never realized. The Republic XF-103 and North American XF-108 Rapier Mach 3 fighters were not built, and the North American XB-70 did not progress beyond the prototype stage. After dramatic advances from 1945 to the end of the 1950s, performance reached a broad plateau. Thirty years after the first Mach 2 aircraft entered service, the most advanced fighters, the F-14, F-15, and F-16, still were performing at the Mach 2 level. Range and altitude dimensions of the flight envelope likewise were similar to those circa 1960. The only true Mach 3 aircraft to enter service was the Lockheed SR-71 Blackbird. Airliners also reached a performance plateau. Jet airliners of 1990 (other than the Concorde) did not fly significantly faster or higher than jet airliners of 1960, although range, capacity, and operating efficiency increased.

Engine progress continued steadily. The P & W J75 and the GE J79 were the first true Mach 2 engines. Airliners and military aircraft were increasingly powered with bypass-flow or turbofan engines, which offered not only increased power through greater airflow but also improved fuel economy. The bypass-flow principle had been pioneered by the British, who were also the first to place the turbofan in service, but once again lost their lead to the United States. The P & W TF33, developed from the J57, and TF30, from the J52, and their civil versions, the JT3 and the JT8, achieved wide application. Allison collaborated with Rolls-Royce from 1958 on the Spey turbofan, produced as the TF41. At the low end of the power spectrum, small turbojets were developed for trainers and business aircraft. The GE J85 and civil CJ610, and the P & W J60 and civil JT12 were the most prominent. The turboprop, while slow to enter wide service, acquired increasingly broad applications, especially in helicopters.

Manufacturing and engineering continued to present challenges to the industry. The unprecedented level of stress encountered in high-performance aircraft and missiles, including extremes of pressure and temperature, required new materials and construction techniques, including adhesive bondings. Development of composite materials, replacing more conventional metals, became more commonly incorporated into production aircraft.

Again, no decade saw more rapid or spectacular progress than did the

1950s. Major developments since have been more in the areas of electronics, engine technology, structures, aerodynamic refinements, and weaponry than in extension of the flight envelope. That rapid progress, however, was not untroubled. Some new concepts, such as the Lockheed and Convair "Pogo" VTOL aircraft and the Convair XF2Y water-based fighter, simply were unsound. Several aircraft, such as the B-66, F7U, and X-3, encountered severe development difficulties, as did some ultimately successful in service, such as the F-100. Wright and Westinghouse exited the engine field due to their failures. Numerous missile and space projects were also troubled, but such problems were inevitable in pushing the frontiers of technology.

The 1960s saw formulation of new military missions and new aircraft to fulfill them, including counterinsurgency (COIN) or limited-war aircraft, battlefield observation aircraft, and combat helicopters. Initially, existing types were converted or modified for new roles, but purpose-designed types followed. The naval A-1 Skyraider was so successful with the air force in Vietnam that in the early 1960s Secretary of Defense McNamara considered reinstating it in production but was dissuaded by the rapid increase in costs since production had ended.[41] Both the piston-engined North American T-28 and jet Cessna T-37 were adapted for the COIN role, and the Douglas Invader, again designated A-26, was modified and also widely employed. The Bell UH-1 was developed as a missile-launching attack helicopter, and the Bell Model 209, the AH-1G HueyCobra purpose-designed combat helicopter, first flew on September 7, 1965. The Grumman OV-1 and North American OV-10 also were purpose-designed.

Politics and Competition

The aerospace industry by 1960 had attained paramount economic importance, provided critical support to foreign policy and defense strategy, and had become the supreme technological industry, employing a majority of the nation's engineering and scientific talent. But it also had become the object of increased scrutiny, due in part to the presumed insidious influence of the so-called military-industrial complex, in which the industry was suspected, in collusion with the military, of exacerbating the Cold War and exaggerating the Soviet threat to enhance business prospects. Related factors included the frustrating Vietnam War, increasing emphasis on domestic social and economic concerns, and continuing concerns over competition and avoidance of excessive concentration in industries.

That a military-industrial complex (perhaps more accurately a congressional-industrial complex: the military can do almost nothing with regard to procurement without the authorization of Congress) existed, and had existed since the Second World War, was not in dispute. Nor was it in dispute that the aircraft/aerospace industry constituted the major part of the defense industry. General Dynamics led aircraft producers into the defense industry model, followed by Lockheed, but all aerospace companies became identified with the defense industry by the 1960s. Yet aerospace included a strong commercial sector, while defense included companies involved in such activities as shipbuilding, land vehicles, and munitions that had no aircraft interests. Even with broadened activities, however, the aerospace industry remained cyclical, and thus highly risky.

The debate, however, was not over definitions but over the influence of the military-industrial complex, how well it served national security and the taxpayers, and how it should be controlled. President Eisenhower, who instilled the term in the national consciousness, was, despite his military background, deeply troubled by the arms race and distrustful of the military establishment. While believing in strong military forces to meet the communist threat, he regarded the defense budget as out of control and faced conflicts with Congress and others in his attempts to restrain spending. He fully realized that large military bureaucracies contained entrenched parochial interests often far removed from those serving national security. Yet like presidents before and since, Eisenhower failed to bring the acquisition process and interservice rivalry under control.

As is the case with the wartime record of the aircraft industry, its Cold War contracting or procurement history has been well researched and widely reported. The issues of the military-industrial complex and the acquisition process have been the object of intensive study within government, academia, and by think tanks. Literature is extensive. It is unnecessary, therefore, to develop and extend the debate here, but rather to summarize the major issues and developments as they related to the aerospace industry. But it should be pointed out that development and cost problems in a major program could possess repercussions beyond the industry and the defense budget. Douglas won the Skybolt air-launched ballistic missile contract in 1959 with optimistic projections both on cost and timing. After the British had committed to that system as their future nuclear deterrent, problems appeared that were so serious that the program became regarded as unfeasible. Some felt that the military-industrial complex withheld information on problems that should have been disclosed earlier.[42] The eventual cancellation resulted in a painful diplomatic crisis with the major American ally.

There was, of course, a long history of official and public suspicion of the aircraft industry, dating from the First World War. But by the beginning of the 1960s aircraft programs were of such magnitude that serious cost overruns carried major budgetary implications and frequently became public controversies. The industry became increasingly criticized for inefficiency and lack of cost consciousness. A contributing factor was that aerospace firms tended to be engineering dominated and generally were not led by strongly financially oriented executives, as was the case in many other major industries. Critics, however, including the economist Stekler and the defense official Fitzgerald, traced well-publicized problems to inadequate competition.[43] Further, with government-financed development, there was no economic penalty to the firm incurring cost overruns.[44]

Congress, the General Accounting Office, and the Defense Contract Administration Service (DCAS) conducted frequent audits and investigations of military program costs. The industry definitely felt pressure to perform, but social and political considerations often played against efficiency. Further, such considerations almost always overrode purely military or national security considerations when questions arose over costs, performance, or military necessity of an aircraft program. Another complication was that ambitious military officers managing major programs defended them against any threat of termination, whatever the reason, feeling that their careers were inextricably tied to the programs. Both military and high civilian procurement officials tended to "sell" their programs to the highest levels of the Defense Department and to Congress, and the flow of information was controlled to that end.[45] Although lobbying by administration officials, including military officers, was prohibited by law, the prohibition was largely ignored or unenforceable. Most disturbing of all, senior military officers unhappy with a reduction or cancellation of a program, rather than obeying civilian authority, often operated in collusion with the industry to lobby Congress to override decisions of the president and the secretary of defense. Eisenhower had fired Secretary of the Air Force Harold Talbott in 1955 over an apparent conflict of interest, yet not one senior military officer was ever fired for contravening his commander in chief in contract matters.

Senior contracting officers frequently could look forward to lucrative postretirement employment with firms they were charged with regulating, hardly an incentive to call firms to account in the areas of costs and performance. The revolving-door relationship of government officials and business, while far from restricted to the defense field, also served to increase public suspicion of the aerospace industry. It was held that the industry employed retired high-ranking officers or civilian officials not so much for their managerial or technical expertise as for their influential contacts in the Pentagon and Congress.

There was no question that the defense acquisition or procurement process was seriously flawed, but there was spirited debate over how to reform the system. The inherent dilemma was that contractors were financially dependent upon government, while the government remained technologically dependent upon a concentrated industry. Even when certain aerospace companies wished to bid on a contract, they often felt frozen out of the competition due to some degree of Pentagon disfavor. If insufficient competition existed, the Department of Defense was as responsible as the industry. Such firms as Martin, Republic, Fairchild, Chance Vought, and Douglas would have preferred to compete for defense contracts as independent firms, but a narrowing business base led them into mergers for survival. Increasing concentration was even more pronounced in certain nonaerospace defense areas. There were, for example, only two submarine builders, two destroyer builders, and one builder of aircraft carriers.

Balanced against the anticompetitive or collusive view was the contention of many in the industry that the military services frequently were unsure of their real requirements, or insisted on features and capabilities that simply were unattainable with existing technology, or otherwise insisted on a last increment of performance that would cost a disproportionate amount to achieve. During the Korean War, Dutch Kindelberger of North American had complained about the frequent turnover of military procurement officials, and that the air force often was unsure about what it wanted.[46] Instability in procurement management personnel persisted.[47] A related problem was that procurement was often regarded by younger military officers as a dead-end career field.[48]

Aerospace firms pointed with some justification to the fact that their supposedly high costs, inefficiency, and cost overruns were not necessarily attributable to their management deficiencies but rather to externally imposed changes and political pressures. The term "gold plating" of weapon systems began to characterize this environment. Contractors frequently were inundated with change orders after development of a new aircraft or other weapon system was under way, inevitably raising costs and lengthening the process. The TFX program was a major example, eventually frustrating both contractors and the military. In other instances the military appeared to be uninterested in securing the lowest-cost weapons. In the triservice Light Armed Reconnaissance Aircraft (LARA) program, potentially a Douglas A-1 replacement, North American won the hotly contested contract with its NA-300 in 1964, designated OV-10A Bronco. General Dynamics, the loser, completed its Model 48 Charger at company expense, which turned out to be less expensive than the Bronco. The Model 48 first flew on November 25, 1964, earlier than the OV-10A, but after a flyoff competition was still rejected by the military because it felt it had too much "invested" in the North American model.[49] The air force felt little enthusiasm for the concept in any event, preferring larger, faster types.

In the 1960s, as defense contractors were charged with becoming more cost conscious and efficient, government also was charged with becoming an informed consumer, mastering the procurement system for the benefit of the taxpayer. A step toward reform began in March 1961 as the air force reorganized its managerial functions with two new commands. Systems Command, at Andrews Air Force Base, Maryland, succeeded the Air Research and Development Command and was responsible for those functions plus Requests for Proposals (RFPs) from the industry. The subordinate Aeronautical Systems Division (ASD), managing the process to production, remained at Wright-Patterson. General Bernard A. Schriever's success in missile development was carried over to other systems. Logistics Command, also at Wright-Patterson, succeeded the Air Materiel Command in the logistics mission.

The navy Bureau of Aeronautics, in existence since 1921, was abolished on December 1, 1959, with its functions transferred to the Bureau of Naval Weapons. Then on May 1, 1966, the Naval Air Systems Command (NAVAIR), roughly the navy counterpart of the ASD and controlling the development of naval aircraft, was established.

Steps toward cost control and efficiency frequently were frustrated by congressionally mandated procurement stretchouts for budgetary reasons, reducing expenditures in a given fiscal year but raising overall program costs and lowering production efficiency. Given the economic impact of major military programs in general, not restricted to those involving aircraft, the competition for large contracts became increasingly politically charged. Aerospace firms felt that they must become politicized simply to survive. Politically aware bidders for contracts focused increasingly on the selection of subcontractors, or location of production facilities, in economically depressed or labor surplus areas, enabling them to point to the salutary effects of their award of a contract on the local economy.[50]

Individual congressmen sought to insure that their districts received or retained their "fair" share of military contracts. Numerous political candidates ran on platforms of cutting or controlling wasteful military spending, except in their districts. Program reductions or cancellations by the military whether for reasons of cost, performance, or changed requirements likewise were strongly resisted for political reasons, especially when one consequence would be massive unemployment. Defense procurement programs became de facto social programs.

The military also could operate a system of political reward and punishment. If a certain senator or congressman questioned a weapon system or other program desired by one of the services, there were sometimes less-than-subtle threats to close military installations in that individual's district, or to steer future contracts to firms located in districts of those more supportive. There could be a heavy political price for conscientious differences of opinion regarding procurement programs or other military matters.

Cost escalation was not restricted to military aircraft. Airliner costs increased relentlessly in the 1960s and beyond. To some degree those cost increases were attributable to government-mandated design and equipment changes for safety and environmental reasons, but airliners were increasingly tailored to the specifications of each customer, increasing unit costs. Inflation also was a factor, but the major increase was linked to the civil equivalent of gold plating, with additions of luxury features and performance-enhancing equipment that may not have enhanced productivity. For airliners, however, market growth and productivity gains more than offset higher costs, so that the critical factor of cost per seat-mile generally declined, and air fares remained stable.

While the broad aerospace industry felt increasing pressure to become more politically aware in the difficult procurement environment, Grumman remained one of the least political firms. In part due to the apolitical Roy Grumman and in part due to its special relationship with the U.S. Navy, Grumman consciously avoided the political maneuvering in which other major firms engaged. And to avoid any possibility of jeopardizing its relationship with the BuAer, Grumman long resisted joining the Aircraft (later Aerospace) Industries Association as being too political, with Clinton Towl finally agreeing to join only in 1961.[51] Grumman gradually altered its approach, and it was believed that the company located its Gulfstream factory in Georgia in part to gain the favor of Senator Richard Russell, powerful chairman of the Senate Armed Services Committee.[52] Grumman also felt that it had lost Super Tiger orders in Japan and Germany because it lacked the political sophistication of Lockheed. But it was the TFX that finally politicized Grumman.

During the 1960s the aerospace industry was by far the largest industrial user of government as well as the leading user of private R&D funding. The relatively few firms in a concentrated industry accounted for a not inconsiderable share of all industrial technological progress.[53] Further, industry R&D knowledge became essentially a public good, since the bulk was government funded and was employed for broad public purposes. In addition to aeronautical advances and the development of the space program, the industry pioneered the new science of systems analysis, with potential solutions to technological problems beyond those of aerospace. Then as R&D accounted for an increasing proportion of total business, a broad decline in production, traditionally the foundation of profits and stability, occurred. R&D spending in the industry exceeded production spending for the first time in fiscal year 1963.[54] That new dominance, with a resulting majority of scientific and engineering manpower, and combined with the special problem of jet airliner development costs, made the industry more risky than before. A new wave of industry mergers, comparable to that initiated by the British, was seen as imperative.[55]

Required resources for major programs, particularly R&D, were beyond even the largest firms, increasing dependence on government funding. In addition, resources of many firms were increasingly concentrated on a single project, requiring government financing or order guarantees to proceed. The researcher Phillips concluded that the increasingly concentrated structure of the industry was due more to science, which overrode any influence that the industry structure played in determining patterns of industrial change.[56] Otherwise stated, the level of aeronautical and space sciences dictated the structure of the industry. This finding seems supportable in that R&D expenses, technological complexity, and financial risks eliminated all but the largest and strongest firms, and even those became increasingly vulnerable. Further troubling was that the combination of increased costs and a narrower production base also meant that the production breakeven point rose steadily.[57] Research and development costs became an increasing percentage of total program cost, which included production, maintenance, and updating over the service life. Timing and performance were major concerns, but cost proved to be the most persistently troublesome.

A factor very much in the favor of the American industry throughout the 1950s and 1960s, but which also gave support to the charge of insufficient competition, was that it was effectively insulated from significant foreign competition. This applied even to such close allies as Canada, Britain, France, and Italy, which in certain instances produced highly competitive aircraft and missiles. The Buy American Act of 1933, which provided that U.S. firms could price up to 50 percent higher than foreign bidders and still win a contract, together with predictable protests from congressmen representing contractors in their districts who would be adversely affected by any foreign aircraft or weapons purchase, provided almost total protection from foreign competition.

Meanwhile, the industry continued to enjoy international market domination in commercial and military aircraft. This domination was of such a degree, moreover, that firms began to run afoul of foreign governments with their aggressive marketing tactics. The industry's efforts were supported and supplemented by the Defense Department's International Logistics Negotiations (ILN) office, a deliberate euphemism for the arms export program. Henry J. Kuss Jr., deputy assistant secretary for ILN, known as the Pentagon's super arms salesman, heightened resentments among affected allied countries. Lockheed, following a major success with the F-104 for NATO, sought to head off the Franco-German Transall tactical-transport program in 1963 and to persuade the German government to order the C-130 instead. Lockheed's tactics even extended to lobbying the Bundestag, the national parliament. Defense Minister Kai-Uwe von Hassel became so incensed that he ordered Lockheed's salesmen from

TABLE 19. Aerospace Exports, 1950–67 (in millions of dollars)

Year	Total	Commercial	Other[a]	Percentage of U.S. Total
1950	$242	$40	$202	2.4
1951	301	13	288	2.0
1952	603	18	585	4.0
1953	881	79	802	5.6
1954	619	93	526	4.1
1955	728	81	647	4.7
1956	1,059	133	926	5.6
1957	1,028	179	849	5.0
		Military	**Civil[b]**	
1958	1,398	713	685	7.9
1959	1,095	557	538	6.3
1960	1,726	637	1,089	8.5
1961	1,653	775	878	8.0
1962	1,923	1,013	910	9.4
1963	1,627	895	732	7.1
1964	1,608	844	764	6.1
1965	1,618	764	854	6.0
1966	1,673	638	1,035	5.6
1967	2,248	868	1,380	7.2

Source: *Aerospace Facts and Figures, 1968,* 18.

a. Includes government and military as well as private/general aviation.

b. Represents change in statistical record-keeping by the Aerospace Industries Association.

his office, forbade further contact, and demanded and received an apology from Courtlandt Gross.[58]

Beginning with the F-104 fighter sale to NATO, major foreign sales gained such importance that they became as critical as Pentagon contracts to the futures of many aerospace firms. Competition for foreign orders was keen among U.S. firms as well as from leading European firms such as Dassault. Political, economic, and diplomatic leverage was applied by all competitors. Firms sought to enlist influential individuals for assistance not only with the Defense Department but also with foreign governments, and the process soon would grow out of control.

A recurring accusation in defense contracting was that of so-called contract roulette, in which new contracts were supposedly allocated to those most in need rather than on merit. If the practice ever existed, it was evidently ineffective, since the number of active aircraft producers declined steadily, and most experienced financial crises threatening their survival at some point.[59] The criticism, of course, was that the happenstance of a lack of business was no reason to be awarded new business. An opposing view was that measures taken to keep the maximum number of firms active and competing promoted cost efficiency and preserved a larger industrial base for rapid expansion during national emergency. That duality for defense contractors was such that North American in particular became widely regarded as a quasi-government subsidiary, which could not operate with conventional corporate financial goals.[60]

In the Kennedy administration, Secretary of Defense McNamara abandoned the idea of preserving a defense mobilization base, believing that it was irrelevant in the age of nuclear warfare, and that the military would have to fight with forces in being.[61] Since the mobilization base concept stressed maintenance of excess capacity, deemphasis of this policy inevitably increased pressures on military contractors. In any event, contracting decisions for major programs continued to be made on a case-by-case basis, with no assessment of the impact of individual decisions on the structure or economic health of the overall industry.[62]

Firms also attempted to gain a competitive edge by anticipating future requirements. Thus advanced development was often undertaken at company expense in an attempt to forge ahead of the competition, although at the risk of severe losses if a requirement never materialized or if the eventual contract was lost to a competitor. Boeing, for example, anticipated air force interest in a variable-wingsweep combat aircraft before the TFX specification, and began development in 1959, a year ahead of its eventual rival, General Dynamics. Firms also aggressively competed for new contracts as they were formulated since winning meant keeping abreast of new technology and remaining technically qualified for later contracts.[63] Losing meant falling behind and presented the added danger of losing the most highly qualified engineers to other firms.

Even in the face of strong attacks on the military-industrial complex, a 1963 survey by the *New York Times* showed that no major defense firms had plans for switching out of defense and moving into nondefense areas.[64] They were determined to remain in defense production even with declining contract opportunities, the danger of cost overruns, and all attendant risks. While there may indeed have been inadequate competition in the aerospace and defense industry, a firm still could face imminent ruin or forced exit with loss of a major contract or failure to win further contracts. The greatest competitive pressure on a defense contractor was the fear of contract termination.

The TFX and the New Defense Environment

The protracted and controversial TFX program probably is the most instructive case study with which to close the era of aerospace/defense convergence. The TFX illustrated more dramatically than any other program the procurement complexities of the day, involving issues of military doctrine and missions, interservice rivalry, technological challenge, the military budget, control of the acquisition process, and, as always, politics. The program would have long-term implications for the military aircraft futures of Boeing, General Dynamics, Grumman, and others and for the economies of several states. It was also among the first of sufficient magnitude to impact the broad aerospace industry.

In 1959 General F. F. Everest, commander of Tactical Air Command, developed the preliminary TFX specification, for an aircraft that would possess performance capabilities not only exceeding anything anticipated from the Soviets, but also would free the air force from undue dependence on vulnerable forward military bases, especially in Europe.[65] The new aircraft, in addition to being capable of speeds up to Mach 2.5, was to possess unrefueled transatlantic ferry range, STOL capability, ability to operate from rough or unprepared airstrips, and to deliver tactical nuclear weapons. The design would advance the state of aeronautical knowledge, requiring new engines, radar, and electronics. One unprecedented requirement was that the TFX was to be capable of sustained supersonic speeds "on the deck," or near ground level, requiring variable wingsweep to reconcile STOL with supersonic capability. It also was the first aircraft designed at the outset for afterburning turbofan engines.

The current F-105 was a capable fighter-bomber but was restricted to highly vulnerable long runways. While in one sense the next-generation successor to the F-105, the TFX's expanded capabilities were viewed in some quarters as further blurring the strategic and tactical air missions. One source of tension in the military establishment, in addition to interservice rivalry, was the existence of intraservice rivalry, and some in the Strategic Air Command regarded the TFX, with its range and nuclear capability, as a threat to its strategic role and to the B-70 bomber. Such blurring of missions had already occurred, since the F-105 was capable of long-range nuclear missions, while the B-52, with its conventional bomb capacity, was capable of tactical missions.[66] In fact, the B-52 was employed as a tactical bomber in Vietnam. There were concurrent tensions in the navy between the strike carrier force and the ballistic missile submarine force, as earlier between the so-called carrier admirals and battleship admirals.

Research with variable wingsweep, beginning with the Bell X-5, had validated the basic concept but also had demonstrated the need for extensive development. John Stack of NASA led major advances and assured the air force that the requisite technology was feasible.[67] The most daunting problem was that varying the sweep also changed the center of gravity, causing major stability problems. Stack placed the wing-pivoting mechanism outside the fuselage, so that only the outer wings pivoted, maintaining the center of gravity.[68] Accordingly, Specific Operating Requirement (SOR) 183, soon known as the TFX (Tactical Fighter Experimental), was issued in July 1960, and RFPs were sent to the industry. The magnitude of the program's technological and financial risks led to widespread teaming among manufacturers to make the risks manageable. Boeing, Lockheed, and North American entered alone, while General Dynamics teamed with Grumman, Republic with Chance Vought, and McDonnell with Douglas.

In the meantime, the 1960 elections had brought into office the Kennedy administration, with its widely heralded new thinking about defense matters, and a new secretary of defense, Robert S. McNamara. Cost-effectiveness, weapons commonality, and acquisition reform became management hallmarks of the coming era. A key aspect of the reforms was that procurement decisions, heretofore largely autonomous for the services, would become increasingly centralized. The Office of Systems Analysis, which applied sophisticated quantitative techniques to weapons evaluation, symbolized the new approach. A new strategy focusing on missions rather than preservation of traditional service roles was to be developed. Both became regarded by the services as threats to their prestige and autonomy. The doctrine of flexible response, emphasizing conventional forces and deemphasizing tactical nuclear capability, became exemplified by the TFX.

The development that more than any other was to bedevil the TFX, however, had originated in the final year of the Eisenhower administration. Eisenhower had suspended initial TFX development to consider a combination with the navy's emerging long-range fleet air-defense requirement and the close-support requirement of both the navy and Marine Corps. Duplication of major weapon systems had peaked in the Eisenhower years, with the army Chrysler Jupiter and air force Douglas Thor ballistic missiles, the army Nike and air force Bomarc interceptor missiles, and others. By 1960 the air force had no fewer than six supersonic fighters, the F-100, F-101, F-102, F-104, F-105, and F-106, in service, by five different manufacturers. Thus while the determination was already in place, the new administration became even more determined than the last to avoid system overlap and mission duplication.

McNamara reviewed the TFX specification and began its redevelopment to fulfill three major missions: the original air force mission, navy-Marine close support, and long-range fleet air defense.[69] The navy soon opted for a smaller and simpler aircraft to fulfill its close-support mission, issuing the VAX specification in May 1961 that resulted in the Vought A-7. The McDonnell F4H still was not regarded by the navy as the answer to its long-range fleet air-defense mission, leading to a development contract awarded on July 21, 1960, for the Douglas F6D Missileer, a "slow fighter" emphasizing long-range radar, missile carrying, and high-altitude, long-endurance loiter capability rather than supersonic speed. The new Eagle missile contract was awarded at the same time. But the Missileer was canceled in April 1961, apparently in anticipation of a dual-service design.[70]

McNamara, although advised otherwise, believed the basic TFX concept could be adapted to meet the navy requirement as well, with savings to the taxpayers in the range of a billion dollars.[71] The total program cost of a dual-service TFX was projected at $7 billion, whereas development of two separate aircraft would be $1 billion higher. That sum was even more significant when it is remembered that annual federal budgets of the day were less than $100 billion. McNamara's curtailment of the F-105 and standardization on the F4H Phantom II as a fighter-bomber were part of his determination to eliminate duplication. He felt the F4H offered greater mission flexibility consistent with new doctrine, and after some grumbling the air force became highly pleased with the navy-developed design.

The navy reacted unfavorably to McNamara's dual-service TFX, pointing out that the two missions were incompatible and that carrier operation imposed severe size and weight limitations that did not constrain the air force version. Air force chief of staff General Curtis LeMay also lacked enthusiasm for the proposal. But since the air force would take the majority of production, the program was to be managed through the Aeronautical Systems Division, raising further concerns within the navy over attention to its specific requirements. Some suspected that the navy also planned simply to outwait McNamara, aware of the brief average tenure, around two years, of secretaries of defense in the fourteen-year history of that office.[72]

McNamara pressed ahead with a revised specification involving different versions for the two services, but with a high degree of commonality. A vast fleet of 1,726 aircraft, including prototypes, for the U.S. services alone was anticipated, with export prospects extending the production run. The potential for the industry was obvious. The two contract finalists were Boeing and the General Dynamics–Grumman team. Boeing gambled that it could develop a superior design by remaining alone in the bidding and optimizing the design for the air force.[73] Award of the contract also would save Boeing's Wichita Division, where B-52 production was ending. For General Dynamics, with B-58 production winding down in Fort Worth, the program came at a time when it was considering closing that plant and consolidating production in San Diego.[74] General Dynamics felt that teaming with Grumman, a "navy company," would make its design more palatable to the navy.[75] GD also was more favorably disposed to commonality, having suffered earlier with curtailed production of the specialized B-58.[76]

All bidders were tempted to follow the discredited practice of "buying in" on a new contract with an unrealistically low estimate, then later renegotiating to recoup the inevitable cost overruns. The government had inadvertently encouraged this practice by separating development contracts from production contracts, but the firm awarded a development contract inevitably was awarded the production contract, where the major profits were to be made. Given the major unknowns in a weapon system breaking new technological ground, overruns were almost expected. Further, the financial risk was borne entirely by the government.[77] New procedures from 1960, however, were based on incentives to contractors for accurate estimation of costs and performance, and penalties for the reverse.[78] The TFX contract was to be awarded on a fixed-price-plus-incentive-fee (FPIF) basis, which largely shifted the financial risk to the contractor.

By December 1961, the Boeing and General Dynamics–Grumman proposals went to the ASD for review. On January 19, 1962, the combined air force–navy Source Selection Board (SSB) recommended the Boeing design.⁷⁹ Boeing then was asked to switch from an advanced General Electric engine to the Pratt and Whitney TF30, less advanced but further along in development, used by the GD-Grumman entry. The change required some redesign. The Boeing entry also had a lesser degree of commonality between the two versions than did the GD-Grumman entry, possibly as little as 50 percent.⁸⁰ The air force also looked askance at the navy-developed TF30.

The selection process then went into not only a second but a third round as disagreements heightened. The navy preferred GD-Grumman, but the SSB favored Boeing. Vice Admiral Robert B. Pirie, deputy chief of naval operations for air, became so vocal in his opposition to Boeing that it led to his early retirement.⁸¹ The controversy and delays eventually worked to the GD-Grumman team's advantage, as McNamara was drawn to the greater commonality offered and remained suspicious not only of Boeing's cost estimates but of its relative lack of experience in smaller aircraft.⁸²

On July 1, 1962, the competition entered its fourth and final phase, with Boeing and GD—Grumman each asked to make their final submissions as if they were the prime contractors. McNamara was determined above all to avoid winding up with two essentially separate airplanes and ordered both to minimize costs and maximize commonality.⁸³ Each made their final submissions in September 1962. As the competition appeared to be deadlocked, McDonnell entered a proposal for a developed version of the F-4 to fill both TFX missions. Although not possessing variable wingsweep, the developed F-4 offered much lower costs. But the SSB remained focused on the TFX. After a final review still favored Boeing, McNamara overturned the SSB and awarded the contract to the General Dynamics–Grumman team. The virtually unprecedented instance of civilian authority overruling military advice alarmed many, but McNamara resisted what he regarded as military parochialism. He was adamant that his perspective was superior to that of military experts, and that a design that would be satisfactory in both missions, at a lower overall cost, contributed more to national security than two separate designs that optimized performance in their missions, but at far greater expense.

Award of the contract for the new aircraft, designated F-111, to the General Dynamics–Grumman team was publicly announced on November 24, 1962, to celebrations in Fort Worth and Bethpage, where the F-111 would be built.⁸⁴ Among other things, the contract assured that the Fort Worth plant would remain open. For Boeing, suspicious of political maneuvering throughout, reversal of the SSB recommendation did little to allay that suspicion. Both the Texas and New York congressional delegations had actively lobbied for GD-Grumman. Vice President Johnson and Congressman Jim Wright and Secretary of the Navy Fred Korth, both from Fort Worth, joined in the celebration. Congressman George Mahon, Defense Appropriations Subcommittee chairman, also was from Texas. Deputy Secretary of Defense Roswell Gilpatric, an attorney who had negotiated the merger of General Dynamics and Material Services Corporation, was viewed as having a conflict of interest, and Colonel Henry Crown, then General Dynamics' largest stockholder and a Kennedy campaign contributor, also was regarded with suspicion.

Allegations of politics moved Senator McClellan's Permanent Subcommittee on Investigations of the Government Operations Committee to began hearings on the matter in 1963, as the F-111 was being developed. William Allen testified reluctantly, fearing that his protests would harm Boeing's future military contract prospects, but emphasized that his bid was $100 million less than the GD-Grumman bid.[85] No formal investigation of the contract award was ever undertaken by the Justice Department, and the McClellan subcommittee ultimately would prove no political favoritism. Then the Kennedy assassination and the succession of Lyndon Johnson to the presidency effectively ended investigations for the time. Secretary Korth eventually resigned, however, having come under suspicion for his earlier role as a banker in assisting with financing General Dynamics through the 880/990 affair.[86]

The prototype F-111A made its first flight on December 21, 1964, and completed its first variable-geometry test flight on January 6, 1965. The Grumman-built F-111B carrier version flew on May 18, 1965, but proved to be hopelessly overweight for carrier operation and quickly lost support within the navy. Feeling all along that the air force version dominated development, the navy began to press for major redesign of the F-111B, which effectively meant a new program. The air force remained skeptical, however, of the navy's weight concerns, pointing out that a sixty-ton C-130 had been flown off an aircraft carrier.[87]

The F-111A was ordered into production in April 1965 by a letter contract, even before completion of flight testing, and the focus of controversy shifted from politics to design merit.[88] The air force version was overweight, deficient in takeoff distance, range, acceleration, and top speed and suffered from excessive drag.[89] Additionally, it was underpowered with early versions of the TF30 and had no air-to-air combat capability. The F-111A was another example of concurrency, in which the RDT&E (research, development, test, and evaluation) phase overlapped with commencement of series production.[90] But the goal, to speed operational deployment, also reduced flexibility and invited severe operational difficulties. Both services' inflexibility on performance requirements diminished the prospects for a successful dual-service aircraft and effectively doomed the F-111B.[91]

The outcome inevitably reopened the question of whether the Boeing design would have been superior both in cost and performance for what eventually became a single-service airplane. The air force version had been compromised in order to accommodate navy requirements and as a result was less than optimal for air force missions. Although McNamara maintained for some time that the savings of commonality would be realized, the aircraft was already costing double the original estimates before it entered service.[92] One result was renewed demand for "fly-before-buy" competitions in the future as opposed to the then-prevalent practice of buying "paper" aircraft based on contract proposals.

The question of producing an effective and economical dual-service aircraft remained active. There were combat aircraft before and after the F-111 adapted for dual-service operation. The North American Fury series developed from the F-86, the Douglas A-1 Skyraider, the McDonnell F-4, and the LTV (Vought) A-7 are prominent examples. The air force also had considered both the Douglas A-4 and Grumman A-6. Dual-service aircraft could be successful, although most involved originally naval designs. But the major factor in the F-111 failure was not that it was dual service but that its missions were so different as to be irreconcilable.

The operational F-111, despite its designation, was a long-range strike aircraft. In fact, the modified FB-111 version served as a strategic bomber. Further, the nuclear emphasis of the F-111 seemingly contradicted the Kennedy administration's strategy of enhanced conventional capabilities. Then the initial combat record of the F-111 in Vietnam, where it suffered severe losses, raised sharp questions about its effectiveness. That the F-111 was not really a fighter, and that the air force's chief fighter was the naval F-4 Phantom II, led directly to a specification for an advanced pure fighter, which the air force had not had since the mid-1950s with the F-104. The requirement also was spurred by awareness from 1965 of the Soviet MiG-25. The navy formulated a new fleet air-defense specification.

For the Department of Defense, the F-111 case was not unique but all too typical of advanced weapon systems acquisition problems.[93] Proceeding with the dual-service F-111 probably was an erroneous decision. Separate designs likely would have better served both national security and the taxpayers. The dual-service concept was viable, but McNamara never fully appreciated the complexities of its implementation.[94]

For the industry, the F-111 case dramatically illustrated the pitfalls, as well as the potential, of advanced-technology programs subject to changing specifications, risky contracting procedures, and partisan and regional politics. Technical merit and efficiency simply were not enough to assure contract awards, and even when awarded, events could negate the basis of the original specification. Again, it might have appeared rational for a major firm simply to opt out of military aircraft and remain only in those fields in which it might exercise greater control over its fate. But the behavior of firms was that they felt compelled to compete aggressively for any and all new Pentagon contracts. There simply was no alternative strategy for their long-term survival.

Chapter 13 ✈ The Advanced-Technology Era

In many respects 1967 was a watershed year for the aircraft-turned-aerospace industry. Major milestones included the sixtieth anniversary of military aviation, the fiftieth anniversary of American entry into the First World War, which did so much to establish the industry, the fortieth anniversary of Lindbergh's flight, and the twentieth anniversary of the unified defense structure and independent air force. It also was the year in which significant transformations in the industry, involving a wave of mergers and the exits of many older firms from the airframe industry, were completed. The *aerospace* appellation was firmly established.

The State of the Industry, 1967–70

By 1967 aerospace was the nation's leading industrial employer, with 1,484,000 employees, of which 834,000 were directly involved in aircraft. With $27 billion in sales, of which $15 billion were to the government, it ranked second.[1] Profit to sales, at 2.6 percent, remained low, especially compared with the 5.6 percent overall manufacturing average for that year. The Cold War and the war in Vietnam maintained military output at high levels, and airliner and general-aviation production also were strong. Boeing delivered its thousandth jet airliner on June 5, 1967. The industry was well advanced toward success in the Apollo moon-landing program, with a successful first launch of the Saturn carrying the Apollo on November 9, 1967.

The airframe sector of the aerospace industry possessed one major structural difference with the automobile industry, with which it was sometimes classified, in that even after exits and mergers it still consisted of nine major military-commercial manufacturers and five major helicopter producers (one a Boeing subsidiary). The auto industry consisted essentially of the Big Three. Thus the aerospace industry, while oligopolistic, possessed greater competitive potential. Certain firms enjoyed commanding positions in a given sector, such as Boeing in airliners and McDonnell Douglas in fighters, but no firm dominated the industry. Neither automobiles nor aerospace yet faced an international competitive threat, however. The market structure still resulted in a competitive environment markedly different from other industries, with five firms, Fairchild (Republic), LTV (Vought), General Dynamics, North American

Rockwell, and Northrop, reduced to a single major aircraft program. Only Boeing, McDonnell Douglas, Lockheed, and Grumman enjoyed a degree of aircraft product diversity.

Formation of giant industrial conglomerates, promising risk reduction and synergy, was at its height in the 1960s, and aerospace was widely viewed as an attractive field for conglomerateurs. In addition to mergers between aerospace firms, certain firms became components of larger, highly diversified conglomerates. The major mergers from 1960 were McDonnell with Douglas, North American with Rockwell-Standard, Martin with American-Marietta, and the Fairchild-Hiller-Republic and Ling-Temco-Vought combines. Helicopter makers Bell and Vertol were acquired by Textron and Boeing.

By the completion of this transformation most of the major founding figures had departed the scene, either through death or retirement. Major firms now were directed by second-generation family members or professional managers, or combinations of both. Donald Douglas became inactive after the merger with McDonnell, and the retirement of Courtlandt Gross in 1967 ended thirty-five years of Gross leadership of Lockheed. William Allen of Boeing and J. L. Atwood of North American were soon to retire, and James S. "Mr. Mac" McDonnell, while remaining chairman of the board of his company, relinquished active management in 1971.

As had been foreseen in the 1950s, nonaviation companies acquired major stakes in the aircraft field. Rockwell-Standard made several acquisitions in general aviation prior to its merger with North American. Even the largest firms found it increasingly difficult to remain independent. Only Boeing, Lockheed, Grumman, and Northrop remained so, and of these only Boeing never became the object of a takeover attempt or merger speculation. The industry thus became more highly concentrated than at any time since the 1930s, with only giant firms possessing the technical and financial resources to undertake major programs. All had moved into other high-technology fields, and airframe production constituted a decreasing percentage of overall business. A new entry into the airframe industry was practically impossible.

A new Department of Transportation was established in 1967, and the Federal Aviation Agency, renamed Federal Aviation Administration, came within that department. Aeronautical development continued under NASA, but major emphasis was on the space program. All aerospace firms were heavily committed to the space program, and leading defense contractors also tended to be leading NASA contractors.

Despite increased diversification, dependence on the government continued, and the financial risks of aerospace remained daunting. In a study of industry finances as of 1969, the researcher Shin found them to be abnormal compared with other industries, due to major government investments in fixed assets and working capital. Earnings tended to be volatile, indicating high financial risk. Shin concluded that industry earnings simply were inadequate given the level of risk.[2] In addition, capital requirements almost forced firms into high leverage, resulting in a potentially lethal combination of high business and high financial risk.[3]

The degree and even the very nature of risk faced by the industry still was a matter of debate, however. Conventional financial theory held that above-average risks must be compensated by above-average returns. Accordingly, a cyclical industry such as aerospace should be forced to pay a risk premium to attract required capital. However, the premium was found to be quite modest, with industry returns exceeding those required by a comfortable margin.[4] But the question of profitability measurement remained critical. Return on sales could be low and seemingly inadequate, while returns on equity and total capital, more meaningful measures, could be very high.

For the aerospace industry, high returns on investment frequently were due to rather modest internal investment, with government-supplied plant, equipment, and development financing providing the balance. Yet a *Forbes* study showed that defense contractors, of the one hundred largest corporations, suffered a profit decline during the years 1965 to 1970.[5] A GAO study of seventy-four defense contractors covering the 1966–69 period found that while returns on sales were lower for primarily defense-oriented firms than for primarily commercial firms, returns on equity were slightly higher.[6] The GAO study also acknowledged the measurement difficulties due to the degree of government funding. For twelve major firms primarily involved with aircraft, missiles, and space, with 80 percent or more of sales in defense, profits to sales averaged 4.3 percent, but return on total capital invested averaged 12.9 percent.[7] One reason for the differences again was that aerospace firms tended to possess higher debt to equity ratios, or higher leverage, than other industries. In explanation, aerospace firms pointed to large capital requirements for new programs.

The aerospace industry had been created from the convergence of aeronautical and space technologies, rather similar to the later convergence of the computer and telecommunications industries, and thus had largely withstood the "creative destruction" challenge posed by the missile and space programs. McDonnell, for example, was prime contractor for the Project Mercury and Project Gemini manned orbiting vehicles. The largest undertaking, Project Apollo, was a true industrywide effort. Boeing was prime contractor for the Saturn V first stage, North American for the Saturn V second stage and Apollo spacecraft, McDonnell Douglas for the Saturn V third stage, and Grumman for the Lunar Module (LM). North American's Rocketdyne Division developed large launch engines, including the F1 for the Saturn V. After success with the F1, Rocketdyne was chosen on July 13, 1971, to develop the engines for the new Space Transportation System (STS or Space Shuttle), program.

The Manned Orbiting Laboratory (MOL), on which more than $1 billion had been spent, was canceled in 1969, a major loss for McDonnell Douglas. The MOL had been backed strongly by the air force but lacked a clear military mission, and space research was the province of NASA. The later Skylab, developed from the Apollo and operated in 1973 and 1974, included contributions by McDonnell Douglas with the airlock module, North American Rockwell with the Apollo capsule, and Martin Marietta with the docking system.

The McNamara defense era ended on February 29, 1968. With regard to his goals of bringing R&D costs and the acquisition process under control, and of rationalizing missions with improved efficiency and effectiveness, McNamara's record was at best mixed. He definitely deserved credit for developing a more unified strategy and structure and reducing the near autonomy of the services in defining their own missions and weapon requirements. But the F-111 program, originally the showpiece of the new era, instead became a symbol of its failure, widely regarded as another program out of control. The same would be said of the C-5A. McNamara's Total Package Procurement (TPP) policy was superseded by revised policies, but the government still had little practical recourse against contractors that did not perform as expected. McNamara began with a mandate for major reforms but was largely diverted by Vietnam.

McNamara also failed to eliminate interservice rivalry. Yet whatever that rivalry said about military effectiveness and value for the taxpayers, its existence probably had a beneficial effect on the industry and on aeronautical advances.[8] The rivalry also led to greater technological progress than otherwise, and even at the cost of mission overlap created a larger aircraft and missile market for the industry. Despite McNamara's disdain of the defense mobilization base, rivalry meant more programs and more contract opportunities for aerospace firms than otherwise.

The McDonnell Douglas Corporation rivaled Boeing for the distinction of biggest in the industry, but effectively merging the two firms was difficult. There was little immediate rationalization, and the Douglas California operations and the McDonnell Missouri operations continued largely as before. Douglas Aircraft Company was organized into an Aircraft Group and a Missiles and Space Group.

McDonnell's management, controlling the merged firm, knew little of commercial aircraft, and James S. McDonnell, while highly successful in dealing with the Pentagon, did not understand the markedly different negotiating structure with the airlines.[9] Nor, for that matter, did his successor, David S. Lewis. Thus it was not surprising that some questioned the merger for the apparent absence of synergy, usually regarded as the prime financial rationale for mergers. With few overlapping product lines, increased stability was promised, but communication problems between St. Louis and Long Beach persisted. In later years Daniel J. Haughton of Lockheed regretted that he had not entered the bidding for Douglas.[10] Among other things, a Lockheed-Douglas merger would have precluded the ruinous head-to-head competition between the DC-10 and L-1011, and the C-5A might have benefited from Douglas's large transport experience.

Fortunately, Jackson McGowen as Douglas president managed to get along with Lewis, his new superior. Lewis learned the commercial business quickly, and the duo became a successful sales team.[11] Donald Douglas Jr. returned to California to manage the McDonnell Douglas real-estate operations and remained a corporate director. When Lewis left for General Dynamics late in 1970, Sanford McDonnell succeeded him as chairman of Douglas Aircraft as well as president of McDonnell Douglas. The company was strong in military aircraft and in missiles and space, but airliner production was to remain the most intractable problem.

The Rockwell-Standard Corporation of Pittsburgh formulated a goal of becoming a diversified manufacturer with a heavy defense and aerospace content. Concurrently, North American's J. L. Atwood, acutely aware of his firm's vulnerable position in lacking commercial business, had begun an active search for a merger partner in 1965. North American had expanded greatly during the decade with the space program, but the disastrous fire in the Apollo capsule in January 1967, costing the lives of three astronauts, subjected the company to intensive investigation and heavy criticism.[12] In addition, the Autonetics Division incurred severe cost overruns in major programs, including the F-111 avionics suite on which it lost $500 million.

North American possessed the distinction of having produced more military aircraft than any other company, but that record meant little in the environment of 1967. In what may well have been its last chance to survive as a major firm and preserve its renowned technological and production capabilities, NAA agreed to be acquired by Rockwell-Standard in

March 1967. The last of the major aircraft industry mergers in the decade, the deal became effective on September 22, 1967, after being delayed by a Justice Department antitrust review. Rockwell-Standard paid $922 million for North American, effectively absorbing the firm even though it was more than three times Rockwell-Standard's size, evoking the General Dynamics takeover of Convair.[13] Colonel Willard F. Rockwell, age seventy-nine, became chairman of the board, and W. F. "Al" Rockwell Jr. became vice chairman. Atwood held the titles of president and CEO, but control was firmly in the hands of the Rockwells.

The North American Rockwell Corporation (NAR) began to enjoy some degree of countercyclical diversification, given Rockwell-Standard's heavy involvement in automotive products and electronics. But after the merger, Rockwell's management found extensive organizational vagueness and lax cost controls, confirming the popular impression of the defense/aerospace industry at the time. Management consultants were called in, and the company was extensively restructured.[14] Two major groups resulted, the Aerospace and Systems Group, which held most of the former North American aircraft and missile operations, including Rocketdyne, Autonetics, and Atomics International, and the Commercial Products Group, which covered most of the former Rockwell-Standard lines, including general aviation. John R. Moore, formerly executive vice president of NAA, headed the Aerospace and Systems Group, while Robert Anderson, formerly with Chrysler, headed the Commercial Products Group. The restructuring subsequently provided a measure of stability, and North American Rockwell continued to compete for increasingly scarce military contracts.

Ryan, which had successfully diversified into many aerospace lines, was acquired by Teledyne for $128 million in 1968 and became the Teledyne Ryan subsidiary in February 1969. Claude Ryan retired but afterward pursued independent experimental work in aircraft for several years.

Technological Advances

Although aeronautical technology had advanced steadily, even spectacularly, through its history, developments from the late 1960s became known as "advanced technology." Major advances were in electronics, missiles and other weaponry, and, for transports and helicopters, in load-carrying ability. For commercial aviation, advances were in quieter, more economical engines, collision avoidance, and air traffic control (ATC).

The large Pratt and Whitney J58 and General Electric J93 turbojets provided unprecedented power and Mach 3 potential, while the turbofan engine had found wide application in both commercial and military roles. The first-generation military turbofans led to supersonic turbofans. Those designated F100 and higher were air force funded, while designations F400 and higher were navy funded.

In flight performance, there emerged a realization after attainment of Mach 2 capability that ever-increasing speeds did not necessarily mean greater effectiveness or survivability. High-speed maneuverability gained increased emphasis. Other advances were in airfoils, composite structures, and onboard computers, but the era of higher, faster, and farther in flight records was largely over.

The General Dynamics F-111, the Lockheed A-11/SR-71, and the North American XB-70 led to a new generation of advanced-technology combat aircraft. There was a strong linkage among space, missile, and aeronautical technologies, as with the adaptation of large missiles as launch vehicles both for manned spacecraft and satellites. The X-15 linked aeronautics directly with astronautics, as it operated beyond the atmosphere to the fringes of space. The X-15 also attained the ultimate in manned aircraft speeds when it reached Mach 6.72 on October 3, 1967. But the end of the X-15 program also ended hypersonic research for the time.

Richard Whitcomb of the NASA Langley Research Center again led a major advance in aerodynamics. On the problem of drag buildup in the transsonic regime, Whitcomb observed the point at which airflow over the airfoil became supersonic, which led to a decrease in drag, and developed a wing cross-section to enhance the development. The wing was flat on top, rather than curved, with a long taper at the trailing edge, conveying a marked decrease in drag. It was first tested on a converted navy F-8A in 1971. The supercritical wing was applied to business and transport aircraft rather than to combat designs, however.[15]

Steady gains in helicopter power, capacity, and control were made. As even larger models appeared in the Soviet Union, the Sikorsky CH-53E Super Stallion became the largest U.S. production helicopter. Powered by three 4,380 shp GE T64 turboshafts driving a huge seven-bladed main rotor, the CH-53E first flew on March 1, 1974, and was capable of transporting fifty-five troops or thirty thousand pounds of cargo. Bell won the joint military-NASA development contract in May 1973 for the XV-15 tilt-rotor transport, following the earlier XV-3.[16]

Time took its toll on many major industry figures during this period. Allan Lockheed, who had returned as a consultant to Lockheed in his final years, died in 1969, age eighty. Malcolm Lockheed had died in 1958. The outstanding aircraft designer and general manager of Chance Vought,

Rex Beisel, died in 1972. Sherman Fairchild died in 1971, age seventy-five. Fairchild, probably the wealthiest man ever involved in the industry after Howard Hughes, left an estate estimated at $200 million. Some believed his fortune would have been closer to $500 million if he had never become involved in the industry at all. Igor Sikorsky died in 1973, age eighty-four, Alexander P. de Seversky died in 1974, age eighty, and Reuben H. Fleet died on October 29, 1975, age eighty-eight. Lloyd Stearman died on April 3, 1975, age seventy-six, and Grover Loening on February 29, 1976, age eighty-eight. The influential designer Alexander Lippisch also died in 1976. Wellwood A. Beall of Boeing and Douglas died on January 28, 1978, at age seventy-one.

Declining Military Procurement

A growing problem for the industry was that advances in aeronautical and avionics technology had resulted in weapon systems of such high cost and enhanced capability that the era of large-scale production of combat aircraft, extending from 1938, was largely over. During the 1950s and early 1960s the industry had produced some two thousand B-47s, over seven hundred B-52s, over twenty-two hundred F-100s, over eight hundred F-101s, over one thousand of the F-102/F-106, over twenty-seven hundred of the F-84F/RF-84F, over eight hundred F-105s, and comparable numbers of naval fighters such as the F-9 Cougar and F-8 Crusader. By the mid-1960s, however, only the F-4 remained in large-scale production, and the future portended further decreases in variety and quantity. No bombers apart from the FB-111 were being produced, and the only new transport was the extremely troubled C-5A. To a major extent the production decline reflected the strategic shift to forces-in-being, but one result was further excess capacity in the industry, often accompanied by inefficient production rates. Excess capacity in turn was the major contributor to profit drops.[17] The trend only increased the anxiety felt by the major manufacturers.

The industry still wished, of course, to keep operating at or near full capacity, a diminishing prospect from the late 1960s. The government shared industry's concerns, and observers felt that several major contract awards had been decided primarily on that basis. For example, contracts for the supersonic transport (SST), the MOL, and the CX-HLS for the C-5A were awarded almost concurrently in the mid-1960s. Boeing, Douglas and Lockheed competed for all. Boeing won the SST over Lockheed, but Lockheed won the C-5A. Douglas, the loser in both, was awarded the MOL. Thus each firm gained an important new program, with the benefits spread among California, Washington, and Georgia. Although the MOL and SST were canceled and the C-5A became troubled, political considerations in awards were rather transparent.

By the beginning of the 1970s, military procurement and production had begun to decline from the Vietnam War peak. In 1968 4,440 military aircraft were produced, but output declined to only 1,110 by 1974.[18] Military helicopter orders also dropped sharply after 1971 as needs in Vietnam fell, leading to problems for that sector after a decade of rapid growth. Concerns over future contract opportunities were widespread. New programs were formulated, although not with the frequency of previous years. The B-1, F-14, F-15, AWACS, and AMST programs, all technologically challenging and unprecedentedly costly, attracted spirited competition.

The Vought A-7 attack aircraft continued for both services, but design changes effectively eliminated most advantages of commonality. Production switched to the A-7D and A-7E with the higher-powered Allison TF41 from 1968. The air force AX program, for a rugged, attack "tankbuster" aircraft, inspired by the success of the A-1 in Vietnam, began in 1970. With a mission of army battlefield support it was unaccompanied by any great enthusiasm by the air force, however, where emphasis still was on high-speed capability. The AX was relatively slow and of strictly utilitarian appearance, and even though heavily armored, its combat effectiveness still depended on a "permissive" air defense environment. The competition also was the first of a series to be conducted under the revived fly-before-buy policy, dormant since the 1950s, with selection of the winner coming after a flyoff between the two contenders. Earlier it had been felt that a design competition, ordering an airplane "from the drawing board," was far less expensive than two or more complete RDT&E programs, but such troubled programs as the F-111 and C-5A led to a return to the fly-before-buy policy. On December 18, 1970, the Northrop YA-9 and Fairchild's Republic Division YA-10 were selected as finalists. The YA-10A made its first flight on May 10, 1972, and won the production contract on January 18, 1973, but not without strong recriminations from Northrop.[19]

The F-111A entered service during 1967, still with numerous operating shortcomings, especially in excess drag. NAVAIR finally canceled the F-111B on July 10, 1968, after notifying the Pentagon that it could afford yet another delay in its fleet defense requirement.[20] While it may be argued that the cancellation was justified both on cost and operational grounds, clearly the navy never held a strong commitment to the F-111B, since the design did not originate with that service.[21] Interservice rivalry was a factor.

With no naval version and shifting requirements, orders for the F-111, at one time projected as high as twenty-four hundred aircraft including exports, were sharply reduced. But the General Dynamics contract, accounting for $6 billion of the $8.6 billion program total, insured a profit of $300 million for the company.[22] Contrasting sharply with the troubles of North American, Lockheed, and Boeing during that period, the contract protected General Dynamics from any cancellation or penalties for default, and performance "guarantees" were effectively meaningless.[23] Only some two hundred F-111s had been delivered by the end of 1969, and total production, which extended into 1976, was 565 aircraft, including prototypes and 24 for Australia. The total also included the FB-111, which with its performance shortfalls still left the air force without an advanced strategic bomber. Originally projected at $3.2 million, F-111 unit cost ultimately reached some $16 million.

Senator McClellan's Subcommittee on Investigations issued its final report on the F-111 in 1970, which blasted McNamara's entire management of the program, from his basic assumptions underlying the dual-service aircraft to his allegedly misleading the Senate.[24] McNamara had assumed personal direction of the F-111 in 1966 under his informal Project Icarus, coordinating the program among himself, the service secretaries, General Dynamics, and Pratt and Whitney, while denying such activities to the Senate.[25] Although later versions, especially the definitive F-111F with the developed TF30-P100 engines, were successful performers, and the aircraft had its defenders, the overall program was a failure both in concept and implementation. Responsibility, moreover, lay with the top levels of the Defense Department.

Both the air force and navy issued specifications for advanced-technology fighters. The new navy VFX specification was in two stages, to speed development and service entry. The initial VFX-1 was to retain the TF30 turbofans and Hughes radar and missiles of the F-111B and was projected to enter service in 1972, with the VFX-2, with air force–funded advanced turbofans, to follow. Grumman, having already studied new fighter designs, won both the VFX-1 and VFX-2 contracts, the award being announced on January 14, 1969. The TPP contract carried a potential value of $5 billion and was written with severe penalties for shortfalls in performance, weight, and timing.[26] President Lew Evans stated that it was "probably the toughest ever written by the Department of Defense."[27] The VFX, designated F-14, would indeed bring severe problems to Grumman. While the proven engines, missiles, and radar were intended to reduce the development risks, the TF30 was acknowledged to be a serious mismatch with the airframe, and cost problems postponed installation of more advanced engines. The F-14 eventually cost far more than the F-111B, in excess of $20 million per aircraft. The navy still supported the program, however, probably reflecting its origins with that service.

The air force "Fighter Mafia" attached great urgency to the FX program, which resulted in the F-15, as a counter to the Soviet MiG-25, believed capable of Mach 3. Contractors selected through the definition phase were McDonnell Douglas, North American Rockwell, and Fairchild Hiller.[28] The prototype award to McDonnell Douglas reinforced its status as the leading military aircraft manufacturer. The air force counterpart of the F-14, the F-15 was powered by advanced F100 engines but was designed with fixed wings. Development still was troubled, although not to the degree of the F-14. Risks included the concurrent development of airframe, engines, and electronics. The program benefited, however, from its award by the Nixon administration, which imposed fewer contract restrictions than those for the F-14.

Both the F-14 and F-15 were by far the most expensive warplanes the United States had built to that time, costing three to four times the F-4 they would replace, which also had increased steadily in cost despite large-scale production. Both led to sharp controversy beyond development and reliability questions in that they represented only next-generation incremental improvements, but improvements gained at simply too great a cost. In the civil sector, technical advances were considered successful only if they accomplished both improved performance and lower cost.[29]

In the transport field, the C-141 program was completed in 1967, finally retiring the veteran Douglas C-124. The CX-HLS long-range heavy airlift specification, in a sense a replacement for the canceled Douglas C-132, became the next step. The specification was ambitious, including the ability to lift a 125,000 pound payload, and it was hoped that later civil air freight developments would attract orders.[30] Five bidders responded to the RFPs issued on December 11, 1964, and spent a total of $60 million in company funds on their proposals. Finalists were Boeing, Douglas, and Lockheed. The SSB initially favored Boeing on the basis of higher estimated performance, but Lockheed, with the lowest bid, won the contract on September 30, 1965.[31] The program covered five development and 115 production aircraft at a total cost of $2 billion. The TPP contract, while carrying penalties for cost underestimation, provided for increased profits with good contractor cost performance. Total orders were projected to reach two hundred, not including civil versions.

The first C-5A flew on June 30, 1968, but it was then apparent that development had been troubled from the outset. A. Ernest Fitzgerald, deputy for management systems under the assistant secretary of the air force for financial management, detected signs of C-5A cost overruns as early as January 1966.[32] The program cost estimate soon increased to $3 billion, and problems also appeared related to structural durability. Lockheed, as had other contractors in similar circumstances, did not alert program managers to the cost overruns and performance shortfalls, fearing that to do so would lead to cancellation. Further, the air force also did not wish to expose problems. Fitzgerald, among others, noted that career military officers or civilian officials charged with monitoring contractor performance tended to behave defensively whenever a program came under criticism, regarding negative disclosures as reflections on their competence, and to question the patriotism of the critic.[33]

Fitzgerald disclosed major C-5A cost overruns to congressional investigators in 1968. Total program cost then approached $5 billion, and as a result the order was reduced to eighty-one on November 14, 1969. Fitzgerald also reported production deficiencies that would result in unreliable performance.[34] In a notorious case of official retaliation against whistleblowing, Fitzgerald was first demoted, then fired on November 4, 1969, although later reinstated after a long investigation.

While Lockheed's bid unquestionably was too low, McNamara's policies, especially the TPP policy, and the inflationary pressures of Vietnam also bore responsibility. But to the public the C-5A became another illustration of long-standing defense acquisition problems, including the precedence of narrow political and economic considerations over military needs and sound management.

The air force had begun a major push for a new Advanced Manned Strategic Aircraft (AMSA) in 1965 to replace the B-52. Although there were widespread doubts about the military necessity for a supersonic strategic bomber, that mission was central in air force doctrine. The air force regarded the AMSA as its highest-priority aircraft program, even exceeding that for the F-15, and set a goal of service entry in the early 1970s. AMSA was designated B-1 in April 1969, and a requirement for 250 was contemplated. The RFP was issued on November 3, 1969, and NAR won the contract over Boeing and General Dynamics on June 5, 1970. The B-1 also became NAR's only major aircraft program, particularly critical after loss of the F-15 contract.[35]

B-1 development experienced delays and cost overruns, and the complex variable-geometry wings, never before attempted on such a large aircraft, also raised serious questions. The initial program cost estimate of $9 billion escalated sharply, and the B-1 became yet another major acquisition controversy. Both contractor and the air force attempted to conceal problems until the program had reached a point that more money had to be allocated to justify the amounts already expended.[36] Strong opposition to the B-1 arose in Congress and the public, but its potential for job creation, and job losses with cancellation, became major selling points, especially in Southern California, where aerospace employment had declined. Yet even prodefense senators felt that they could not support the B-1 if, as feared, unit cost should reach $100 million.[37]

The air force and NAR expended every effort to extend political benefits of the program. Subcontracts were spread over the forty-eight contiguous states, even at some increase in cost when NAR could have produced the subcontracted components more efficiently.[38] The air force planned to base the first B-1 wing near the home area of Texas Senator John Tower, of the Armed Services Committee, and another would be based in Kansas, represented by Senator Bob Dole.[39]

Still, the post-Vietnam environment was not conducive to expensive new weapon systems, and the B-1 became a campaign issue in both 1972 and 1976. NAR, by then Rockwell International, combined lobbying efforts with the military to defeat a Senate effort to cut off B-1 funding in 1974–75.[40] Bastian J. "Buzz" Hello, previously with the Apollo program, was appointed B-1 program manager and led the lobby. Many in Congress fought not for outright cancellation but for a delay in production authorization, which the air force feared would be just as fatal. President Carter, who had opposed the B-1 in the election of 1976, after a final review canceled the program on June 30, 1977.[41]

Other advanced programs involved electronics more than aerodynamics. The Wild Weasel program, resulting from heavy aircraft losses to surface-to-air missiles in Vietnam, led to modification of existing aircraft specifically to suppress radar and other electronic defenses. McDonnell Douglas received large orders for the F-4 in Wild Weasel configuration, adding to its already illustrious record. Boeing received the contract for the Airborne Warning and Control System (AWACS) on July 8, 1970. AWACS was based on the 707–320 airframe and designated E-3A. While small in numbers, the complex technology made the aircraft the first to break the $100 million barrier. Its mission also was controversial: originally intended for long-range early detection of Soviet manned bombers, that virtually nonexistent threat led to a redefinition of its mission toward NATO and tactical control of theater aircraft. This led in turn to questions of vulnerability, both in terms of radar jamming and in being shot down in a combat zone. The AWACS program eventually involved in excess of $3 billion, which, coming after Boeing's 747 development crisis, earned it the unofficial designation BBO, for "Boeing Bail-Out."[42]

The air force requirement for an Advanced Medium STOL Transport (AMST) to succeed the C-130 did not result in production. Boeing and McDonnell Douglas were selected to build prototypes in 1973, and the Boeing YC-14, which first flew on August 9, 1976, and the McDonnell Douglas YC-15, which first flew on August 26, 1975, represented radically different design approaches. The YC-14 featured a rather small supercritical wing and two large overwing turbofans, while the YC-15 mounted four turbofans in underwing pods. While both had merit, development ended in 1978 due to declining funding, and the capable C-130 continued to serve.

No major additions to the U.S. strategic airlift capability would appear for years after the C-5A, which then required an expensive program of structural strengthening. C-5A weight problems had led Lockheed to save weight from the wings, resulting in inadequate strength. Lieutenant General James T. Stewart, who became commander of the Aeronautical Systems Division of Systems Command in 1970, stated that C-5A production should have been stopped when the structural problems were discovered, but to have done so would have provided further ammunition to critics. Stewart felt the air force needed the heavy airlift capability and opted for continued production and later modification.[43]

The Lockheed AH-56 Cheyenne for the $1 billion AAFSS program also experienced severe performance problems, and the production order was canceled on May 19, 1969, for default. The army continued some R&D on the Cheyenne but ended all work on August 10, 1972. The army then launched the successor Advanced Attack Helicopter (AAH) program, based on Vietnam experience, issuing RFPs in November 1972. This was to be the first purpose-designed combat helicopter to enter service, the Bell AH-1 having been developed from the UH-1. Bell, Boeing Vertol, Hughes, Lockheed, and Sikorsky responded. After a protracted competition the contract went to the Hughes AH-64 Apache on December 12, 1976, winning over the Bell YAH-63 and Vertol YAH-65. Teledyne Ryan was the major subcontractor.

Advanced-Technology Airliners

Truly advanced technology in the commercial sector could be considered to date from 1966, when the Boeing 747 was launched and project definition began for the Douglas DC-10 and Lockheed L-1011 wide-bodied airliners. All were precipitated and facilitated to some degree by the CX-HLS program, and development of large, economical turbofans made such airliners feasible. But commercial programs involved rapidly changing markets and political environments, with no direct government support. Advanced-technology airliners, therefore, carried an entirely different dimension of risk.

Pratt and Whitney, General Electric, and Rolls-Royce developed turbofans in the forty- to fifty-thousand-pound thrust class required. The P & W JT9D became the launch engine for the 747, and Rolls-Royce enjoyed, for the moment, a major coup by winning the Lockheed L-1011 contract with its RB.211. The GE CF6, the third large turbofan, powered the DC-10.

Preceding the large airliners, however, there was enthusiastic anticipation of the supersonic transport (SST). Development of the Anglo-French Concorde had begun in 1962, with the partners hoping to gain a major lead over what they regarded as inevitable U.S. competition. The American industry in fact envisioned an SST even larger and more advanced than the Concorde, if later in timing. Preliminary development began in 1963, and the FAA sponsored a design competition, with Boeing and Lockheed selected as finalists. The Concorde, while a true Mach 2 aircraft, nonetheless was based on engines and structures of the day, but the American SST promised breakthrough advances. For the first time in forty years of airliner development, the effort clearly was beyond the resources of a single company, requiring government funding. The Boeing Model 2707 won the FAA competition on December 31, 1966, but Boeing continued to refine the design, switching at one point from variable-geometry wings to a fixed gull-winged delta. The General Electric GE4 engines, developed from the J93, were projected to develop almost seventy thousand pounds of thrust with afterburning. Major airlines rushed to take options, fearful of falling behind their competitors despite the many unknowns of SST operation. But increasing environmental and cost concerns led the U.S. Senate to cut off funding on March 24, 1971, forcing the FAA to end the program.

The subsonic "jumbo" airliners promised a revolution in air travel. Offering passenger amenities never before possible as well as greater operating efficiency, they were expected to make air travel still more accessible to the masses. All would become technical successes and would enter service reasonably close to schedule but brought major problems to their developers. Apart from technical problems, the dominant concern with commercial programs of such scale and cost was that the market simply could not support two or more models in head-to-head competition, as occurred with the L-1011 and DC-10.[44] A similar competition was less critical in the days of piston-engined airliners.

Boeing built its market dominance during the 1960s with a steadily expanding airliner family. By the mid-1970s, however, only low-volume production of the 707–320 remained of the basic 707. Total production of the 707 airframe was 1,031 over twenty years, of which 878 were to commercial customers. The 727 proved to be a classic case of the right design at the right time, and the short-range 737, which first flew on April 9, 1967, also gained global popularity.

It was the 747, however, which would shape Boeing's future in airliners and as a corporation. The most ambitious airliner undertaking ever, it problems became so great as to threaten Boeing's very survival. Pan Am, as it had done over more than thirty years, solicited development of a new class of airliner, while Boeing's initial interest grew out of its CX-HLS loss. The first discussions between Juan Trippe of Pan Am and William Allen of Boeing took place in the fall of 1965. In addition to financial risks far greater than those of the 707 of fourteen years earlier, the 747 presented major technical risks, in that it stretched both airframe and engine technology, which in both cases did not benefit from any direct military ancestry, and was also committed to an ambitious development schedule. Further, it was to be assembled at a massive new Everett, Washington, factory. Boeing's existing facilities were at capacity, requiring a new plant. Pratt and Whitney, the engine contractor, faced similar risks in delivering a turbofan of requisite power and reliability in such a short time.

Acknowledging the risks, the parties pushed ahead. Pan Am obligated itself for some $500 million, more than its annual revenues at the time, for twenty-five 747s, with half to be paid to Boeing six months before first delivery, and before award of the FAA Certificate of Airworthiness.[45] Juan Trippe was an optimist: his forecasts of airliner needs and airline traffic growth had always proved conservative.[46] Boeing, likewise, favored a new design after rejecting a major stretch or double-decking of the 707 fuselage.[47] Allen and Trippe signed the letter of intent on December 22, 1965, for an airliner weighing 550,000 pounds, capable of carrying up to four hundred passengers at thirty-five thousand feet over fifty-one hundred statute miles, at a speed of Mach 0.88. Charles A. Lindbergh, a Pan Am director and consultant, served as a private liaison between the companies during the early phase of development.[48]

As the program progressed, the specification became regarded as nearly impossible, even with the new large turbofans. When the formal contract for the 747 was signed on April 13, 1966, the performance specification had been cut back, but the anticipated weight had increased to 655,000 pounds.[49] By comparison, the maximum weight of the latest 707–320 was 335,000 pounds. Optimism still prevailed, however, and the 747 mockup became the focal point of the City of Seattle's civic celebration of the fiftieth anniversary of Boeing on July 15, 1966.

The 747 launched the wide-bodied airliner era, but it soon became evident that the project, in addition to being too ambitious technically, was well ahead of the market. Trippe had negotiated a contract, similar to the military weapon system concept, in which Boeing was responsible for power as well as the airframe and could not transfer blame to the engine supplier for performance deficiencies. The 747 program also represented a shift in marketing policy for the industry, in that its price, initially $23 million, was indexed to inflation and other unforeseeable changes rather than established as a fixed, guaranteed price as before.

Despite the leadership positions of Boeing, Pan Am, and P & W, and their determination to maintain those positions, there was extensive second-guessing of the basic concept. The 747 represented a major departure from the historical progression of airliner designs, in which new airliners represented incremental increases in size over earlier models. But the 747, still the largest airliner ever built, preceded development of the smaller three-engined widebodies as well as the twin-engined European Airbus. Further, while designed for the intercontinental market, and thus for the export market, it was originally too large even for that market. Only Pan Am forecast the need for an airliner of its size at the outset. Weight escalation resulted in initial models weighing 710,000 pounds. At that weight and with the forty-one thousand pounds of thrust from the JT9D, it was underpowered, and operating economics were doubtful, raising the prospect of a service delay. Pratt and Whitney, which had not scheduled uprated engines until after three years of service, now was pressured to develop the JT9D beyond the initially guaranteed thrust and before the design had been proven.[50]

With the initial version of the 747 appearing to fall short of contract specifications, a triangular conflict among the three principals developed.[51] Allen and Trippe especially were at loggerheads over changes and development responsibilities, but Pan Am continued to make progress payments. Then management changes occurred at all parties in 1968. Juan T. Trippe retired from Pan Am, and William M. Allen relinquished the duties of president and CEO of Boeing, becoming chairman of the board on April 29, a position vacant since the retirement of Claire Egtvedt. Jack Horner also retired as chief executive of United Aircraft, parent of P & W. To Trippe's successors, first Harold E. Gray and then Najeeb Halaby, and to Allen's successor Thornton A."T" Wilson, fell the task of managing the 747 through to service. Wilson, who had joined Boeing in 1943 and who had led the B-52 and then the Minuteman, soon faced a major corporate restructuring in addition to 747 problems.

Other airlines, despite reservations about the 747's performance and economics, ordered heavily to avoid being left behind in the market. Charles Tillinghast of TWA later stated that if he had been aware of the TriStar and DC-10 at the time he probably would have bypassed the 747 entirely.[52] Critical to new airliner introduction, especially with a design as complex and expensive as the 747, was that major orders were necessary from two or more airlines simply to make introduction costs manageable. Small orders from several buyers made the costs prohibitive for each.[53] If no major orders had followed that of Pan Am, the project might have been canceled, perhaps not to be revived for a decade.

The 747 prototype first flew on February 9, 1969, and testing proceeded smoothly. With initial performance still below specifications, Pan Am refused to make full payment on delivery, and in fact delayed until December accepting its first 747. Final payments were keyed to Boeing's progress in bringing it up to specification.[54] The first 747–100 entered service with Pan Am on January 21, 1970, followed rapidly by service with other airlines, but its debut was marred by frequent serviceability problems with engines and other components. Problems were overcome fairly quickly, however, and the type became firmly established and popular. The 1970 recession, rather than 747 costs, was largely responsible for Pan Am's operating losses in 1970 and 1971.

Financial problems also mounted for Boeing even as technical problems were being addressed. Unavoidable with airliners is the lag in cash flow from development to delivery. Development costs are at the risk of the company, as are initial production costs. Thus cash outlays are greatest during the period just before revenues are realized from customers, requiring a strong capital base and generally heavy borrowing to bring the new model into service. Customer progress payments helped, but they seldom compensated for inadequate capital of the manufacturer. Delays and performance shortfalls brought financial penalties and cancellations, prolonging the period of net cash outflow, forcing increased borrowing and potentially a financial crisis. This was precisely the situation faced by Boeing, threatening its survival as well as those of Pan Am and Pratt and Whitney. Even Boeing's world market dominance and the fact that it had an effective monopoly with the 747 did not make the situation more manageable at that point.

Further, Boeing lacked a large military base other than missiles. It had won no aircraft prime contracts for years and remained embittered by loss of the TFX and by what it perceived to be the often arbitrary, eccentric nature of contract awards. Thus Boeing's future was tied to the 747 program, on which it had gambled that

- It could develop the 747 within time, cost, and performance estimates
- P & W could do the same with the new JT9D engine
- Others would follow Pan Am in ordering the 747
- Overall economic growth and airline traffic growth would support the economic basis of the 747
- It could bring a completely new manufacturing plant into operation concurrently with production

It would fall short on all projections except the third, and even then orders dropped after the initial rush. Orders for more than 150 were booked before the first flight, but from January 1969 through September 1970 only 25 more 747s were sold. The cash flow situation forced a major retrenchment and severe cost cutting to save the company.

Of the three airliner builders, Lockheed faced perhaps the most daunting risks. Another characteristic of the market is that airlines tend to be heavily influenced by the company as well as by a specific airplane in ordering decisions. When it decided to develop the TriStar, Lockheed had been out of the airliner field for several years. Its reputation had declined with the troubled Electra program, and its military capabilities did not directly transfer into the commercial sector. But Dan Haughton, who became chairman and CEO in 1967, decided that Lockheed must have a future in the commercial field, and that to do so it must build the most advanced of the new airliners.[55] The L-1011, while benefiting to some extent from the C-5A, broke new technological ground and in fact was regarded as more advanced than the Boeing and McDonnell Douglas designs. Further, with engine selection becoming as hotly competitive as the airliner design itself, Haughton also was determined to offer the most advanced, highest-performance engine. This led to the fateful negotiations with Rolls-Royce.

Rolls-Royce maintained that its new RB.211, with its compact dimensions, turbine blades of advanced composite materials, and a higher power-weight ratio than competitors, would offer a distinct advance. Rolls-Royce, moreover, knew that it must penetrate the North American market to have a future in large jet engines. The TriStar was launched with announcement of the selection of the RB.211 engine on March 29, 1968, to great celebration in Great Britain as the largest single export order in British history.[56] Simultaneously, TriStar orders by Eastern and TWA were announced. Selection of a British engine over American competition brought predictable protests from the American industry, but Haughton, in what was regarded at the time as a masterstroke, announced an export order, however vague, from an organization called Air Holdings, Ltd., for fifty TriStars. Air Holdings, controlled by London merchant banks, was not an operating airline, and the "order" was not really a firm order, but rather an agreement to market the TriStar in Europe. But Haughton, the master salesman, had largely defused protests by promising a net financial return to the United States with strong export sales of the TriStar.[57]

The prototype L-1011 TriStar first flew on November 16, 1970, and entered production at the Palmdale factory. By 1971, however, as it was attempting to bring the TriStar into service with Eastern, Lockheed was in the throes of the most severe crisis in its history. Lockheed had suffered heavy financial losses in 1969 and 1970, in large part due to cost overruns on the C-5A contract, for which it had underbid competitors. The company quite simply needed a major bailout. It requested an additional $750 million recovery on the C-5A but compromised on $400 million, still leaving it with a severe deficit in addition to soaring TriStar expenses. But as the C-5A matter was stabilizing, Lockheed received what was potentially a fatal blow.

Heretofore unknown either to Lockheed or the British public, Rolls-Royce was financially past the point of no return, principally due to RB.211 development expenses. The Heath Conservative government was less inclined to bail out a failing company than the previous Labour government would have been, and Rolls-Royce declared bankruptcy February 4, 1971. Lockheed by that time had exhausted its bank credit lines, had major unsecured creditors, and had already rejected as unfeasible the replacement of the RB.211 with the GE CF6.[58] Financially strapped airlines could not be induced to provide higher advance payments. Edward Heath nationalized military engine operations into a reorganized Rolls-Royce (1971) Ltd., as essential to security, but civil engines were to fend for themselves in the market.

Dan Haughton, accompanied by Kelly Johnson and others, flew to Great Britain in desperation. Intensive negotiations followed. The British would not continue the RB.211 without assurances from the U.S. government that Lockheed would survive, and it was rumored that President Nixon did in fact make such assurances.[59] The British still stated that RB.211 development would end unless a guarantee was in place by August 8.[60] The provisional agreement then negotiated included both higher prices for the RB.211 and higher prices for the TriStar. Armed with the agreement, Lockheed then turned to the Congress for a guaranteed loan of $250 million to preserve the company until the TriStar could generate revenues. The controversial proposal, which revived the old debate over the relationship between government and private enterprise, narrowly passed both houses of Congress on July 2, 1971. One inescapable conclusion was that, in addition to politicized military procurement, large airliner programs, especially with increasing internationalization, also had become highly politicized.

Quite apart from Rolls-Royce's problems, Lockheed had seriously underestimated costs of the TriStar program, and as a result projected a low breakeven figure. In fact, the entire financial estimation process was unrealistic: Lockheed probably calculated a breakeven quantity of only 195–205 TriStars.[61] Further, the time of peak development expense almost coincided with the refusal of the Department of Defense to absorb all cost overruns on Lockheed's military contracts.[62] Another problem was that with the RB.211, Lockheed had ordered an unproven "paper" engine. Composite blade development had to be abandoned, leading to greater engine weight and power shortfalls. The contract price to Lockheed of $840,000 per engine, unrealistic originally, subsequently became ludicrous. Although the loan guarantee, specifically for the civil program, was more controversial, the $400 million recovery for the C-5A actually was more beneficial, providing cash when the L-1011 was draining the company more than the C-5A.[63]

After the financial rescue, Lockheed was able to sign contracts with airlines and with Rolls-Royce to proceed with TriStar deliveries and service. But another complication arose when the world financial crisis of August 1971 led to the imposition of a 10 percent U.S. tariff surcharge. The surcharge raised the cost of three RB.211 engines by some $280,000 over the renegotiated higher price. After intense negotiations, Lockheed was able to pass the fees on to customers.[64] Initial TriStar customer delivery took place in April 1972, somewhat later than originally scheduled, and was accompanied by operating restrictions.

The RB.211 engine gradually attained its specification, however, and the TriStar became regarded as superior to its direct competitor, the DC-10. But the L-1011 still suffered a prolonged order drought, and the Air Holdings marketing arrangement also proved to be sluggish.[65] There were no L-1011 sales in 1975, and then only thirty-one from 1976 through 1978. Implication in the international bribery scandals and shaky finances did little to enhance Lockheed's sales prospects. Increased costs pushed the breakeven quantity probably to well above 400, a figure never to be achieved.[66]

The original Douglas interest in a large airliner, or airbus, also sprang from its CX-HLS loss. American Airlines, long associated with Douglas, formulated a specification for a twin-aisle, 250-passenger medium-range airliner for its major routes. Douglas, in turn, developed its proposals around the twin-engined widebody configuration, as did Lockheed, but other airlines with overwater routes favored three engines.[67] Twin engines for such a large airliner then were regarded as too risky, although the later success of the Airbus series would dispel such doubts. Ironically, the original American/Douglas proposals were almost identical to the Airbus A300, and if proceeded with would likely have eliminated Airbus as a major competitor.[68]

In any event, Douglas was too weak financially by 1966 to undertake a new venture, but the postmerger McDonnell Douglas Corporation, with access to the necessary capital, approved a larger three-engined design as the DC-10. John C. Brizendine, who earlier had headed the DC-9 and DC-8 programs, was named program manager for the DC-10. The DC-10 was officially announced on April 25, 1968, less than a month after the TriStar, and the first prototype flew on August 29, 1970. American, United, and Northwest provided sufficient orders to launch the program.

In the meantime, DC-9 production continued at a high rate, with more than three hundred delivered by mid-1968, including the military C-9A Nightingale. The DC-9, incidentally, was one of the first airliners to widely employ international production sharing, as major components were produced in Canada and Italy for assembly in Long Beach. DC-8 production was ended May 17, 1972, with 556 completed over fourteen years. McDonnell Douglas felt that resources devoted to the DC-8 were needed for the DC-10, but the DC-8 Super Sixty series presented at the time a more economical alternative to the huge 747, and many more likely would have been sold.

Although the DC-10 did not experience a development crisis comparable to those of the 747 and L-1011, market problems still resulted in severe losses. Positioning the DC-10 directly against the TriStar split the market and insured that neither would ever attain profitable volume.[69] TriStar service delays definitely worked to the advantage of the DC-10, which outsold the TriStar, but competition between expensive large airliners in a market of financially pressed airlines resulted in heavy pressure to cut prices in order to secure orders. Dan Haughton, understanding the pressures, probably was more ruthless in price cutting, while James S. McDonnell resisted. That each airline wanted a custom-tailored version for its particular needs also increased costs and increased pressure on the manufacturers' learning curves. Then the broad recession of 1970 and 1971, depressing air travel, slowed all new airliner sales.

Regardless of problems experienced by their builders, American airliners continued to dominate the world market. Manufacturers continued to refine their current models and even considered new designs. A broad range of Soviet jet airliners was produced but could attract no orders outside the Soviet sphere of influence. Early British and French jet airliners simply were behind the American models, and the Concorde was a market failure. The Concorde partners, aware of the poor record of the European industry for on-time development and delivery, had publicly committed to a firm first flight date of February 28, 1968. That date passed, as did others, and the first flight did not in fact take place until a year later. While the partners could hardly be faulted for caution with such a critical program, the delays did little to enhance the credibility of the project.[70] A target service date of 1972 also fell by the wayside, by which time soaring fuel costs and environmental concerns had ended the Concorde's viability. It finally entered service in 1976, but only with the government-controlled airlines.

The eventual, credible threat to American dominance was Airbus Industrie. France had studied the airbus concept, for a high-density, medium-range airliner, for several years, and a formal specification, influenced by the original American Airlines proposal, was developed in 1966.[71] Led by France and Germany, a multinational development agreement, including the initially reluctant British, was signed on September 26, 1967. Airbus Industrie (AI), the manufacturing consortium owned by aerospace firms but with heavy government backing, was not formally established until December 1970, and included Britain, Spain, Belgium, and the Netherlands. With a view toward the North American market, the Airbus A300 was powered by American engines and contained a heavy proportion of American electronics and other equipment. The prototype, capable of carrying more than three hundred passengers, flew on October 28, 1972. But orders were slow, as the initial model was regarded as too large for the market, and the American industry did not regard Airbus as a serious threat for several years.

Another significant change in the airliner market was that airframe and engine orders became largely separate decisions. After selecting an airliner, the ordering airline then specified an engine. Engine manufacturers thus competed fiercely throughout the airliner production cycle. Large airliners, with the exception of the Lockheed TriStar but including the Airbus, offered two or more engine options.

Major manufacturers strove to maintain their positions in the commercial market, even as technical and financial risks threatened ruin. Large jet airliners became the most expensive mass-produced products in history. Although stricter noise, environmental, safety, and energy/fuel consumption regulation posed new challenges, these factors actually served to enhance demand. Airlines could no longer afford to operate noisier, fuel-gulping airliners, requiring replacement with new models meeting the new standards.

At the opposite end of the commercial spectrum, third-level service, or commuter carriers, grew steadily, and with the expansion came demand for specialized, efficient small airliners. Those seating no more than nineteen passengers were exempt from major airline regulations, but the CAB, which officially established the commuter category in 1969, after 1972 commonly permitted third-level carriers to operate aircraft carrying up to thirty passengers. By 1978 this was expanded to include up to sixty-seat aircraft.[72]

Chapter 14   An Industry under Stress

The new Nixon administration included the team of Melvin Laird as secretary and David Packard as deputy secretary of defense. Both were determined to end the abuses and failures of the acquisition process and to reform the system. Reacting to excessive centralization of R&D and procurement of the McNamara era, Laird and Packard moved to decentralize, to improve decision making, and to increase efficiency. Yet the military still resisted efforts toward simpler, less costly weapon systems, the justification being that American fighting men must have the absolute best to counter the Soviet threat. In application this often meant electronic and weaponry additions, or gold plating, which in turn denigrated the performance of the original design. While individually insignificant, design changes collectively resulted in increased weight, development delays, and cost escalation. But pressures for improved management increased, given the decline in new programs and the financial magnitude of those remaining. The air force's lightweight fighter and the navy's less expensive multimission aircraft programs are especially instructive in this regard.

In 1972 Deputy Secretary Packard, heading the R&D and procurement functions, allotted $200 million to the services to present proposals for a new weapon system for which two competing models could be built. The air force, wanting the funding as much to keep it from the navy as for its own purposes, issued a specification to the industry for a lightweight fighter (LWF) possessing the technology of the F-15 but smaller, lighter, less costly, and more easily maintained. While the air force expressed no operational requirement, it was anticipated that a production LWF might be adopted by NATO as a replacement for the Lockheed F-104G.

General Dynamics, Northrop, Boeing, Lockheed, and LTV submitted proposals, Lockheed with a development of its F-104 Starfighter and LTV with a development of its F-8 Crusader. After evaluation the air force awarded contracts to General Dynamics and Northrop for prototypes, termed "technology demonstrators," for a flyoff competition.

Northrop, following industry practice of anticipating official specifications, had been developing an advanced P-530 design since 1966. Among other things, it hoped to position the P-530 for the eventual NATO F-104G replacement order. General Dynamics, on the other hand, having lost out on the B-1 and F-15 competitions, was desperate for a production successor for the F-111, then winding down at Fort Worth. The LWF thus represented possibly its last hope in aircraft.[1] Emulating Lockheed, GD set up its own version of the Skunk Works to develop the Model 401, which became the YF-16, in record time.[2] The F-16 project office represented a sharp departure from the bloated F-111 organization, and the YF-16 first flew on January 20, 1974.

Northrop had not built a P-530 prototype, but the military-funded YF-17 development first flew on June 9, 1974. The YF-16 proved the clear winner in flyoff tests and demonstrated strong operational potential. But air force priorities remained with the more complex and expensive F-15 and with the B-1. General Stewart, ASD commander during peak development of both the F-15 and F-16, was frustrated at the degree of internal contention between the systems: "I was trying to keep the F-15 and the F-16 SPOs (System Program Offices) from competing and fighting with each other, and trying to remind them that the enemy for the F-15 was not the F-16—it was the MiG-23 or something like that."[3] The programs also exemplified the continuing argument between smaller numbers of a more sophisticated system versus larger numbers of a less sophisticated system.

In a move toward operational application, the LWF was renamed the ACF (Air Combat Fighter) program in 1975, and new secretary of defense James Schlesinger won air force support by allotting an absolute increase in combat wings.[4] The F-16 then entered production, with an initial order for 650 in January 1975 against a total requirement for 1,388 aircraft. Attention then turned to NATO. General Dynamics had limited experience in military exports, and its Canadair operation was facing difficult times, making a NATO order especially critical.

Northrop also waged an aggressive campaign for the NATO order with its F-17, and in fact held a 20 percent interest in Fokker at the time. But the F-16, with an unprecedented level of government backing, won not only over Northrop but also over French and Swedish competition, on June 7, 1975.[5] Orders for Norway, Denmark, Belgium, and the Netherlands totaled 348 aircraft, to be built by a Dassault-controlled plant in Belgium and by Fokker in the Netherlands.[6] Although the air force would have preferred to sell the F-15, the NATO victory led to a strong worldwide market for the F-16.

Northrop, although losing both the LWF and NATO competitions, was far from finished. The navy had issued a new VFAX specification for an aircraft to replace both the F-4 Phantom II in the fighter and the A-7 Corsair II in the attack missions. But the VFAX program was terminated by Congress, mindful of the problems of the F-14, in August 1974, with a directive to adapt an existing design with resulting cost savings. The successor Naval Air Combat Fighter (NACF) thus inspired adaptations of both the YF-16 and YF-17. McDonnell Douglas teamed with Northrop and became the prime contractor on the NACF. Northrop agreed to the arrangement due to the larger firm's experience with carrier-based aircraft, and awarded the production license on June 27, 1975. In 1976 McDonnell Douglas–Northrop won over the General Dynamics–LTV team, which had entered an F-16 development. Although pressured to adopt a version of the F-16, the navy resisted, maintaining that the twin-engined F-17 was better adaptable to carrier operation.

McDonnell Douglas developed the F-17 into a somewhat larger version that would be produced as the F-18 for the fighter mission and the A-18 for the attack mission. The advanced GE F404 engines promised long range and high performance. The F-18 prototype flew on November 18, 1978, and the two variants were soon combined as the F/A-18 Hornet. Eventual navy and Marine Corps requirements were for 1,377 aircraft. The navy hoped export orders would control unit costs by extending production, but in an all-too-common occurrence, the F/A-18 became unprecedentedly costly, as well as escalating in weight and complexity.

This history also was shared by the F-16, as the air force transformed the original air superiority fighter into a "gold-plated" multimission aircraft, with nuclear-weapons delivery capability and ever-increasing weight. The cost also became a political issue in NATO, as General Dynamics had given the Europeans a not-to-exceed unit price of just over $6 million, an unrealistic figure originally, but one that became totally subsumed in the reality.

Controversy and Scandal
The major structural, financial, and competitive problems of the aircraft/aerospace industry, which had always existed but had been studied and debated intensively since the early 1960s, still were unsolved. The aerospace/defense industry also was marked with the unsolved problems of the procurement or acquisition process. Neither the government nor the industry appeared to be capable of solving issues of contract procedures, terms, costs, competition, technological risks, and finances. Cost overruns and development problems seemed to be the norm.

Against that backdrop, two major threats to the status of the industry emerged almost concurrently. The first, a continuation of controversies over contracting problems and the industry-government relationship became more critical and urgent by sharp declines in defense spending and in new programs after 1970. The second involved widespread and damaging disclosures in the export field, then rising in importance. The environment thus became even more volatile.

The acquisition process already was widely viewed as a costly failure, exemplified by the F-111, C-5A, F-14, and Cheyenne, as well as by several missile, shipbuilding, and army land weapon programs. The NAR Autonetics Division was a particular target for its technical problems and major cost overruns with the F-111 avionics suite and the Minuteman guidance system, and it was realized that the air force could have purchased two privately developed 747s for the price of one C-5A. Industry bore the brunt of criticism, but organized labor also shared in the finding of excessive costs. Fitzgerald noted that aerospace wage increases were higher than those for industry overall, leading to locked-in cost increases.[7] Yet the system still resisted change.

One major flaw, increasingly evident with the decline in major weapons programs, was that the relationship of the industry with the military was at something less than arm's length. Industry influence in development of military RFPs was viewed with increasing suspicion. The management problem remained that the military and the industry were charged with cooperating closely while still maintaining that arm's-length relationship. General Stewart, during the eventful 1970–76 period in which he addressed major problems with the F-111, C-5A, and the Short-Range Attack Missile (SRAM), as well as managing the A-10, F-15, F-16, and AMST programs, regarded his inability to reform the source selection process as his major failure.[8] He considered RFP procedures to be unduly complex, riddled with special provisions for small business, minority contractors, and others. In particular he deplored that he would read fifty pages of an RFP and still not understand what the air force wanted in a new system.[9] He regarded the military bureaucracy, with its excessive input from all functional areas, rather than any undue influence of industry, as his greatest frustration.[10]

To many, cost overruns were the product of the failed system, not its cause. The Laird-Packard regime, abandoning the TPP policy in 1971, reemphasized contractor cost reimbursement but attempted to achieve a more equitable distribution of risk between contractor and government. New procedures by no means reduced pressures on contractors, however. The government had long recognized a symbiosis in that it was dependent on the industry both for technology and production, while at the same time holding the power of life or death over individual firms. But with fewer, yet larger and riskier, programs, this relationship resulted on balance in a more threatening industry environment, in fact moreso than at any time since before the Second World War. David Packard acknowledged that poor Pentagon management and contractor overoptimism had combined to encourage contractors to undertake more than they could deliver.[11] Packard left his post in 1972 after a valiant attempt at reform, frustrated by the intractability of the system, and later expressed regret that he had permitted the services greater autonomy in developing weapon systems requirements.

The aerospace industry became a particular target of those more oriented toward government programs other than defense. The familiar allegations of inefficiency and lack of cost control were increasingly accompanied by allegations of unsavory political influence and unethical conduct. To many, it appeared that firms were earning huge profits on costly, unreliable products. Countering this view was the fact that those same firms were constantly undertaking major diversification to reduce their dependence on those defense contracts on which they were allegedly earning such profits.

Aerospace problems certainly were less than unique when considered in the wider context of the troubles of the entire nation. The 1970s brought the Vietnam debacle, the Cambodian genocide, the Watergate scandal, the energy crisis, high inflation and significant economic dislocation, an increasing Soviet military threat and a perceived broad decline in American power and influence, accompanied by declining industrial competitiveness. Yet while it might be argued that aerospace/defense was not more culpable than other sectors of the economy, it was still held responsible for cost overruns, delays, and other inefficiencies.

Some still felt, however, that the industry should be held accountable for the same level of efficiency as were other, nondefense businesses. Thomas V. Jones of Northrop, although enmeshed in ethical problems at the time, was almost alone among aerospace executives in advocating a largely free-market, entrepreneurial approach in the marketplace. In a 1976 statement, in noting that some 80 percent of major military programs suffered cost overruns, he advocated holding defense firms to the same standards of efficiency as other firms and felt that they could be in fact as efficient.[12] Later, in testimony before the Joint Committee on Defense Production, Jones called for greater budget discipline by the government and establishment of more realistic goals in pursuit of procurement reform.[13]

Cost overruns, while associated in the public mind almost exclusively with military programs, affected many areas of government. Urban mass transit, the DC Metro system, energy, the Veterans Administration, and highway programs suffered cost overruns to a greater degree than did most defense programs.[14] The military budget and contract procedures, however, forced management to focus almost solely on cost overruns, with little consideration of future markets, industry structure, or overall strategy, hampering efforts to attain greater efficiency.

It was also evident that the impact of contracting problems, including cost overruns and reactions to their disclosure, extended well beyond the government and the industry. Societal reaction to threatened programs could become even more visceral. In a widely reported episode, the veteran Lockheed engineer Henry Durham resigned his position at Marietta in 1968 and went public with allegations of mismanagement and production deficiencies in the C-5A program. With his charges coming at a time when the C-5A already was under fire in Congress, Durham immediately found himself a pariah in the community, the isolation even extending to the Durham family church. His family was shunned and received numerous threats of violence and even death.[15] It became clear that his neighbors and former Lockheed associates reacted to his charges not as taxpayers or citizens but only as individuals who felt their jobs and the local economy threatened. The motivation to cover up or ignore problems extended beyond those directly responsible.

The industry had always been vulnerable to overall economic, political, and national security conditions, shifts in defense strategy, and changes in technology. Thus it experienced periods of high activity and other periods of retrenchment. Further, the industry was cyclical in all sectors, military, space, and commercial. As sales and employment dropped in 1970, the consensus was that capacity was excessive, that too many firms existed for anticipated business, and that the industry must contract.[16] Most pronounced was the drop in military procurement after the Vietnam War: the defense budget declined in real terms from 1973 to 1979.[17] Apollo 17, the final moon mission in December 1972, signaled a long decline in the space program. Aerospace thus was more vulnerable to downturns than other, more commercial, industries.

The defense economist Gansler formulated eight salient characteristics of the defense (as opposed to specifically the aerospace) industry.

1. The extremely cyclic nature of defense procurement
2. Lack of structural planning (perhaps due to confidence in the free-market economy)
3. Inadequacy of industrial-preparedness planning
4. Lack of actual industrial readiness
5. Importance of technology and research in defense
6. Differences among the industries that make up the industrial base
7. High concentration within industries
8. Dependence on international military assistance[18]

These characteristics applied to aerospace firms, as major defense contractors, to a high degree. The immediate problem in the post-Vietnam era was that, in addition to the pressure on overall defense spending, rising personnel and R&D costs were at the expense of procurement.[19] The industry was pressed to make up the shortfall and found a partial salvation in military exports. The Foreign Military Sales (FMS) program rose from $1.5 billion to over $14 billion in the first half of the 1970s.[20] The industry strongly supported the FMS program, of course, since exports enhanced both international competitiveness and domestic stability, spreading R&D costs and manufacturing overhead. Further, without such sales, foreign competitors would make inroads in markets traditionally dominated by Americans. In 1968 exports constituted only 10 percent of overall U.S. military procurement but rose to 60 percent by 1976.[21]

Yet while stabilizing the industry, military exports grew increasingly controversial. The 1970s marked the height of a global arms race, drawing competition from the Soviet Union, Britain, France, and newer exporters such as Israel and Brazil. But the U.S. government, while stressing arms control, especially during the Carter administration, frequently worked at cross-purposes, with arms transfers included in many international agreements, as between Egypt and Israel. The International Logistics Negotiations (ILN) office was succeeded by the Defense Security Assistance Agency (DSAA) as the chief military export agent.

In the volatile environment of the 1970s, aerospace and military exports involved not only the Departments of State, Defense, and Commerce, but also required congressional approval. Certain transactions were made contingent on human-rights progress or foreign-policy stances in the recipient country and could be proscribed due to concerns over advanced technology falling into hostile hands. American combat aircraft of Iran, including the F-14 exported during the regime of the shah, came under the control of the ayatollah in 1979. At one time Congress banned exports of combat aircraft to Latin America entirely. American firms looked on in frustration as French and British companies booked orders instead.[22] President Nixon then rescinded the ban, to the relief of the industry, on June 5, 1973. But President Carter later reimposed restrictions. Battles ensued over F-15 and AWACS sales to Saudi Arabia, which drew vehement opposition from Israel, and over a proposed fighter sale to Taiwan, which ended due to protests from China.[23]

Efforts to develop combat aircraft specifically for countries with modest financial resources met with mixed success. The Northrop F-5 Freedom Fighter and the improved F-5E Tiger II enjoyed strong export orders with government backing, but recipient countries increasingly insisted on the latest models adopted by U.S. forces rather than less sophisticated and less costly designs, which they tended to consider second-rate. Prestige became a stronger motivation in weapons purchases than cost-effectiveness. Further, when those aircraft were not adopted by the originating country, marketing seldom proved successful. Lack of a French air force requirement was a negative for Dassault with its turbofan-powered Mirage development for the NATO order.

The industry viewed the international arms trade as a commercial opportunity on one hand but as a formidable economic threat on the other, in that it could lose its position not to free-market competitors but to vagaries of policy. Arms exporters, including aircraft firms, defended their efforts with the axiom "If we don't sell, someone else will," which while widely derided, was never proved invalid. An added imperative for the industry was that, other than Boeing, it was heavily dependent on the military market. Accordingly, it was never realistic to expect individual firms to restrain arms exports voluntarily. Controls on the arms trade could only succeed by international agreements, with effective enforcement, and most obviously including the Soviet Union. But such progress was not then in prospect.

With a single-buyer, or monopsony, structure, the aircraft/aerospace industry could never operate in a free-market environment. The issue was how the Department of Defense could best utilize the structure that existed.[24] Illustrating its influence over the industry, some one-third of manufacturing space, much major equipment, and all repair depots remained government owned. The defense industry was a "sick" industry in the sense that it was permanently in excess capacity, largely due to occupancy of government-owned plants, and had little flexibility to reallocate resources to other, faster-growth areas. In the mid-1970s the aircraft industry was operating at approximately 55 percent of one-shift capacity.[25] Further, while the industry was tagged with the inefficiency label, an efficient contractor with high financial returns would find itself facing criticism as well. Contract negotiators tended to believe they protected the public interest when contractors' financial returns were low.[26]

More disturbing, American defense contracting bore certain similarities to the widely disparaged European system of "lemon socialism," in which high-cost, inefficient private-sector companies were bailed out or acquired by their governments simply to preserve employment and the industrial base. Losses thereby were socialized, while profits remained privatized. In effect, inefficiency was rewarded, while efficient firms were penalized in that they did not receive the financial and political assistance extended to the failing firms. Thus government rescues of defense contractors were deeply troubling. There was indeed spirited competition by major firms for new contracts, but if all were high cost and inefficient, no cost saving would result. In addition, after contract award there was no alternative for a given aircraft line. For example, there was only one B-1 and only one firm to produce it.[27] After unsatisfactory performance, the government could not switch to a different aircraft or a new supplier. Further, a major contractor usually could negotiate a bailout, as with Lockheed's "restructured" C-5A contract announced on May 31, 1971.[28]

Socially, giant defense contractors were "too big to fail," as the largest commercial banks became regarded in the banking crisis of the 1980s. Consequently, cancellation power was seldom exercised by the Department of Defense, and renegotiated contracts suggested that failure was being rewarded. Defense would request more money from Congress; accommodation with existing contractors was the order of the day.[29] The contractor possessed strong leverage in that the military seldom would face the calamity of outright cancellation and of beginning the process again. Further, contract incentives, while well intentioned, could themselves become counterproductive: an improvement made during development could conflict with the incentive formula, requiring renegotiation.[30] Contractor leverage was not unlike the situation faced by banks when a major borrower threatened a default, which could in turn pull the lender into insolvency. In the competition award phase, the Defense Department was stronger, but afterward the contractor became stronger as the sole source.

Even as reformers deplored the system, others viewed the level and nature of defense spending, efficient or not, as beneficial to broad economic prosperity. Companies experiencing development difficulties continued their behavior of stall, delay, and denial until the program could reach a stage where it could not easily be canceled. Then the resulting call for a financial bailout would be justified more on economic grounds than military necessity. Fitzgerald became convinced on just such grounds that inefficiency was de facto national policy.[31] Aerospace contractors could hardly be blamed for taking that approach, since it had proved effective. In the debates over the C-5A and the L-1011 loan guarantee, senators from both California and Georgia openly argued the case on the basis of the thousands of jobs at stake. Representative L. Mendel Rivers, chairman of the House Armed Services Committee and procurer and protector of numerous military installations in South Carolina, also was a defender of the C-5A up to his death in 1970. The first squadron was to be based in his district, further reinforcing his campaign slogan, "Rivers Delivers."

In pursuit of greater efficiency, Gansler recommended emphasis on maintenance of efficient production rates, probably with fewer types in production.[32] Debate continued on the structure of the industry and its relationship to the government. While there was a virtual continuum of alternative structures for the defense industry and defense market, major alternatives appeared to be

- The free-market oligopoly extreme
- Limited competition in a concentrated industry
- The nationalization extreme[33]

Regarding industry structure, Gansler concluded,

> Clearly, traditional theory of economic behavior in a free market does not apply to the defense industry, and thus many of the actions that have been taken under the assumption that the theory is applicable have not had the desired effects.[34]

A subsequent question concerned the relationship between design and development and series production. Some advocated separation, since a contractor had a vested interest in a program's continuation regardless of cost and technical problems or changing military requirements. Even the Soviet model, with a powerful Ministry of Aviation Industry, the Central Aerodynamics and Hydrodynamics Institute (TsAGI) for research, several independent design bureaus, and unaffiliated state-controlled production plants, seemed to have merit.

Increasing the frustrations felt by those who attempted to increase efficiency and deliver value for the taxpayers was another counterproductive practice of the congressional-industrial complex: the sheer inability of the military to terminate a production program even when it wished to do so. For years, as the F-111 and A-7 programs ran down, the defense budget excluded requests for further procurement. But Congress persisted in restoring funds for both types, both built in Texas, for low-volume production, only twelve per year. Thus production extended years beyond planned completion, to 1982 in the case of the A-7, and in excess of requirements. Ostensibly motivated by the need to preserve the production base so that in the event of war output could be expanded rapidly, in reality the extensions reflected local political pressures. Congressional hypocrisy reached its height in this area, in that many who called for industry efficiency and denounced military waste fought to keep unneeded programs affecting their districts alive when scarce funds could be spent more effectively elsewhere.[35] The pork-barrel approach transcended party and ideology.

While achieving consensus on reforms was as difficult as ever, proposals were developed that were intended to benefit all, the military, the industry, and the taxpayer. Gansler recommended several changes.

- Reduced reliance on detailed performance requirements in developing specifications
- Greater separation of development and production, with better testing before a production decision
- Reduced emphasis on exotic subsystems
- An increase in autonomy and flexibility of program management[36]

The acquisition system unquestionably tended toward excessive centralization, which almost inevitably led to unintended consequences. The major problem of procurement management thus was organization. The obvious solution was to decentralize, but that remained unlikely given the billions at stake in each major program.

The magnitude of defense contracts, with their long-term economic implications, meant that the acquisition process became even more politicized, with contract battles carried to the halls of Congress and even to the president's office. Aerospace prime contractors knew that while winning a contract also meant acceptance of high risks that might doom the firm, failure to win a contract would mean total withdrawal from the field. Consequently, firms still competed fiercely on the chance that they could survive and prosper for another aircraft generation. One conclusion emerging from procurement reform efforts and closer scrutiny of defense contractors was that the reforms did not change the *behavior* of the participants.

The second threat to the American aerospace industry's traditionally strong competitive position emerged almost concurrently with its domestic stresses. The export market, in which the American industry had long enjoyed overwhelming advantages and to which it attached increasing urgency, became increasingly competitive, and in commercial as well as military markets. The principal damage to the industry, however, came not so much from any competitive losses but in the industry's behavior. Most aerospace firms responded, especially from the beginning of the 1970s, at some point and to some degree, in a manner inconsistent with acceptable marketing practices and ethical standards. Those included certain practices that were illegal under U.S. law, and others that led to laws making them illegal thereafter.

While many companies engaged in government contracting and many industries were faced with intense international competition, aerospace firms felt more seriously threatened than most, due in no small part to their relative lack of diversification. While the pressures were understandable, the industry's responses became indefensible either ethically or legally, seriously damaging its reputation.

International scandals were preceded by exposures of illegal corporate contributions to presidential campaigns by many aerospace firms. There were also exposures of improper lobbying of influential officials, including questionable entertainment expenses included in contract billings. The Watergate investigations from 1972 revealed a pattern of violations. As with other violators of campaign-financing laws, the aerospace industry felt these contributions necessary to buy or retain influence in Washington, both for future defense contracts and for arms exports.

Illegal campaign contributions involved such a broad spectrum of American business that aerospace was not singled out for opprobrium, but far more damaging were the exposures of questionable or illegal payments in connection with foreign sales. As with the record of the industry during the Second World War and with the military-industrial complex and procurement issues from the 1960s, the payments scandals of the 1970s have been exhaustively researched and analyzed. Rather than retrace the entire history, it will suffice to summarize key events with a view toward understanding the environment in which the industry operated and the impact of the scandals.

International business practices defined by the United States as corrupt, chiefly involving bribery and extortion in large commercial transactions, many directly involving governments, extended over many industries and product categories. In fact, the scandal first came to public attention with the suicide of United Brands chairman Eli Black on February 3, 1975, as questions were raised concerning that food company. The Securities and Exchange Commission (SEC) then required those public corporations with international sales to conduct internal investigations to detect misapplications of funds.

While payment of foreign bribes was not then illegal under U.S. law, questionable payments nonetheless were of concern to such agencies as the Federal Trade Commission (FTC), the GAO, and the IRS, as well as the SEC, especially if involving government-guaranteed funds. These concerns spurred widespread investigation, first by the Senate Subcommittee on Multinational Corporations, under Senator Frank Church, which began in June 1975. Questionable or improper foreign payments by many American corporations were disclosed, but aerospace firms were particularly prominent, and the reputation of the industry suffered more than most. Payments included inducements to foreign governments to order aircraft from a particular company and inducements to order aircraft they would not otherwise have ordered. Investigation also disclosed that, rather than involving exceptional cases in extraordinary circumstances, there was a well-established, pervasive pattern of questionable foreign payments, sometimes referred to as "grease."

The juxtaposition of the international arms race with the disclosures of widespread international corruption put the aerospace industry in a still more unfavorable light. Further, bribery for orders in poorer countries resulted in egregious misallocations of resources. Countries that might better have emphasized constructive economic development instead built unnecessary military machines, largely with imported equipment, and often in regional competition.

Of all the firms besmirched by the bribery and extortion disclosures, Lockheed, regarded as a "worldly" company, was the most severely damaged. Investigation revealed that improper or questionable payments by Lockheed to foreign officials extended back to the 1950s and were a factor in the F-104 sale to NATO. Within Lockheed, as in others, the presumption existed that securing an export contract almost necessarily involved a payoff somewhere in the process. Lockheed had first cultivated Prince Bernhard of the Netherlands during the days of Robert Gross. KLM Royal Dutch Airlines, on whose board Bernhard served, was a major customer for the Super Constellation. Lockheed was aware that Bernhard, despite the wealth of the Dutch royal family, did not possess a personal fortune and was in fact on a quite restricted allowance.[37] Further, he had a taste for high living, including certain tastes not known to the Dutch public. Lockheed's payments to Bernhard, through a Swiss bank, began in 1960. Disclosure of payments totaling some $1 million and their linkage to the Dutch F-104 order led to a major scandal within the Netherlands and added to Lockheed's troubles.[38] Lockheed employed other well-connected European consultants, but nothing was ever proved, although widely suspected, of illegal payments to the largest F-104 buyer, West Germany.

Some payments were defined as extortion, in which an influential foreign official would demand payment in order to insure a sale or prevent its cancellation, raising the question as to who was corrupting whom.[39] Lockheed's notoriety was due in no small part to the prominence of the individuals involved, which included Prime Minister Tanaka of Japan as well as Prince Bernhard. The disclosures also came after Lockheed's contract difficulties with the C-5A and only four years after its survival had been preserved by the government loan guarantee, which was still in effect. That juxtaposition triggered widespread anger.

Further, there was a generally unrepentant attitude among top executives, including chairman Daniel J. Haughton, who emphasized that such payments were not illegal at the time, and that Lockheed also had been fighting for its very survival. But bribery had also occurred in cases in which foreign competition was not a factor. The TriStar sale to All Nippon Airways in August 1972 occurred when the United States held a monopoly on wide-bodied airliners, and Lockheed's competition was not foreign but the McDonnell Douglas DC-10. A political sidelight was that Prime Minister Edward Heath also promoted the Nippon sale, because of his country's stake in the TriStar's RB.211 engines.

Haughton had entrusted most international business to president A. Carl Kotchian and in 1972 ordered Kotchian to Japan, to remain as long as necessary and do whatever was necessary to gain an L-1011 order. What proved to be necessary was some $14 million in payments to Japanese agents and officials.[40] Kotchian regarded such payments in the same context as insurance, as a business expense with return on investment as the decisive criterion.[41] The eventual disclosures carried another high cost to Lockheed, however, as Japan canceled a large order for the P-3C.

The most severely damaged company after Lockheed was Northrop. Investigation indicated that Northrop was even more aggressive than Lockheed in offering payments for orders and did not merely respond to solicitations or extortion. Further, Northrop was financially sound during the period in question, unlike Lockheed, and had enjoyed the strong support of the Defense Department in its exports of the F-5 series. That program led to healthy profits but reinforced Tom Jones's view that the key to future success and survival was political connections.[42] Jones made the investment in Fokker due to the presence of Prince Bernhard on the board and became a Fokker director himself. Northrop actually led the era of scandal in aerospace with its conviction in May 1974 for illegal contributions, totaling $476,000, to the Nixon campaign.[43] Although Jones surrendered his position as president under pressure after the conviction, the most troubling aspect of the affair was that he survived as chief executive, only to generate later ethical controversies. Boards of directors, stockholders, and creditors in other corporations, including Lockheed, demanded stricter accountability, and other companies affected by the bribery scandals dealt with executives severely. Boeing, McDonnell Douglas, and General Dynamics were also revealed to have engaged in questionable payments, although not to the extent of Lockheed and Northrop.

Again, political corruption and international scandals involved a broad cross-section of American business, but the involvement of major aerospace and defense contractors perhaps was more troubling than others. Such firms were widely regarded as quasi-public utilities, as North American in particular had been earlier, and thus were felt subject to tighter regulation and higher standards than other firms or industries. Yet in another sense, as with the problems of the acquisition process, it was argued that the conduct of aerospace firms, both in defense contracting and export, was a manifestation of a system that they had not created. The government has remained the most influential factor in the development, structure, and policies of the industry.

Attempts to rationalize improper payments abroad emphasized that refusal to do so simply indicated American naïveté in other environments, and that such payments were necessary if the markets were not to be surrendered to foreign competitors. One ironic conclusion of the Senate investigations, however, was that most payments probably were not cost-effective. Many contracts might have been won or preserved without such payments, and many American aircraft had no direct foreign competitors. But willingness to pay bribes or extortion represented a basic insecurity on the part of the aerospace firms. Executives and salesmen were not necessarily predisposed to corruption, but the pressures were similar to those of the campaign contribution scandals, in that management was too fearful of the consequences of *not* making such payments.[44]

With hindsight the entire bribery-corruption scandal need not have occurred, or not to the extent to which it did, at least in the aerospace industry. While certain contracts might have been lost, over the long run the export position of most firms would have remained fairly strong without payments. A difference with other industries, however, was in the sheer clumsiness of payment procedures: aircraft firms generally lacked discretion and finesse. Finally, the atmosphere conducive to such payments could be linked to the industry structure. With few active programs and few major contract prospects, firms felt highly vulnerable with the loss of any potential order. None felt secure enough to simply shrug off the loss of a major military or commercial export order and go on to the next. But many lost such orders and still survived. Yet the structure precipitating that problem was not by design of the government or the industry; as the researcher Phillips had developed, the structure was dictated by the nature of science and technology.[45] Science and the engineering effort required in turn dictated fewer, larger firms, and fewer, more expensive aircraft programs.

The legislative remedy was the Foreign Corrupt Practices Act (FCPA), signed into law by President Carter on December 19, 1977. The law defined corrupt practices, imposed detailed reporting requirements, and provided criminal penalties for violations.[46] The law did, however, recognize that in certain countries facilitating payments were necessary to enable previously negotiated contracts to be fulfilled and were therefore exempted.

Global extension of U.S. law (extraterritoriality), which had not previously dealt with foreign bribery, caused considerable controversy among legal experts as well as within the aerospace industry. Many pointed out that fees or commissions to foreign sales agents and "consultants," or outright bribery, was long accepted. It was also feared that imposing American standards of conduct would serve only to concede future orders to European competitors, unrestrained either by law or principle, further damaging exports. The purpose of the FCPA was not to enhance competitiveness, however, but rather to upgrade the image and reputation of American business, already seriously damaged by Watergate revelations. While beyond the scope of this book to debate its effectiveness, one respondent to a survey of business executives about the Act stated the question succinctly:

> To the extent the Act sought to reform business practices around the world, it is a naive failure. To the extent it seeks to make U.S. executives aware of their special responsibilities and immense influence in the international business community, it is a success.[47]

Industry Survey, 1967–79

Despite its unquestioned importance and spectacular technical achievements, especially in the space program, the aerospace industry endured a highly troubled period from the later Vietnam era. It ended the 1970s with a declining aircraft production base and as the object of undiminished public suspicion. Yet fortunes varied widely among firms.

Fairchild Hiller continued attempts to compete in combat aircraft through its Republic Aviation Division. An R&D contract for a supersonic VTOL fighter, in cooperation with the German EWR consortium led by Messerschmitt-Bolkow-Blohm (MBB), was awarded in 1966. The aircraft was to have two propulsion engines and four retractable lift engines but was canceled by outgoing secretary of defense McNamara in February 1968, ostensibly because the U.S. lacked a requirement.[48] The firm then was a finalist for the F-15 program. While the loss also was a heavy blow to NAR, Fairchild Hiller believed that its loss was primarily due to political considerations. If it won the contract, meaning production at Farmingdale, the military's two main fighters would be built on Long Island, believed to be politically unacceptable as well as militarily vulnerable.[49]

The firm became Fairchild Industries, Inc., in 1971, after the death of Sherman Fairchild. Edward Uhl continued as chief executive. The Republic Division, which became the Fairchild Republic Company in 1973, continued to focus on military aircraft. Success came with the A-10 that entered production in 1975, ending a ten-year gap at Farmingdale.

With its VTOL experience Fairchild also studied small VTOL airliners but did not proceed with development. Fairchild acquired 90 percent of Swearingen Aviation Corporation of Texas in 1971, believing that it would be able to capitalize on its experience with local-service airliners. The Metro II, developed from earlier Swearingen models, was certificated under FAR 23 for third-level airline operation and achieved a strong market share.

Production of the Fairchild F-27/FH-227 series ended in 1968, although the last delivery was not made until 1972. Total production by Fairchild was 128 F-27 and seventy-eight FH-227. The end of the program ended aircraft production in Hagerstown and also ended American efforts in the local-service field, as Fairchild decided in June 1968 not to develop the FH-228 twin-turbofan airliner based on the Fokker F28. Fairchild sold the Hiller operation and design rights back to Stanley Hiller in January 1973 after production of some 250 FH-1100 civil models. The A-10 and Metro kept Fairchild in aircraft, but subcontracting was a major part of total business. Uhl proposed a merger with Bunker-Ramo Industries of Chicago in 1979, but the proposal encountered resistance, and Fairchild remained independent.

Reflecting its current activities, Grumman reorganized in 1969 as the Grumman Corporation, with four operating divisions. With continued diversification, the number of operating divisions rose by 1973 to include important positions in allied industries, including boats and truck bodies, data systems, and ecosystems.[50] It acquired American Aviation of Cleveland, a successful builder of small general-aviation aircraft, on January 2, 1973, and created Grumman American Aviation Corporation (GAAC). Military production remained active, as the A-6 and E-2 underwent significant updating: both designs dated from the late 1950s. The definitive E-2C Hawkeye first flew on January 20, 1971.

Grumman won the VFX competition over strong competition from McDonnell Douglas, General Dynamics, North American Rockwell, and an LTV-Lockheed team. Grumman's reputation with the navy was undamaged by the F-111B affair, and it appeared poised to regain its position as leading fighter supplier. The F-14 prototype first flew on December 21,

1970, ahead of schedule, but a looming crisis finally tarnished Grumman's reputation for quality and efficiency.

Grumman's troubles with the F-14 were those of a declining business base over which to allocate company overhead, an inflation rate far higher than that built into the contract, and technical and performance problems with the TF-30 engines. By the early 1970s the F-14 represented the bulk of Grumman's sales and was also the most complex and advanced project the company had ever undertaken. Losses in 1971 and 1972, with post-Vietnam annual sales falling from $1.2 billion to $800 million, left it weakened.[51] By mid-1972, with costs escalating, Grumman informed the navy that fulfilling the contract for production lots 1–5, covering 134 aircraft, would severely cripple the company, and that production of the full requirement of 313 F-14s would result in a cumulative loss of $400 million, far more than the company's net worth.[52] Grumman had lost some $1 million on each F-14 delivered to that point at a fixed price of $11.5 million and then stood to lose over $2 million on each lot 5 unit. Unit cost was estimated to have risen to $16.5 million. The situation reached an impasse. Grumman simply refused to build lot 5, forecasting a loss of $105 million in addition to the $85 million already incurred on lots 1–4, covering 86 aircraft.[53] Said Clinton Towl: "We will close our doors if we have to. We can't proceed."[54]

Secretary of Defense Laird was critical of the F-14 contract, written under TPP procedures, but the navy still threatened Grumman with breach of contract on lot 5.[55] An extended and painful renegotiation preserved the program but resulted in the F-14 becoming by far the most expensive naval aircraft ever built. Grumman emphasized during the renegotiations that the F-15 contract did not contain restrictions as had been applied to the F-14, but the F-14 became the focus of another debate over contracting and industry cost controls. The navy and Grumman shared the total blame, as both were too eager for the program, undertaken with overoptimistic assumptions. The TPP system also was then in disrepute, with the realization that it was impossible for any contractor, no matter how cost conscious or competent, to forecast development costs accurately for a system breaking new technological ground.

As renegotiations neared completion, resulting in still higher prices, Grumman president Lew Evans died of a heart condition on July 9, 1972. Clinton Towl was forced to take over the burden of renegotiation. John C. Bierwirth, with long financial and industrial experience, succeeded Evans as president. Bierwirth then became chief executive on December 31, 1974, with Towl continuing as chairman. Joseph G. Gavin Jr., who had led the LM program, later became president.

Grumman returned to profitability in 1973 as the F-14 entered service. The order for eighty F-14s from Iran in January 1974, gained with the personal support of President Nixon, represented a timely financial coup for Grumman.[56] Grumman still was far from stable financially, however, as banks had not restored credit lines suspended in 1972, and the company effectively became a ward of the U.S. Navy.[57] Advance production payments were cut by the Senate, and a proposed loan from the Iranian state bank was vetoed by the navy, forcing the firm to seek added private financing.[58]

Since the VFAX-2 version of the F-14 could not proceed due to costs, the F-14A with its troublesome TF30 engines continued. Cost escalation of both the F-14 and its Phoenix missile also continued, with inefficient production rates adding to the continuing inflation and overhead problems.[59] But once beyond its F-14 contract problems, and with private financing restored, Grumman regained its status as a successful aircraft producer. Bierwirth established a goal of 25 percent of sales from civil lines of business, but diversification would still frustrate the company.[60] In another strategic shift, the general aviation subsidiary, including the successful Gulfstream, was sold on September 1, 1978. Grumman continued to subcontract on the Gulfstream.

TABLE 20. Largest Aerospace Manufacturers, 1979

Industry Rank	Firm	Fortune 500 Rank	Sales (in millions)	Employees
1	United Technologies[a]	26	$9,053	197,700
2	Boeing	29	8,131	98,300
3	LTV Corporation[a]	31	7,997	68,125
4	Rockwell International	45	6,466	114,452
5	McDonnell Douglas	54	5,279	82,736
6	Lockheed	82	4,070	66,500
7	General Dynamics[a]	83	4,060	81,600
8	Textron[a]	100	3,393	64,000
9	Northrop	204	1,582	28,800
10	Grumman	216	1,476	28,000
11	Cessna	301	939	18,956
12	Fairchild	354	718	13,082
13	Beech	396	608	10,533
14	Gates Learjet[b]	593	302	4,471

Source: "The Fortune Directory," *Fortune*, May 5, 1980, 274–99. (© 1980 Time Inc. All rights reserved.)
  a. Diversified. Included for airframe-manufacturing activities.
  b. From *Fortune* Second 500 Tables.

Boeing, while enjoying a dominant commercial position, and with major stakes in other aerospace sectors, carried out a major internal restructuring from 1968. Major divisions resulting were the Aerospace Group (responsible for the Minuteman ICBM series and the Saturn launch vehicle), the Commercial Airplane Division, the Vertol Division, and the Wichita Division. Boeing had eliminated its Military Airplane Division after loss of the CX-HLS contract, reviving the Wichita Division. But in 1969 a new Military Aircraft Systems Division was formed, largely to support the large B-52 and KC-135 fleets. The Commercial Airplane Division, under the management of E. H. "Tex" Boullioun since 1967, was the largest in the company.

Chief executive T Wilson faced a major financial and operating crisis linked to the 747. Early orders by airlines fearful of being left behind competitively required a high initial production rate for the 747, in turn requiring a massive investment. Company profits declined to only $10 million in 1969. The order drought from 1969 to 1972, continuing weight growth, and pressure to install uprated engines were further cost complications. In addition, 727 orders dropped sharply, while the 707 backlog disappeared entirely, although the type was still available.[61]

Despite a strong overall position, Boeing had suffered major reverses with loss of the TFX, C-5A, and SST programs. Further, there was no question that the company, with a 1968 employment level of more than 120,000, was overstaffed and inefficient, and that major downsizing would be necessary.[62] Wilson eliminated more than twenty-five thousand jobs in 1969, followed by forty thousand in 1970, causing a major economic crisis in Seattle, where Boeing employment, which reached a high of 101,400 in mid-1968, was reduced to 38,000 by June of 1971.[63] Recovery began in 1972, but the downsizing was clearly necessary to save the company. The long-term success was that Boeing became an even more efficient producer.

Ironically, Boeing's loss of both the CX-HLS and SST ultimately worked to its benefit. The C-5A resulted in a major crisis for Lockheed and the SST would have had no market even if successfully developed. In addition to escaping almost inevitable major problems, Boeing was forced to concentrate its resources on more productive and remunerative endeavors. Contracts for the AWACS and for the SRAM also greatly aided Boeing's recovery.

A further reorganization completed on December 19, 1972, resulted in the formation of the largely autonomous Boeing Commercial Airplane, Boeing Aerospace, and Boeing Vertol companies. A new eight-hundred-thousand-square-foot plant opened in Winnipeg, Manitoba, for production of 747 components. William M. Allen retired as chairman of the board on September 29, 1972, after twenty-seven years of leadership. T Wilson became chairman while continuing as CEO, and Malcolm Stamper became president. For years Boeing had ranked among the most admired American companies, but its reputation was besmirched in the international scandals. Allen had long resisted the employment of foreign agents or consultants, but Wilson and others began to view them as necessary, resulting in damaging disclosures.[64]

A new Boeing Military Airplane Company was formed, including the Wichita Division, on October 23, 1979, for production of subassemblies and still providing support for the aging B-52 and KC-135. Boeing Aerospace remained responsible for E-3, E-4 and E-6 developments. A total of thirty-four E-3 AWACS aircraft was exported to NATO for joint operation, and to Saudi Arabia, Great Britain, and France.

General Dynamics continued the dubious history of managerial instability and strategic upheavals that had characterized its aerospace component since the Convair acquisition. Henry Crown, supported by Nathan M. Cummings of Consolidated Foods in Chicago, controlled 18 percent of GD common stock by the end of 1969. Initially Crown attempted to negotiate a working agreement with Roger Lewis but insisted that he relinquish the office of president. Lewis tentatively agreed but resisted giving up executive authority and as a result was removed as CEO by Crown in October 1970.[65] Crown sought a more commercial orientation for the company, but his original choice as CEO, Semon "Bunkie" Knudsen, formerly president of Ford and son of wartime production chief William S. Knudsen, declined. David S. Lewis, unrelated to Roger Lewis, then was recruited from McDonnell Douglas as CEO despite his strong military orientation. It was believed that David Lewis insisted on relocation to St. Louis as a condition of his acceptance, and General Dynamics announced the move on February 10, 1971.[66] Aircraft production remained centered in Fort Worth. Lewis's experience at McDonnell Douglas was an advantage in solving production problems at Convair, a key subcontractor on the DC-9 and DC-10 airliner programs.

By December 1970, General Dynamics was divided into the Convair Aerospace Division and the Electro Dynamic Division. Although the firm recognized its vulnerability in being so dependent on defense, it actually increased its dependency with the acquisition of the Quincy shipbuilding operation and later the Chrysler tank plant. After increasing nationalistic pressures, GD sold money-losing Canadair to the Canadian government in 1976 for $38 million.[67] With F-111 production finally ending in November 1976, the air force LWF program and the later F-16 production contract enabled General Dynamics to remain at the forefront of the aircraft and defense fields. Eventual domestic and export orders were to push the F-16 program past four thousand aircraft, making it the largest and most successful aircraft program for General Dynamics (Convair) since the Second World War.

Chairman Daniel J. Haughton and president A. Carl Kotchian led Lockheed during its most troubled period. During the financial crisis of 1971 Haughton explored merger possibilities with such firms as Ford, General Electric, and U.S. Steel. Textron emerged as the strongest prospect but eventually withdrew, and Lockheed continued to struggle as an independent firm.[68] With the C-5A settlement and the loan guarantee, Lockheed's finances eventually stabilized, and its credit lines with twenty-four banks were restored.

In 1972 the Lockheed organization consisted of the headquarters in Burbank and four major divisions, Lockheed California and Lockheed Georgia for aircraft, Lockheed Air Service Company, and Lockheed Propulsion Company. Lockheed Shipbuilding and Construction, Lockheed Air Terminal, Lockheed Electronics, Lockheed International, and Lockheed Missiles and Space were subsidiaries. Lockheed Propulsion, never highly successful, was eliminated in 1975.

While the C-5A cost overrun coverup was damaging for Lockheed, a possibly mitigating factor was that it could not have taken place without the complicity of high defense officials. But it was the airliner program that almost pulled the company into bankruptcy. Naval programs continued successfully, and the YP-3C Orion, the definitive patrol bomber model, first flew on October 8, 1968, and remained in continuous production for over twenty years. Lockheed also won the contract for a next-generation antisubmarine aircraft on April 8, 1969, with the S-3 Viking. The S-3 first flew on January 21, 1972, and was another success.

Even with improving finances, Lockheed remained heavily indebted to its banks, which began to insist on new management. In the board meeting of February 1976, Haughton and Kotchian were forced to resign. Haughton, later credited with having saved the company, departed quietly, but Kotchian protested vehemently, feeling himself a scapegoat.[69] Both were charged by the SEC with violating financial-disclosure rules, but not convicted. Lockheed director Robert W. Haack, formerly head of the American Stock Exchange, became chairman, charged with stabilizing management. Chief financial officer Roy A. Anderson became vice chairman, and Lawrence O. Kitchen became president and chief operating officer, although both were slightly tainted by the payments scandals.[70]

While only an interim chairman, with no aircraft background, Haack was later credited with restoring confidence in the company.[71]

Kelly Johnson had retired at age sixty-five in 1975, the recipient of two Collier Trophies and the Medal of Freedom. Several far-reaching changes emerged from the board meeting of September 1977. Roy A. Anderson succeeded Haack as chairman and CEO, and Courtlandt Gross retired as a director. A name change to Lockheed Corporation reflected its diversification. Perhaps of equal significance, however, was that the government loan guarantee expired at that time, and Lockheed became financially independent again.

Illustrative of the overall decline in aircraft production, Lockheed completed fewer than eighty new aircraft in 1978, with only low-volume production of the L-1011, P-3, JetStar, and C-130 remaining. C-130 production continued, however, with exports and more than one hundred sold to commercial users. Lockheed won a contract to modify the C-141 fleet with provision for in-flight refueling and a fuselage extension from 1975, somewhat alleviating the volume restrictions imposed by the rather slim cross-section. The entire fleet was modified during 1978–82.

Sales of the TriStar gradually improved but still offered no prospect of breakeven. Production of 187 S-3A Vikings ended in mid-1978, but the P-3C Orion continued at Palmdale and under license by Kawasaki in Japan. Lockheed ended the JetStar II in 1979 and began production of the TR-1 development of the U-2. Although highly secret for a decade, Lockheed won the contract to develop the first stealth technology aircraft in 1976. Small test models began flying late in 1977, leading to the operational F-117A. Despite its well-publicized financial and ethical problems, Lockheed's technical capabilities were undiminished.

David S. Lewis became chief executive officer of McDonnell Douglas Corporation in 1970, succeeding James S. McDonnell, who remained chairman. Lewis was to hold that position only briefly, however, moving to General Dynamics at the end of the year. With the departure of Lewis, Sanford N. McDonnell became president and chief executive. He was regarded as more democratic and less autocratic than his uncle, the company founder.[72] John Brizendine, previously DC-10 program manager, became president of Douglas Aircraft on July 17, 1973, succeeding Jackson McGowen. The old Douglas Santa Monica headquarters complex was razed in 1975.

The major program for McDonnell Douglas during the 1970s was, of course, the F-15. Preliminary work on an advanced fighter had preceded the official specification, and the contract award was announced on

December 23, 1969. But the F-15 contract, while incorporating Laird-Packard reforms, still did not eliminate concurrency, and numerous delays and cost escalation resulted. But the supersonic turbofans of the F-15 made it the first American fighter to possess a thrust-weight ratio of more than one to one, enabling it to accelerate while climbing vertically. The F-15 first flew on July 27, 1972, and entered production rapidly.

McDonnell Douglas also held the U.S. license for the British Harrier attack fighter. The advanced AV-8B version was developed under a defense contract, and 342 were ordered by the Marine Corps. Operationally a short-takeoff vertical-landing (STOVL), rather than a true VTOL, aircraft, the first converted YAV-8B flew November 9, 1978.

In part due to efficiency efforts, but principally due to declining production, overall employment declined from almost 140,000 at the postmerger peak to about 57,000 in 1976, although it began a long rise afterward. While maintaining its aerospace focus, McDonnell Douglas continued to diversify into other nonaerospace areas, including solar power, information systems, and commercial real estate development. Commercial aircraft at Long Beach remained important, but unprofitable and troubled. By 1978 major operating units were McDonnell Aircraft, Douglas Aircraft, the Astronautics, Automation, and Electronics companies, the Tulsa company, and the International company.[73]

The two longest-running McDonnell Douglas military aircraft programs, the F-4 Phantom II and the A-4 Skyhawk, finally ended. A-4 production ended on February 27, 1979, with total production of 2,960. The last of 5,057 F-4s was delivered on October 26, 1979. Mitsubishi built another 138 in Japan, bringing total production to 5,195.

Thomas V. Jones gave Northrop the dynamic leadership and strategic direction it needed to become a strong aerospace and high-technology firm. Although never ranking higher than seventh among airframe manufacturers, Northrop built a strong reputation in electronics, missiles, and component production. Commercial diversification lagged, however, the major activity being the manufacture of Boeing 747 center sections, which began in 1966. In some respects Jones evoked earlier aviation entrepreneurs such as Douglas and Fleet but regrettably led the firm into disrepute.

Jones still dominated the company after the disclosures regarding Watergate and international bribery, but a 1974 shareholder lawsuit resulted in restrictions on his authority, addition of four outside directors, and a policy that 60 percent of board members must be outsiders.[74] Later investigations implicated Northrop in bribes paid in Iran, Indonesia, and Saudi Arabia in connection with aircraft sales, whereupon board member and industry veteran Richard Millar replaced Jones as chairman of the board, although not as chief executive. But Jones's dominance was such that he regained the post after a year, with Millar stepping down to vice chairman until his retirement.

Exports of the F-5A Freedom Fighter built up from 1965; then the more powerful and developed F-5E Tiger II flew in August 1972, and also won export orders. F-5 variants were license-produced in Canada, Switzerland, Spain, Korea, and Taiwan. Production of the T-38/F-5 series ultimately totaled 3,805, excluding prototypes. Northrop still bid aggressively for military aircraft contracts against larger competitors, and its originally company-funded P-530 design ultimately would show results as the McDonnell Douglas F/A-18 Hornet.

Northrop established an Aircraft Group in October 1976, largely succeeding the Norair Division. The Aircraft and the Aircraft Services Divisions were subordinate. Although secret at the time, Northrop also developed a stealth technology design for an operational strike aircraft but lost the contract to Lockheed in 1976.

At North American Rockwell, military aircraft remained in the Aerospace and Systems Group. The Columbus Division continued the T-2B, RA-5C, and OV-10A, and the Los Angeles Division developed the B-1. Production of the OV-10A for export continued at a low level through the 1970s. The A & S Group also encompassed the Atomics International, Space, Autonetics, Rocketdyne, and Tulsa Divisions. Willard F. Rockwell retired, passing the title of chairman to his son. President J. L. Atwood retired in January 1970, succeeded by Robert Anderson, but remained on the board until 1977.

After loss of the F-15 contract and with the Apollo program approaching completion, NAR implemented major organizational changes, largely deemphasizing the aerospace component of the company. NAR felt that perceived weaknesses in management structure had biased the air force against its F-15 proposal. John R. Moore was transferred to a staff vice presidency, and former North American operating divisions now reported to Anderson.[75]

The aircraft production future of NAR rested with the B-1. Employment declines at the Los Angeles Division also gained the attention of Governor Ronald Reagan and Senator George Murphy.[76] The NAR bid of $1.35 billion for the RDT&E phase proved critical in winning the contract. After some delay, the B-1 prototype flew on December 23, 1974, by which time controversy was intense, and the B-1 was also announced as negotiable under the Strategic Arms Limitation Talks.

Rockwell also held a contract for the XFV-12A V/STOL naval fighter, to serve aboard the new class of Sea Control Ships (SCS) being developed. Employing the augmentor-wing concept and powered by a single vectored-thrust engine, the XFV-12A was scheduled to fly in 1977, but funding delays and the fact that earlier augmentor-wing developments in the 1960s, such as the Lockheed XV-4, had not proved promising prevented this innovative design from being realized. When the SCS program was canceled the navy ended support of the XFV-12A.

NAR began investing in Collins Radio Company, another aviation-related activity, in 1971, and acquired complete ownership in 1973. The subsidiary became Rockwell Collins. North American Rockwell was renamed Rockwell International on February 16, 1973, after a merger with the previously independent Rockwell Manufacturing Company, also begun by Colonel Willard Rockwell. A further reorganization on February 15, 1977, split the Aerospace and Systems Group into the North American Aircraft Group and the North American Space Group. The Industrial Products Group, which controlled the General Aviation Division centered at Bethany, Oklahoma, was unaffected. The Aircraft Group then encompassed the B-1, Los Angeles, Columbus, Tulsa, and Atomics International Divisions. The Sabreliner Division relocated to St. Louis. The Space Group included the Rocketdyne and Space Divisions. Bastian J. "Buzz" Hello, who led the B-1 program and later headed the Aerospace and Systems Group, gained a wide reputation for his aggressive management and for his defense of the B-1 in Washington.

The 1977 cancellation of the B-1, on both cost and strategic grounds, was a major blow to the company's aircraft fortunes, but supporters managed to preserve some R&D funding. With the cancellation, a September 1978 reorganization combined the Aircraft and Space Operations Groups into a single North American Aerospace Operations Group, controlling the Aircraft Operations Group, the Space Transportation Systems Group, and the Rocketdyne Division.

Willard Rockwell Jr. retired as CEO of Rockwell International in February 1979 and was succeeded by Robert Anderson. Anderson and Rockwell had openly disagreed on the direction of the company, with Anderson favoring greater diversification from military contracts. The board backed Anderson, and Rockwell departed to pursue new ventures. Donald R. Beall succeeded Anderson as president.

Ling-Temco-Vought experienced numerous financial difficulties in the late 1960s, accompanied by a massive increase in debt. James J. Ling, having earlier lost much authority, was forced to step down as chairman in May 1970. The new chief executive was W. Paul Thayer, formerly head of Vought. The firm became LTV Corporation, following the fashion in corporate names, on May 5, 1971. Remaining shares of Chance Vought held by the public were acquired by LTV. Under LTV Corporation was LTV Aerospace, which housed Vought Aeronautics Company, Vought Helicopters, the marketing subsidiary for French Aerospatiale helicopters, and other units. LTV Electrosystems eventually became the independent E-Systems, important in defense electronics.

Continuing reorganization reflected attempts to refine a conglomerate strategy. Vought units were combined as the Vought Systems Division in 1972, still under LTV Aerospace, and Vought Helicopters was resold to Aerospatiale in 1974. Then, on January 1, 1976, LTV Aerospace was renamed Vought Corporation, and the A-7 again carried the Vought nameplate. Although built at a low rate in later years, A-7 production totaled 1,545, including exports, when completed in 1982. By 1977, LTV operations consisted of Jones and Laughlin in steel, Wilson in food processing, and Vought. Vought appeared to be the most consistent performer of the divisions, yet while holding several major subcontracts, it lacked a successor to the A-7. In general, the record of large conglomerates with aerospace acquisitions was at best mixed. It did not appear that Vought with LTV, Ryan with Teledyne, or Bell with Textron had benefited, nor had their acquirers.[77] Aerospace diversification remained an imperative, however.

In helicopters, Bell output continued at a high level, with the five thousandth UH-1 delivered during 1967. The first production AH-1 HueyCobra was delivered in June 1967, and total Bell helicopter production reached twenty thousand on April 23, 1974. Although Bell by that time had been in helicopters for some thirty years, most had been delivered since 1964. Bell won the reopened army LOH competition on March 8, 1968, with the OH-58 development of its Model 206 JetRanger, and twenty-two hundred were ordered. The XV-15 twin-tilt-rotor experimental transport made its first hovering flight on May 3, 1977, then made its first in-flight transition on July 24, 1979. It showed sufficient promise to lead to further development. Bell also made major advances in rotor vibration reduction.

For the civil market, Bell introduced the Model 212, known as the Twin Two-Twelve, based on the UH-1N, in 1971. The Model 222, which became popular both for executive use and offshore oil operations, first flew on August 13, 1976. The advanced Model 206B JetRanger III emerged in 1977, and the Model 412, an updated Model 212, followed in 1979.

Sikorsky continued its S-61 and S-65 models in several military versions. S-62 production ended, but the CH-54 flying crane began series production. It also developed the S-67 Blackhawk attack helicopter as a private venture, followed by the experimental S-72 with the advancing rotor blade (coaxial) concept and compound power, under an army contract from 1972.

New United Chief Executive Harry J. Gray, arriving from Litton Industries in 1971, stressed greater diversification, acquiring Otis Elevator and Carrier Corporation. Sikorsky continued as a major subsidiary of the renamed United Technologies and regained its former preeminence when it won both the advanced LAMPS III and the Utility Tactical Transport Aircraft System (UTTAS) competitions with developments of its new-generation S-70 design. The high degree of commonality between the army and navy airframes was advantageous for both the military and the manufacturer. Sikorsky's UTTAS entry, the YUH-60A Black Hawk, first flew on October 17, 1974, and won over the Boeing Vertol YUH-61A on December 23, 1976. The LAMPS III airframe contract was awarded in September 1977, and the resulting SH-60B Seahawk first flew on December 12, 1979. IBM was designated prime contractor for Mark III avionics, indicative of the increasing importance of that field in new weapon systems. Sikorsky also developed the twelve-passenger S-76 commercial model in 1975 and the S-78 commercial model of the UH-60A.

Boeing Vertol continued production of the CH-47 and variants of that design. Loss of the UTTAS competition was a major blow, and Vertol also suffered reverses in diversification attempts. Development of light-rail vehicles for urban rapid transit, for which Boeing felt it could capitalize on its technological expertise, instead led to serious operating problems and financial losses. Boeing Vertol withdrew from the transit and subway field in 1979, but still subcontracted to the Commercial Airplane Company.

Kaman, with no follow-on orders for its SH-2 SeaSprite, ended production in the late 1960s with 184 delivered. From 1965 Kaman diversified successfully, transferring its rotary-wing expertise to guitars and other precision devices. It thus was able to maintain sales and employment in the face of declining government business.[78] Kaman grew to five plants and maintained employment at around thirty-five hundred. A parent Kaman Corporation was formed to hold Kaman Aerospace Corporation among other operations. Then Kaman managed a return to helicopter production in 1971. Although LAMPS II was found to be beyond the capabilities of the SH-2 airframe and was canceled, and the successor LAMPS III program went to Sikorsky, the original LAMPS continued. The first SH-2D LAMPS prototype, converted from an existing airframe, flew on March 16, 1971, and the first new-build LAMPS SH-2F model was delivered in 1973. Both conversions and new production SH-2Fs carried Kaman through the decade in helicopters.

Hughes OH-6A production ended in August 1970 after 1,434 had been delivered. The reopened LOH competition was won by Bell, but Hughes still developed military versions of the civil Model 500. One outcome of the confusion and litigation over Howard Hughes's estate after his death in 1976 was that the Toolco Aircraft Division became Hughes Helicopters. Hughes Helicopters in turn became a division of the Summa Corporation, formed in 1972 to hold Hughes's extensive interests. Hughes Aircraft remained separate. Most important for Hughes was the army attack helicopter program, won by its AH-64 Apache. The prototype flew on September 30, 1975, but production would encounter numerous delays.

International Competition and Collaboration

Another major impact on the industry's structure and strategy, though not necessarily a threat, came from internationalization. While the United States possessed the largest and most technically capable aerospace industry in the world, political pressures and development costs of advanced projects created incentives for risk sharing internationally. The fragmented European aerospace industry had first initiated such arrangements, as smaller firms and governments simply lacked the resources to develop advanced-technology programs alone. Great Britain, after a disastrous 1955–65 decade of false starts, confusion, and cancellations in aircraft development, decided never again to undertake major military aircraft programs alone.

Other than the brief Fairchild-EWR VTOL fighter project, the United States did not collaborate on military aircraft. The American strength was, of course, directly related to strong military R&D and a large military market, and emphasis remained on exports and licensed production. Most technical development had been government funded, but major firms also possessed impressive R&D capabilities to maintain their advantage.[79] This powerful combination translated into civil-market dominance, which with export assistance made government policy resemble the Japanese system of industrial policy.[80]

It was understood on both sides of the Atlantic that Britain had not lost out in the commercial field solely because of the Comet disaster; American airliners also had suffered crashes due to design flaws but had recovered in public confidence. The British lead rather had been lost by a series of unfortunate commercial and technical judgments in the industry, generally involving an unwillingness to take risks of the scale that Boeing had taken.[81] Britain's future commercial efforts also would be collaborative.

European collaboration presented the prospect of effective competition for the American industry for the first time. The initial threat was felt not so much in the domestic market for military or commercial aircraft, but in international markets, long American-dominated. Foreign aircraft, in some cases of a single nation but increasingly from collaborative ventures, began to make market inroads. European firms indeed became aggressive exporters, and collaboration also enabled those nations and certain of their trading partners to better withstand pressures to buy American, particularly military aircraft, where the ILN and DSAA programs had been so effective.

American firms entered international collaboration somewhat reluctantly, with no permanent commitment as in Europe. One reason was that no single office within the government regulated or set policy for such arrangements.[82] American airliner builders instead explored collaboration on a case-by-case basis, mindful that even with their market dominance future programs would involve greater risks than before. Boeing entered into an agreement with Aeritalia of Italy in 1971 on the projected Quiet Short-Haul Aircraft (QSHA), keeping Aeritalia out of the Airbus program while helping Boeing financially.[83] Boeing terminated the program abruptly in 1976, however, to the dismay of Aeritalia.[84] Boeing also

TABLE 21. Leading Exporters among Aerospace Companies, 1988

Company	Total Sales (in billions)	Exports (in millions)	Rank among All Industrial Exporters	Exports as a Percentage of Total Sales
General Motors[a]	$121	$9,392	1	7.8
Ford[a]	92	8,822	2	9.6
Boeing	17	7,849	3	46.2[b]
McDonnell Douglas	15	3,471	8	23.1[b]
United Technologies	18	2,848	10	15.8
General Dynamics	9.5	1,597	18	16.8
Textron	7.1	1,127	23	15.9
Lockheed	10.7	686	34	6.4
Rockwell International	11.9	620	39	5.2

Source: Edward Prewitt, "America's 50 Biggest Exporters," Fortune, July 17, 1989, 51. (© 1989 Time Inc. All rights reserved.)
a. Included for comparison.
b. This high percentage is attributable to exports of commercial airliners.

studied smaller airliner proposals under the provisional 7J7 designation, signing a Memorandum of Understanding (MoU) with Japan in 1973 for collaboration, but did not proceed with any.

McDonnell Douglas ended discussions with Dassault on its Mercure short-range airliner but seriously studied in the mid-1970s the ATMR (Advanced Transport Medium Range) program, provisionally designated DCX-200. In many respects a twin-engined DC-10 with a supercritical wing and larger engines, the design closely compared to the Airbus A300 and would have become the DC-11 if produced. Needing financial support, it discussed Japanese participation in the ATMR, but James S. McDonnell in particular was skeptical after the DC-10 and ended development in the summer of 1978.[85]

The market structure also forced internationalization. Orders, whether to governments or government-owned airlines, often hinged on other than free-market factors. Inclusion of locally manufactured components and equipment, job creation, technology transfer opportunities, offset sales, and the impact of an import on a nation's balance of payments, were more likely to be decisive. Firms thus found it necessary to produce components locally or include local equipment to make a sale politically acceptable, regardless of its financial or technical rationale.

The most impressive collaborative progress was in engines. Both Pratt and Whitney and General Electric collaborated with the French firm SNECMA, GE on the CFM56 advanced "ten-ton" (twenty thousand pounds of thrust) turbofan on a fifty-fifty basis, to succeed the first-generation civil jet engines.[86] This also marked the first collaboration in aerospace between an American company and a foreign state-owned firm. The issues, however, were far more complex than simply those of private versus public sector.[87] Among other things, the U.S. government blocked transfer of CFM56 core technology to SNECMA since it was derived from the large F101 military engine for the B-1. The impasse delayed the program by one year but was resolved by allowing the core to be exported as a self-contained "black box" due to the modular design of the engine.[88] The CFM56 became a major success, inspiring further efforts, and a direct benefit was that GE won adoption of the CF6 engine for the Airbus.[89]

The procurement changes and international developments of the 1970s led to numerous reform attempts in the aerospace industry, but its inherent instability was perhaps most dramatically illustrated by the almost constant restructuring that every firm carried out. Seemingly this represented a continuing if frustrating search for the structure that would maximize chances for survival in aircraft, or otherwise the optimal diversification strategy for long-term stability. Industry was as much concerned with restructuring as it was with defense procurement, technical development, exports, or collaboration. Acquisitions of both aviation and nonaviation companies for diversification were commonplace, but just as commonplace were reversals of diversification efforts undertaken only a few years earlier.

Chapter 15   The 1980s and Beyond

The end of the 1970s could be considered another low point for the aircraft/aerospace industry. New aircraft and missile programs, the space program, and airliner orders had all declined. Contracting and corruption scandals and stronger international competition were also damaging. While these problems were neither unprecedented nor unique, their magnitude and concurrency were. Further, the decline was in all sectors. NASA space outlays were substantially lower than in the boom days of the 1960s.[1]

The industry saw a turnaround, or even salvation, with the Reagan administration. President Reagan, in addition to undertaking a major military buildup, also promised to revitalize the economy and to lighten the regulatory burden on business, potentially strengthening commercial and general aviation as well as the military sector. Yet the era would be marred by further scandals and new investigations.

Advancing years claimed several pioneering names, five in the space of two years. James S. McDonnell, probably the only aircraft industrialist to build and retain a major personal fortune, died on August 22, 1980, age eighty-one. Donald Douglas died on February 1, 1981, age eighty-eight, and John K. Northrop two weeks later, on February 18, 1981, age eighty-five. Northrop, in poor health for the last years of his life and unable to speak due to a stroke, was visited by defense officials at his hospital and told that the secret Northrop B-2 bomber then under development would be of the flying-wing configuration.[2] Northrop died feeling that his design concepts had been vindicated. Claude Ryan died September 11, 1982, age eighty-four, and Leroy Grumman on October 4, 1982, age eighty-seven.

In July 1982, Courtlandt Gross and his wife were slain in their home by an intruder. William M. Allen died October 29, 1985, at age eighty-five. Jerome Hunsaker, never an industrialist but whose efforts greatly influenced the industry, also died in 1985, at age ninety-eight. Clinton Towl, who retired as chairman of Grumman in 1974 after a forty-five-year career, died on January 25, 1990, at age eighty-four. Possibly the two greatest military aircraft designers, Kelly Johnson and Ed Heinemann, died within a year of each other. Johnson died on December 21, 1990, age eighty, and Heinemann on November 26, 1991, age eighty-three.

Airliner Developments

Commercial air travel had grown to the point that two-thirds of the world airliner market was outside the United States, but American manufacturers still dominated. Further, given the importance of economies of scale, production under license generally was unfeasible. Component production was increasingly internationalized, however, extending even to China. Domestically, the majority of airliner production by value was subcontracted. The Boeing 747, for example, was only assembled at the Everett plant. Competitive challenges increased with penetration of the American market by Airbus Industrie, followed by Fokker and several third-level airliner manufacturers. Their success gave rise to increasing concern and protests by American manufacturers.

U.S. airlines were deregulated by legislation in 1978. Entry into the industry became easier, acquisition of routes was eased, and airlines were free to set fares and schedules as never before. The CAB was abolished after forty years. It was anticipated that such enhanced free-market competition would lower fares and serve to expand and upgrade air travel, and that demand for new airliners would rise over the long run. Another significant development was the Civil Aircraft Agreement of 1979, concluded within the framework of overall international trade negotiations. It largely eliminated tariffs on aircraft and parts, expanding the market and diminishing incentives toward internationally licensed civil-aircraft production.[3]

Major firms had observed that lighter regulation of international arms sales from the Nixon era had led to more bribes and kickbacks, and those practices spilled over into commercial markets. Again, risking the company on a new venture, and the desperation felt at the loss of a major order, led to corruption. Lockheed was the first company exposed and fined, but others were implicated. McDonnell Douglas settled a case involving improper payments and misapplication of funds in January 1979. Then on November 9, 1979, four executives, including James S. McDonnell III and John Brizendine, were indicted for fraud in connection with airliner sales, since such sales were financed through the Export-Import Bank. The company defended on the grounds that the charges referred to acts in the 1972–77 period, before the Foreign Corrupt Practices Act, and the case was resolved by plea bargaining in September 1981 with no convictions.[4] One outcome was that the McDonnell Douglas board was required to have an outsider majority for the first time.

In December 1981 Lockheed announced that it would end marketing and production of the L-1011 TriStar, thereby exiting the airliner field. Lockheed took an immediate financial write-off of $730 million due to its heavy unrecovered development costs. The last of 246 production TriStars was completed on August 19, 1983, although deliveries continued from unsold stocks. Chairman Roy A. Anderson reported that Lockheed lost a total of $2.5 billion on the TriStar.[5] The loss dwarfed that incurred by General Dynamics in the 880/990 debacle twenty years earlier.

Despite its technical merit, the TriStar remains a case of unrealistic planning. In the words of one researcher,

> One may doubt that there existed in 1968 a feasible price-sales combination for the *Tri Star* [sic] at which the program could have been expected to generate a positive net present value, even if one projected nonrecurring costs of only $800 million, a cost-of-capital rate of 5 percent, and an annual traffic growth as high as 10 percent per year![6]

The exit of Lockheed left only two American airliner builders. Boeing's airliner family covered virtually the entire spectrum, so that a large airline had the added inducement of commonality to standardize on Boeing models.[7] Boeing's marketing staff also had gained a strong reputation

TABLE 22. Transport Aircraft Shipments

Year	Total
1973	294
1974	332
1975	315
1976	222
1977	155
1978	241
1979	376
1980	387
1981	387
1982	232
1983	262
1984	185
1985	278
1986	330
1987	357

Source: Aerospace Facts and Figures, 88–89, 32.

Note: Includes U.S.-manufactured fixed-wing aircraft over 33,000 pounds empty weight, including all jet transports plus the four-engined turboprop-powered Lockheed L-100. Shipments are actual deliveries to customers, including those aircraft produced in an earlier year.

with airlines for understanding and representing their needs to senior executives.[8] McDonnell Douglas, like Lockheed, had incurred heavy losses and offered only two basic designs. As Airbus Industrie became a credible competitor, Boeing especially protested against the subsidized Airbus as constituting unfair competition against companies operating in a free-enterprise environment. British Airways, however, became Boeing's largest customer in Europe, repeatedly ordering Boeing even though Great Britain was a partner in AI, and wags represented that BA meant "Boeing Always." AI countered criticism by pointing to heavy U.S. military R&D spending as effectively subsidizing development of the 747 and other airliners, and by emphasizing the heavy U.S. content of its products. The Airbus also captured third-country markets previously dominated by American firms. U.S. airlines, while generally expressing loyalty to the American industry, still were somewhat divided, since the existence of a third strong competitor worked to their advantage in holding down prices.

An unpleasant reality of the commercial sector was that offering a competitive, economical, correctly sized and correctly timed airliner was not enough. In addition to political considerations, major orders often hinged on financing, involving such factors as trade-ins of older airliners, lenient payment terms, low interest rates, and even countertrade arrangements with unrelated products of the customer country. American protests also included financing subsidies. Europeans countered that the Export-Import Bank (Eximbank), established in 1934, had become an equivalent financing subsidy for American competitors, providing credit for foreign countries and airlines for U.S. purchases. Eximbank frequently was called "Boeing's Bank," since its largest activity was financing Boeing airliner sales. Eximbank assistance indeed posed a policy dilemma in that, for example, it financed exports of the Boeing 757 with its Rolls-Royce engines, while it would not finance sales of the Airbus with its GE engines and other American equipment. Thus airframe manufacturers were helped, but engine manufacturers were likely to be injured.[9] A political concern was that the Reagan administration was felt to look askance at the Eximbank, regarding it as inconsistent with market principles.

In 1978 Boeing launched development of an advanced and risky fourth generation of jet airliners, a step McDonnell Douglas could not match. After studying future airliners for several years under provisional 7X7 and 7N7 designations, Boeing announced the concurrent development of the Models 757 and 767. The 767 offered the latest twin-aisle configuration with advanced technology engines, but its development costs alone, around $2 billion, were more than Boeing's net worth.[10]

There were widespread assessments that the new airliners were the wrong size and at the wrong time.[11] Their development, beyond the expected financial outlay, evoked the 747 case in that after strong launch orders no significant further orders were placed for some time. Airlines were in an especially shaky condition at the end of the 1970s, made worse by the oil crisis, which had sharply increased fuel costs, causing an industry recession.

The basic Airbus A300 model, like the TriStar, suffered a sales drought during 1976 and 1977, but in 1978 and 1979 captured significant worldwide orders. The American breakthrough was an order from Eastern after an initial lease arrangement, overcoming its resistance of being the first to place into service a new foreign design. Then the downsized, more efficient A310 directly competed with the Boeing 767.

The 767's stablemate, the Model 757, was originally conceived as a narrow-bodied 727 replacement and began with a lengthened basic 727 fuselage and advanced-technology wings. Originally projected for the 150-seat category, the design quickly escalated to a capacity of 185–220 seats, leading many to believe it was too large for its intended market. But orders from BA over Airbus competition and from Eastern in August 1978 provided strong launch support. Then Delta's order for sixty 757s in November 1980, the largest ever to that time, appeared to secure its future. The 757 was offered with either RR RB.235 or P & W PW2037 engines, leading GE to drop out of the 757 engine market.[12]

The American industry still lacked a new 150-seat design, which by the mid-1980s had emerged as the next growth segment. Boeing instead developed further advanced versions of the popular 737, while McDonnell Douglas continued to develop the DC-9. But AI developed the completely new A320, optimized for the 150-seat market and promising greater efficiency than earlier designs and won strong orders, including many from the United States.

McDonnell Douglas, having decided against a new large-capacity airliner, conceded important market segments to the Boeing 757 and 767. But it felt that its offerings were the most efficient positioning feasible, with the improved MD-80 sized just below the 757, and with the DC-10 between the 767 and 747. The overall market structure remained troubling, however, in that only Boeing offered broad coverage.

The first advanced DC-9, the stretched Super 80 version of 1976, could carry up to 160 passengers. Redesignation of the DC-9–80 as the MD-80 occurred in 1982. Designed to operate within the environmental, energy, and economic constraints of the 1980s, the MD-80 series became a strong seller.[13] An MD-90 designation initially applied to the older and smaller DC-9-30 was soon dropped, to be reapplied to developments with advanced-technology engines. The milestone of one thousand DC-9s was attained on September 3, 1981. The MD-87 short-fuselage version was introduced in 1985. The new MD-90, powered by international V2500 engines, was launched in November 1989 with an order from Delta.

Four major crashes within a relatively short time in the late 1970s damaged the DC-10's reputation and market prospects, although none was attributed to any design flaw.[14] The DC-10 program ended in the spring of 1989 with a disappointing total of 386 produced, including military variants, ending the DC series after some fifty-five years. Incidentally, the Douglas family opposed dropping the DC designation for airliners, but to no avail.

McDonnell Douglas reinforced its commitment to the airliner field in July 1985 with announcement of the MD-11. Based on the DC-10, the MD-11 was more than eighteen feet longer and was extensively reengineered to eliminate all shortcomings of the original. It also was to be powered by three new-generation PW4460, CF6–80C2, or RR Trent 650 engines, each of more than sixty thousand pounds of thrust. Takeoff weight could exceed six hundred thousand pounds. The MD-11 gained launch orders in December 1986 and first flew on January 10, 1990.

Boeing progressed rapidly with the Model 757, which first flew on February 19, 1982, and with the Model 767, which followed on September 26. Both succeeded in operation, but slow orders led skeptics about the programs to believe they had been right.[15] Sales of the 767 eventually gained momentum, and development of ETOPS (extended range twin-engined operations) made transatlantic service permissible.

The veteran 727 finally ended production in 1984, with a record 1,832 being completed. The 747 and 737 continued to enjoy strong demand, with Boeing booking some 250 737 orders in 1984 alone. The thousandth 737 was delivered on December 9, 1983, and by 1988 deliveries had exceeded those for the 727. The first advanced 737–300 model, with CFM56 engines and increased seating capacity, was delivered in November 1984. Another milestone was delivery of the five thousandth Boeing commercial jet on August 17, 1986. In almost thirty-five years of continuous development and production, Boeing had built more than 60 percent of all jet airliners. Boeing completed the thousandth 747 in 1990, by which time only the advanced 747-400 remained in production. The 747–400 suffered major delays, in part due to the 1989 IAM strike, but regained momentum after entering service later in 1989.

Oddly, the American industry had no position in the growing third-level segment, for those models carrying thirty to eighty passengers.

Grumman discontinued a thirty-two-seat commuter version of the Gulfstream I, and Beech did not proceed with its 1979 studies for a twin-turboprop commuter in the twenty-five-seat to forty-seat range. After Fairchild withdrew from its agreement with SAAB on the Model 340, there was no American-built entry between the nineteen-seat commuters and the smaller versions of the Boeing and McDonnell Douglas twins, with capacities in excess of one hundred passengers. Reluctance of U.S. builders in this segment was believed to stem from Fairchild's F-27 experience.

New international engine collaborations focused on the growing twenty- to thirty-thousand-pound-thrust commercial market. The International Aircraft Engine (IAE) venture to develop the V2500, a "ten-ton" engine competitive with the CFM56, was signed in February 1983. The Justice Department approved the program even though only three companies then controlled 100 percent of the large civil-engine market. The V2500 program structure was highly complex in that it consisted of two major partners that were themselves international ventures. One partner was composed of Rolls-Royce and the Japan Aircraft Engine Company (JAEC), while the other consisted of Pratt and Whitney, Fiat, and MTU of Germany. Management was based in Hartford and engineering was based in Derby, England.[16] Rolls-Royce and GE also signed an agreement in February 1984 for risk and revenue sharing on advanced developments of their RB.211 and CF6.[17]

Increasing costs and risks of new airliners, and the emergence of credible international competition, renewed consideration of collaborative ventures by the U.S. industry. As in other fields, collaboration tended to be regarded as a necessary evil, forced by a combination of competitive, technological, and political variables.[18] Although increasing its international subcontracting, Boeing regarded its airliners as company programs with foreign participation rather than as full collaborations. But after entering a risk-sharing agreement with the Japan Commercial Transport Development Corporation (JCTDC) consortium, formed in 1978, for the 767, Boeing became more committed to collaboration. Aeritalia entered the agreement as well. Then in March 1984 Boeing joined with Japan Aircraft Development Corporation (JADC, successor to the JCTDC), for a new 7J7. But after Boeing delayed the launch, the project finally ended in August 1987.[19] Boeing also proposed a British Aerospace production line for the 757, but BAe rejected the deal, fearful of being reduced to a subcontractor role.[20]

McDonnell Douglas entered into a tentative arrangement with Fokker in May 1981 to develop the advanced F100 version of the F28 into the

MDF100 for the 150-seat market. A market slump, conflicts, and McDonnell Douglas's commitment to the MD-80 led to termination in March 1982, underscoring the inherently fragile nature of international agreements.[21] The F100 later became an export success with Dutch government backing. McDonnell Douglas finally became established in collaboration with an agreement with China in April 1985, following six years of discussions, for coproduction of the MD-80. Initially twenty-five MD-80s would be assembled in China.[22]

Lockheed, with declining production at the Georgia division, had seemed a logical American partner for airliner collaboration with Europe. It discussed in 1987 license production of the new-generation Airbus A330 and A340, and in 1988 of the smaller A320, for the North American market. But Lockheed's financial resources were too limited at the time, and no agreements were reached.[23]

The overall record was that collaboration in airliner development and production was more discussed than consummated. The difficulties, in language, management styles, logistics, and finances, were formidable. Political considerations often dominated commercial judgments, especially when involving a government-owned partner. Yet the rationale for collaboration remained compelling.

Airliner sales boomed during the late 1980s. Perhaps most important, the industry appeared better prepared to meet the development challenges than during the near-disastrous late 1960s with the 747, DC-10, and L-1011. The market boom was aided by a fundamental change in the airline competitive environment, with deregulation becoming global, with larger airliners required due to space limitations at most world airports, and large numbers of aging and inefficient models needing replacement. Many jet airliners remained in service more than twenty years. The rapidly growing Pacific Rim nations especially constituted a major growth market, and a global trend toward privatization promised more equitable competition, with orders less influenced by political factors. The fall of the Soviet bloc and its airline monopolies also held market potential.

Industry Trends

In 1979 and 1980 civil aircraft sales exceeded military sales for the first time in the modern era.[24] This was due as much to flat military procurement as to strong commercial growth, however. There were also projections that earnings of the aerospace industry would decline as a percentage of total earnings of U.S. manufacturing.[25] Then with the Reagan defense buildup, rising military procurement and flat airliner business early in the decade resulted in a military-civil sales ratio of greater than two to one. The R&D proportion of total aerospace business rose steadily and was recognized as increasingly vital to the industry's future. It was speculated that Lockheed's Skunk Works, predominantly R&D, was one of its most profitable divisions.

Regardless of the changing domestic mix of military, civil, and space business, exports of aircraft, engines, and parts consistently led all U.S. manufacturing exports. Exports of complete aircraft were valued at some $24 billion in 1987.[26] Even with continuing emphasis on missiles and space, aircraft constituted the majority of total aerospace industry sales. Of total sales of $112 billion in 1987, for example, military and civil aircraft sales combined were some $60 billion, while missiles, space, and related products and services constituted $52 billion.[27] Aerospace remained one of the world's biggest businesses.

Risks continued to increase for defense contractors. Continuing reforms included further cost shifting to bidders, in that finalists would not have all expenses reimbursed but would share development costs with the government. Thus the loser in a competition would suffer a major write-off of company-funded expenses, while the winner still would not be assured of recovery of all expenses.[28] More stringent audits

TABLE 23. U.S. Aircraft Production, 1969–87

Year	Total	Military	Civil
1969	17,795	4,290	13,505
1970	11,796	3,720	8,076
1971	11,072	2,914	8,158
1972	13,106	2,530	10,576
1973	16,530	1,821	14,709
1974	16,839	1,513	15,326
1975	17,030	1,779	15,251
1976	17,747	1,318	16,429
1977	19,047	1,134	17,913
1978	19,958	996	18,962
1979	19,297	837	18,460
1980	14,681	1,047	13,634
1981	11,978	1,062	10,916
1982	6,244	1,159	5,085
1983	4,409	1,053	3,356
1984	3,935	936	2,999
1985	3,602	919	2,683
1986	3,258	1,107	2,151
1987	2,999	1,199	1,800

Source: *Aerospace Facts and Figures, 88–89*, 30–31.

of allowable expenses in R&D and production programs also imposed a burden on contractors.

Generally to the industry's benefit was the Defense Advanced Research Projects Agency (DARPA), independent of the three military departments. DARPA formulated and supported long-term, high-risk projects in advance of an operational application, in effect acting as a venture capitalist. Further, DARPA held no vested procurement interest in programs, as did the military services. The most promising DARPA-sponsored projects, such as stealth technology, eventually could move into the development phase against a military requirement. DARPA contracts became a significant activity for the industry. Related and also benefiting the industry were so-called black programs, for which even their existence was secret. The Lockheed F-117A became perhaps the best-known example. Program budgets were hidden in other areas, in much the same manner as the CIA's budget, preventing even congressional scrutiny and thereby protecting programs from domestic and international political pressures. Because of their secrecy, the impact of black programs on the industry cannot be assessed, but it may be expected that they would afford more R&D latitude than would be the case otherwise.

Also, regardless of views of NASA's effectiveness or importance, no other American industry had a government agency devoted to R&D for its support. Such key industries as steel, chemicals, automobiles, and computers were not recipients of government support for generic technologies that could be employed in the private sector.[29]

The military likewise continued to emphasize the importance of the aerospace industrial base. An Aeronautical Systems Division study during fiscal year 1984, entitled *Blueprint for Tomorrow: A Joint USAF/Industry Assessment,* defined the structure and status of the industrial base, to identify constraints to more efficient peacetime production and to suggest ways of improving efficiency.[30] But a cohesive national-security strategy remained an unattained ideal. An industry desperate to retain and attain new programs also found itself criticized for placing its interests ahead of broad national-security goals in advocating new requirements. That practice often meshed with the desires of the military, however.

In the major example of the B-1 program, General Stewart, ASD commander during the height of the controversy, conceded that the B-1 should have been canceled on technical grounds but justified its later resurrection as being the only way in which the air force could be assured of a new bomber.[31] Effectively outmaneuvering President Carter, who favored stealth and cruise missile technology, the air force revised the

B-1 into a Cruise Missile Carrier Aircraft (CMCA) in 1979. Then the CMCA was replaced by a Long-Range Combat Aircraft (LRCA) specification, combining missile launching and bombing, which the B-1 could fill.[32]

The revolution in Iran and the Soviet invasion of Afghanistan strengthened political support for a new strategic aircraft. Further, Ronald Reagan had asserted in his 1980 campaign for the presidency that the country was entering a "window of vulnerability" in which retaliatory forces were becoming inadequate. Reagan's window of vulnerability soon was shown to be as illusory as the bomber gap and missile gap of earlier campaigns, but he succeeded in reviving the B-1 as the much-modified B-1B.

The B-1B deemphasized supersonic performance, carried a variety of weapons, and featured a lowered radar signature. Delayed by a decade, the B-1B was by then threatened by the new Advanced Technology Bomber (ATB, or stealth) program, originated in the Carter administration. The issue became that of pushing the B-1B or waiting for the ATB, designated B-2. Rockwell and the air force advocated early production of the B-1B, as the B-2 was not projected to enter service until the 1990s. In addition, by 1980 some $5.5 billion had already been spent on the B-1 program. Cost had escalated to the point that $20 billion would be required for a force of one hundred B-1Bs, as compared with the earlier estimate of $15–$19 billion for a fleet of 240 of the B-1. But T Wilson of Boeing and Paul Thayer of LTV, whose firms subcontracted on the B-1B but also had major stakes in the B-2, began to lobby for the B-2 over the B-1B, creating internal conflicts for those firms as well as angering Rockwell.

Many were alarmed that a program the magnitude of the B-1B would be followed by a B-2, then projected to cost in excess of $60 billion. Nonetheless, on October 3, 1981, President Reagan announced a massive new weapons program involving one hundred B-1Bs, followed by the B-2, one hundred MX missiles, and a fleet of Trident II SLBMs. Although under questioning Reagan appeared to be unsure why the nation needed such a costly mix of new strategic weapons, his program was supported by Congress.[33] One reason was that the old coalition opposing the B-1 had become fragmented.

With an ultimate program cost of $29 billion, and with the air force accused of devoting more effort to covering up B-1B problems than to solving them, the acquisition process appeared to have failed again. To critics, the B-1B was yet another example of an unneeded system for an air force defining itself by the manned bomber when other weapons might be more cost-effective, and another confirmation of political considerations gaining primacy over national-security needs. Further, troubled or questionable programs such as the B-1 and B-2 continued to be redefined by their sponsors in order to maintain their relevancy in new environments. During the 1991 war against Iraq, as the B-2 was under attack in Congress, the air force maintained that a fleet of B-2s would be valuable in such limited wars as well as for strategic nuclear missions.[34]

For the few major contracts opportunities appearing, teaming among prime contractors prevailed, both to share the risks and to maintain credibility with the Pentagon. Teaming became the military counterpart of commercial international collaboration. While not new, extending to the General Dynamics–Grumman TFX program, teaming dominated from the 1980s and resulted at times in firms cooperating in one program and competing in another. For example, Boeing and Bell teamed on the XV-22 Osprey, but Boeing teamed with Sikorsky on the LHX program, for which it competed against Bell teamed with McDonnell Douglas. In the ATF program General Dynamics, teamed with Lockheed, competed with McDonnell Douglas, teamed with Northrop, but for the ATA program, McDonnell Douglas and GD were teamed. But major companies had become so large and compartmentalized that separate resource groups could pursue specific projects without the precise knowledge of others within the company.

Closer contract scrutiny and more stringent reporting requirements from 1970s reforms did not end the era of scandal in the industry. There were numerous exposes of improper contract billing by companies, most prominently involving General Dynamics. Contracting scandals erupted again in 1985 and 1986, when Operation "Ill Wind" exposed improper contacts between high Defense officials and the industry with regard to specifications being developed. Certain firms gained an advantage over competitors by learning specifications before RFPs were issued, or by being provided with details of a competing bid. Investigations resulted in numerous convictions of defense firms and millions in civil and criminal penalties. Several high officials, including Melvin Paisley, an undersecretary of the navy, also were convicted. Even more intensive scrutiny and a heightened adversarial relationship resulted.

The Blue Ribbon Commission on Defense Management in 1986, chaired by David Packard, investigated and recommended changes. Packard noted at the outset that the environment had not improved from that he had battled fifteen years earlier. The commission conducted a public opinion survey that disclosed, among other things, that the public believed defense contracting was rife with waste and fraud.[35] The commission concluded, however, that waste was due more to inefficiency than dishonesty and was symptomatic of underlying problems.[36] That finding was consistent with the Grace commission, which studied overall spending in the early 1980s and found that two-thirds of all federal waste occurred within the Defense Department.

The Packard commission did acknowledge the declining authority of contracting officers, and the lack of strong coordination of programs within Defense, but emphasized that the process should be self-governing. One outcome was the new office of undersecretary of defense (acquisition), to control the process throughout. The first three individuals to serve in the post, however, left in frustration at the intractability of the system. One promising development after the "Ill Wind" scandals was that the Defense Industry Initiative on Business Ethics and Conduct was established.[37] The consensus view, however, was that the government still did not effectively control the acquisition process, and prospects for improvement were dim with defense downsizing and fewer supervisory officials.

The declining NASA space budget meant fewer industry contracts, but the Space Transportation System program led a substantial revival. STS contracts were awarded to Rocketdyne for the launch engines, Thiokol for the boosters, Martin Marietta for the main fuel tank, and Rockwell for the Orbiter, which had an aerodynamic configuration for landing. The STS became the major U.S. space activity for the 1980s.

Company Developments

Fairchild Industries in 1980 consisted of the Fairchild Republic, Fairchild Aircraft Service, Fairchild Space and Electronics, and Swearingen Aviation companies. A 1984 reorganization created a Fairchild Aerospace Company to hold both the Fairchild Republic Company and Fairchild Aircraft Corporation, renamed from Swearingen in September 1982. Emanuel Fthenakis became president and chief executive in May 1985, then chairman on January 1, 1986, upon Edward Uhl's retirement. The A-10 program was finally completed in 1984 with delivery of 713 aircraft. Fairchild pursued studies for potential successors for both the A-10 and F-15 but did not succeed in winning development contracts.

The company nonetheless made a vigorous attempt to remain a significant aircraft manufacturer. First, in a return to the airliner market, Fairchild in January 1981 entered into a collaboration with the Swedish firm SAAB-Scania to develop, produce, and market the SAAB 340 thirty-four-seat twin-turboprop regional airliner, dropping its own studies in that category. A partnership of technical and financial equals, Fairchild felt that the SAAB-Fairchild SF340 represented a logical step up from the

Metro nineteen-seat commuter and would give it a lead in a segment not entered by American producers.[38] SAAB felt that it would benefit from Fairchild's name recognition in the field. The SF340 was to be coproduced by Fairchild Republic at Farmingdale.

Fairchild also won the New Generation Trainer (NGT) competition to replace the Cessna T-37B. The fixed-price development contract for its T-46A was awarded on July 2, 1982, and the prototype first flew on October 15, 1985. The T-46 design was influenced by the SAAB collaboration, owing much to the SAAB-105 trainer of the 1960s. Fairchild thus appeared well positioned in aircraft with the SF340, the T-46A, and the Metro. Problems loomed, however, as SF340 development costs proved higher than forecast, and the T-46A program became troubled, creating overhead allocation problems. Fairchild could not afford to develop both and suffered financial losses and layoffs in 1984 and 1985.

Late in 1985, Fthenakis announced a strategic decision to withdraw completely from civil-aircraft production.[39] The coproduction agreement for the SF340 was ended on November 1, 1985. Fairchild reduced its status to that of subcontractor, and component production was completely transferred to Sweden by the end of 1987.[40] The outcome probably was more a failure of Fairchild than of SAAB, as the redesignated SAAB 340 went on to production success in Sweden. Fairchild then decided in 1986 to sell its commuter aircraft subsidiary and rights to the Metro.

Problems with the T-46A led eventually to its cancellation. The air force severely criticized Fairchild's management rather than any inherent design flaws, and Fairchild may have suffered from the constraints of its "buy-in" bid to win the contract. T-46 development costs also diverted resources away from the SF340. Even while trying to defend the T-46 with the air force, Fairchild sought a buyer for the program. Boeing considered buying Fairchild Republic, already a large subcontractor on the 747. Grumman also considered the acquisition contingent on continuation of the T-46, an assurance the air force would not give. Both companies ended negotiations at the end of 1986.[41] The T-46 cancellation became official on March 13, 1987, ending more than sixty years of Fairchild aircraft manufacturing. Late in 1987 Fairchild terminated the remaining thirty-five hundred employees at Farmingdale and closed the plant, bringing to an end the long history of that site in the aircraft industry.

Fairchild Industries, retaining significant activities in space, electronics, and subcontracting, became an acquisition candidate, perceived as possessing a higher value broken into its components than as a continuing entity.[42] A proposal by the Carlyle Group investment firm in 1988 did not materialize, and Fairchild was then acquired by Banner Industries of

Ohio, a metals firm, on August 18, 1989. That arrangement soon required considerable financial assistance, however. The merged firm was renamed Fairchild Corporation on November 15, 1990, with the Fairchild name surviving, but as a diversified manufacturer having only a peripheral connection to the aerospace industry.

Grumman experienced problems in its diversification efforts. It suffered serious losses on its Dormovac freight refrigerators and Ecosystems environmental management and research ventures. The Flxible transit bus operation, acquired from Rohr in 1978, was sold in 1983 after incurring $200 million in losses.[43] Grumman became the target of a hostile takeover attempt by LTV in 1981, which bid for 70 percent of Grumman's stock. But LTV underestimated the loyalty of employees to the company and its determination to remain independent. Grumman's strategy maintained that the move would be anticompetitive in that Vought also was a major naval contractor.[44] Even Leroy Grumman at age eighty-six was called from his Florida retirement to lead a rally against the attempt. The

TABLE 24. Largest Aerospace Manufacturers, 1989

Industry Rank	Firm	Fortune 500 Rank	Sales (in millions)	Employees
1	Boeing	15	$20,276	159,200
2	United Technologies[a]	17	19,765	201,400
3	McDonnell Douglas	25	14,995	127,926
4	Rockwell International	29	12,633	108,715
5	General Dynamics[a]	44	10,053	102,200
6	Lockheed	45	9,932	82,500
7	Raytheon[a]	52	8,796	77,600
8	Textron[a]	61	7,440	58,000
9	LTV[a]	72	6,362	38,000
10	Martin Marietta[b]	84	5,814	65,500
11	Northrop	98	5,200	41,000
12	Teledyne[b]	109	4,664	43,200
13	Grumman	129	3,559	28,900
14	Kaman	385	802	6,424
—	Banner Industries[c]	480	579	6,115

Source: "The Fortune 500," *Fortune*, April 23, 1990, 371. (© 1990 Time Inc. All rights reserved.)

Note: No firms listed include the term *aircraft* in their corporate names.

a. Highly diversified. Aircraft constitutes a minority of total sales. United Technologies includes Sikorsky, General Dynamics includes Cessna, Raytheon includes Beech, Textron includes Bell.

b. Does not manufacture aircraft, but included for related activities.

c. Acquired Fairchild on August 18, 1989. Included for comparative purposes.

bid was finally defeated after a federal court rejection on competitive grounds.[45]

Grumman restructured again in 1984 to position itself as a modern aerospace firm, holding Grumman Aerospace, Allied Industries, Data Systems, and the International company. John O'Brien, president of Grumman Data Systems, became president of Grumman Corporation in 1986, CEO at the beginning of 1988, and finally chairman on August 1, 1988, with the retirement of John C. Bierwirth.

Grumman possessed an extraordinary distinction by the end of the 1980s in that it had remained in full production without a totally new design for over twenty years, the most recent design contract being that for the F-14 in 1969. Grumman became involved with the F-111 again in the early 1980s with a contract for conversion of forty-two F-111As into EF-111A Raven electronic countermeasures aircraft, giving the air force a modern and badly needed capability. The F-14 finally flew with the originally planned GE F110 engines on September 29, 1986, and entered low-volume production as the F-14A+, later F-14B. The definitive F-14D, with F110 engines and new avionics and radar, followed in 1990. Only seventy-five F-14Ds were ordered, however, and production finally ended in 1992.

The advanced A-6F Intruder II of 1987, which combined advanced GE F404 engines with a new Boeing-built wing of composite materials, was intended to strengthen the naval strike force until an Advanced Tactical Aircraft (ATA) could be brought into service. But some within the navy feared that the A-6F would drain funding away from the ATA program, and production was canceled. Production of the older A-6E, also with Boeing composite wings, continued until 1992.

The Joint Surveillance Target Attack Radar System (Joint STARS) program for the air force, based on the Boeing 707–320 airframe and designated E-8C, was highly important. But Grumman, teamed with Northrop and LTV, lost the ATA contract to a McDonnell Douglas–General Dynamics team, raising further doubts about its aircraft future.[46] Grumman actually felt threatened as a prime contractor by its strengths, specifically electronics and electronic-warfare aircraft. Such capabilities were coming under increasing attack by reform-minded politicians, who regarded the electronics on such systems as the EA-6B, the EF-111, the E-2C, and the Joint STARS as unduly costly and unreliable.

Grumman's declining aircraft production resulted in steady employment drops and continuing questions over strategy, and after the board meeting of July 1990 it was tersely announced that John O'Brien had resigned as chairman and CEO. He was succeeded by Renso L. Caporali, a

Grumman veteran, who pursued further restructuring and downsizing. O'Brien later became the highest-ranking industry official to be convicted under the "Ill Wind" investigations.

Boeing remained unique among major aerospace firms in being primarily commercial, with government sales in 1979 only 17 percent of the total. But military and space business remained important and increased during the 1980s. In March 1980 Boeing won the important Air-Launched Cruise Missile (ALCM) program, regarded as an alternative to the then-canceled B-1, over General Dynamics.[47] Coincidentally, Oliver C. Boileau, chief of military airplanes, had moved to General Dynamics as president the previous December.

Boeing entered the computer and electronics fields in 1985, acquiring the small ARGOSystems firm after failing to acquire Hughes Aircraft. Unfortunately for Boeing, its military and space efforts proved to be far more troublesome than its commercial side. Overhead on military contracting was inherently higher than on civil programs due to contracting regulations, and Boeing was felt to have pursued many programs in which it possessed little expertise. Compounding its problems, Boeing pled guilty in 1985 to illegal trafficking in contract information as part of the "Ill Wind" investigations and was heavily fined.[48] Then, separate from the "Ill Wind" charges, Boeing pled guilty in November 1989 to obtaining classified contract data, which it had used as a basis for lobbying the Pentagon as well as to gain advantages over other contractors. Boeing was fined $5 million, and an employee went to prison.[49]

Frank A. Shrontz, a Boeing veteran who had served in senior Defense Department posts in the 1970s, and as head of the 707/727/737 division, succeeded Tex Boullioun as president of the Commercial Airplane Company in 1984. Shrontz then succeeded T Wilson as president of the Boeing Company on February 25, 1985, and became chief executive on April 28, 1986. Wilson retired as chairman of the board on January 1, 1988, and Shrontz also succeeded to that title. Continuing growth brought employment to 164,500 at the end of 1989, including all subsidiaries, a record for an aerospace firm. Sales rose from the $10 billion level in the slump year of 1984 to more than $29 billion by 1991.

Boeing acquired de Havilland Canada, producer of a successful fifty-seat commuter airliner, on January 31, 1986, for $112 million, after the Canadian government reversed policy on foreign ownership in its aircraft industry.[50] Boeing felt that the acquisition would be an effective diversification move and would enable it to take advantage of the large U.S. market for commuter airliners.[51] But despite heavy investment, de Havilland Canada continued to incur heavy losses. Boeing acknowledged its error and resold de Havilland to the Canadian firm Bombardier in 1992.

In another restructuring announced on October 30, 1989, in the aftermath of the major IAM strike, Boeing combined most of its noncommercial activities into a Defense and Space Group.[52] Formed on January 2, 1990, the group included Aerospace and Electronics, Advanced Systems, Helicopters (formerly Vertol), Military Airplanes, and the new ARGOSystems. Boeing held no prime contracts on military aircraft but was teamed on the F-22, B-2, and RAH-66. Commercial Airplanes evolved into the Boeing Commercial Airplane Group. With its commanding position in the commercial market, Boeing more than any other aerospace firm became defined as a Blue Chip company and became a component of the Dow Jones Industrial Average. Significantly for an industry seldom characterized by managerial stability, only two men had served as Boeing chief executive more than twenty years after the Allen era.

General Dynamics seemingly benefited more than any other defense/aerospace firm from the Reagan buildup of 1981–87, with the high-volume F-16 program, continuing shipbuilding and tank production, and exports. The thousandth F-16 was delivered on July 8, 1983. David S. Lewis, with a strong managerial reputation, dominated the company.[53] In 1984 General Dynamics recreated four aerospace divisions, Convair in San Diego, GD–Fort Worth, GD-Pomona, and GD-Electronics. The Convair Division, while no longer producing aircraft, continued with Atlas-Centaur launch vehicles and subcontracting on Douglas airliners. The general aviation firm Cessna was acquired in 1985 as a subsidiary. Cessna reflected the company's desire for more commercial business, although some suspected that another motivation was to gain favor with Kansas senators Dole and Kassebaum.

The GD land and sea systems, the ballistic missile submarine and the M-1 tank, soon became embroiled in intense controversies, and the F-16 remained the star of the company. David Lewis testified before Congress in the spring of 1985, defending his company against charges of contract improprieties and attributing questionable billings to bookkeeping errors. But further damaging disclosures led in September 1985 to a brief suspension of General Dynamics from contract bidding, and the company's reputation was further weakened. Lewis, as a result, became a liability and retired at the end of 1985 at age sixty-eight, although remaining on the board.[54] He was succeeded by Stanley C. Pace, formerly CEO of TRW, who had joined GD as vice chairman on June 1, 1985. President Oliver Boileau was passed over for the top position by the Crown family, which preferred an outsider to distance itself from the frequent controversies occurring under Lewis.[55] The Pace appointment also signified that any strong move toward commercial markets was not yet in prospect. Boileau later moved to Northrop to head the B-2 program. William A. Anders, the former astronaut and previously with Textron, joined GD in 1989 as vice chairman and heir apparent to Stanley Pace.

Despite the company's problems, selection by the air force of F-16 developments for air defense and close-support missions assured its continued production. A new Space Systems Division was added to the aerospace functions. GD's aircraft future appeared even more promising when, teamed with Lockheed, it won the F-22 advanced-fighter program, and, teamed with McDonnell Douglas, won the ATA contract. That contract, however, awarded on January 13, 1988, for the stealthy A-12 naval strike aircraft, imposed a fixed $4.8 billion development budget and a rigid timetable. But the program experienced serious delays and developmental difficulties, leading to cancellation by the secretary of defense on January 7, 1991, for default. Undersecretary of Defense for Acquisition John A. Betti was fired for failing to keep senior officials and Congress fully informed on the mounting A-12 problems, and both GD and McDD faced huge potential liabilities with the requirement to return advance payments.[56] Layoffs resulted in both St. Louis and Fort Worth, and GE also lost the F412 engine program. The navy sought immediate recovery of $1.35 billion from both firms, possibly pushing at least one into bankruptcy, and the contract terms made the survivor liable for all contract claims if the other should go bankrupt.[57]

By 1982 Lockheed Corporation consisted of five major divisions, Lockheed Missiles and Space in Sunnyvale; Lockheed California in Burbank, the traditional aircraft production center; Lockheed Georgia; Lockheed Shipbuilding and Construction of Seattle; and Lockheed Electronics of Plainfield, New Jersey; plus international operations. Sanders Associates of New Hampshire, in defense electronics, was acquired in 1986 and became Lockheed Sanders. A major reorganization on September 3, 1987, included creation of the Lockheed Aeronautical Systems Group, combining the California and Georgia operations. The Skunk Works was renamed Lockheed Advanced Development Company in 1989.

The C-130 was produced in greater numbers than any Western transport since the Second World War and set a new record for continuous production of a basic design. In a major modification contract, Lockheed began rewinging all 77 C-5As in 1978, and the project was completed by 1987. Lockheed also proposed an improved, strengthened C-5B to the air force at a fixed price, in part to head off a possible order for military Boeing 747s. The air force ordered fifty C-5Bs, which were delivered between 1986 and 1989.

Lockheed moved its headquarters from Burbank to outlying Calabasas in 1986, then announced the phaseout of its main Burbank factory in 1988, ending sixty years of unbroken production at that location.[58] The problem was excess production capacity, and of the major facilities in Burbank, Palmdale, and Marietta, Burbank was the oldest and least efficient. The highly secret F-117A, which the Skunk Works had developed in the 1970s, was not officially revealed until November 1988, by which time production was almost complete. Lockheed also had largely completed production of the C-5B, P-3, and TR-1A by 1988. The Georgia operation still was seriously underutilized.

Lockheed led a team, with Boeing and General Dynamics, competing for the important ATF program for a successor to the F-15. RFPs were issued in September 1985, and the specification required stealth characteristics with further advances in engine technology, including thrust vectoring and supersonic cruise capability. Lockheed's YF-22 was pronounced the clear winner in a flyoff competition with the Northrop McDonnell Douglas YF-23. Lockheed received the development contract in April 1991 but immediately faced the prospect of postponement, reduction, or cancellation due to the improved world situation and budgetary pressures.

Both the Lockheed and Northrop teams had committed substantial capital to the risky ATF program, with Northrop as the loser suffering an immediate writeoff. Even in the production phase, however, the winner might not find the production contract profitable.[59] Defense contractors were forced to pay far more attention to economics than in the past.

Indicative of the increasingly hard line taken by Defense, the LRAACA (long-range air antisubmarine warfare capable aircraft) program, resulting in the P-7 to replace the long-serving P-3, was canceled in November 1989 after engineering problems and major cost overruns. Lockheed had anticipated that the P-7 would have perhaps 40 percent commonality with the earlier design, but events largely negated that expectation.[60] The contract was held to be in default, rather than canceled for the convenience of the government, and Lockheed was forced to absorb losses rather than gain reimbursement.

Lockheed began yet another major restructuring in April 1989, specifically to position it for anticipated future markets. The four major groups resulting were Missile and Space Systems, Aeronautical Systems,

Technology Services, which included airport operations and many NASA contracts, and Electronic Systems. It was projected that Missile and Space Systems would constitute 50 percent of the company's business in the future, while Aeronautical Systems, even with the F-22 program, would decline to only 20 percent.[61] A major implementing change was the move of LASC from California to Georgia, announced on May 8, 1990. Remaining Palmdale work would be transferred, and all future aircraft production, principally the F-22, would be in Georgia.[62]

Lockheed figured frequently in takeover speculation. In the early 1980s there were recurring rumors of a bid by Ford, then felt to be interested in a major aerospace acquisition, but an attempt never materialized. Daniel Tellep, formerly head of the Missiles and Space Systems Group, upon becoming CEO in January 1990 recognized that possibly Lockheed's greatest challenge was to remain independent.[63] Tellep immediately faced a major battle for control, as more than 40 percent of Lockheed was owned by institutional investors, who demanded stronger results. Billionaire Texas investor Harold Simmons, through his NL Industries controlling more than 18 percent of Lockheed stock, organized a proxy battle to oust many board members in favor of his nominees, pointing to the poor financial results after the 1989 restructuring. Tellep prevailed in a closely contested election in March 1990, but at some cost to managerial autonomy. The situation calmed, although disrupting continuing restructuring.[64]

While vulnerable to all environmental difficulties, by the mid-1980s McDonnell Douglas alone enjoyed the product line diversity possessed by several companies in the 1950s, with the F/A-18, F-15, AV-8B, and airliners. The company won a low-priority C-X contract over Boeing on August 29, 1981, for the C-17 intratheater transport. Largely replacing the earlier AMST program, the C-17 followed the configuration of the YC-15. Other active programs included the T-45 Goshawk naval advanced trainer based on the British Aerospace Hawk, the A-12 teamed with General Dynamics, the YF-23 teamed with Northrop, and the LHX teamed with Bell. The VTXTS contract for the T-45A was awarded on November 18, 1981, winning over new designs from Beech-Grumman and Northrop-Vought teams.

British collaboration, established with the AV-8B and T-45A, progressed with an agreement on ASTOVL (Advanced STOVL) development. The slowing DC-10 program received a boost with the air force Advanced Tanker/Cargo Aircraft (ATCA) requirement for a more capable replacement for the KC-135. The KC-10 version of the DC-10 won over a variant of the Boeing 747. The KC-10A Extender first flew on July 12, 1980, and the last of sixty was delivered in September 1988. But major setbacks followed as the LHX went to Boeing Sikorsky, the YF-23 lost, and the A-12 was canceled.

Airliner production remained financially troublesome even as the market expanded in the 1980s. High costs and continuing losses forced a major restructuring at Long Beach, and international collaborations appeared critical for the future. Upon becoming president of Douglas Aircraft in February 1989, Robert H. Hood Jr. implemented a Total Quality Management (TQM) program for the MD-11, although it was later felt to have resulted in overhiring. Organization changed from functional to product line, as Douglas also managed the C-17, the T-45, and KC-10 programs. But the role of the commercial field had never been resolved in the merged corporation. Hood later stated,

> When the merger came, the companies never really merged. There was a St. Louis faction and a West Coast faction and the Midwest guys made all the money and they weren't going to stick it into the commercial aircraft business.[65]

Hughes Helicopters, with its important AH-64 attack helicopter contract, opened a new factory in Mesa, Arizona, in July 1981. Although a strong producer, the company was placed up for sale to provide funds to pay various claims on the estate of Howard Hughes and was acquired by McDonnell Douglas as a wholly owned subsidiary on January 6, 1984. It was renamed the McDonnell Douglas Helicopter Company on August 27, 1985, with operations in Culver City and Mesa. McDonnell Douglas felt that the acquisition would strengthen it both in diversity and growth prospects.[66]

Sanford "Sandy" McDonnell, nephew of the founder and president since 1971, became chairman as well as CEO in 1980. John C. Brizendine, the veteran executive who led jet airliner development, retired in 1983 after ten years as president of the Douglas Aircraft Company, although he later joined Lockheed. Control of the company remained firmly in McDonnell hands, as Sandy McDonnell upon retirement was replaced by John F. McDonnell, son of the founder.

Despite its diversification efforts in aerospace as well as other fields, McDonnell Douglas became increasingly troubled financially. Airliner losses persisted, and many military programs experienced delays and cost overruns. The company suffered perhaps more than others as its several programs began to be stretched out for budgetary reasons. Employment, which peaked at over 132,000 in 1990, began declining sharply.[67] Facing a potentially devastating liability with the A-12, both McDonnell Douglas and General Dynamics began a lengthy petition to have the cancellation changed from default to convenience of the government.

By 1991 McDonnell Douglas was experiencing a crisis in cash flow, exacerbated by cost overruns on the C-17 program. The Pentagon attempted to speed up progress payments on existing contracts, and John F. McDonnell even petitioned the undersecretary of defense for acquisition for what amounted to a $1 billion loan, a highly controversial action.[68] At the time the production future of McDonnell Douglas looked uncertain.

The Rockwell B-1B was pushed into production rapidly, in part to avoid being overtaken by the B-2 but also to counter mounting criticism of its cost, reliability, and performance. At peak production in 1986, the B-1B was responsible for forty thousand jobs and contributed two-thirds of Rockwell's profits.[69] The first B-1B squadron was declared operational, with restrictions, on October 1, 1986. Rockwell attempted unsuccessfully to gain an order for a further one hundred B-1Bs, and with completion of the original order in 1988 had only subcontract work remaining. The company then did not aggressively pursue new aircraft prime contracts due to the risks. Anderson eventually succeeded in his quest to diversify, since after completion of the B-1B program only 25 percent of sales were defense. Diversification, however, left the company without a clear identity.

Facing a declining and overcrowded market, Rockwell sold its remaining general-aviation operations to Gulfstream on February 3, 1981. Sam F. Iacobellis, who succeeded "Buzz" Hello as director of the B-1 program in 1982, became executive vice president in charge of Space Systems and Aircraft. Donald R. Beall then succeeded Anderson as chairman and CEO in February 1988, and the corporate headquarters also moved from Pittsburgh back to El Segundo. Earlier Rockwell had formed the Space Transportation Systems Division (STSD) and delivered the four STS Orbiters. The first Shuttle Orbiter spaceflight came on April 14, 1981. When the *Challenger* was lost on January 28, 1986, Rockwell eventually received a contract for a replacement as well. Entering the 1990s, the aerospace organization consisted of North American Aircraft, Rocketdyne, and Space Systems.

Northrop was largely out of production of complete aircraft by 1980, with no immediate successor to its T-38/ F-5 series. Northrop, principal subcontractor to McDonnell Douglas for the navy F/A-18, was to be prime contractor for the F-18L export version. The forty-sixty production sharing ratio was to be reversed for export production, with McDonnell Douglas becoming major subcontractor. McDonnell Douglas then negotiated directly with Canada in 1979 for the F/A-18A. Northrop objected, then sued in October 1979, claiming violation of the agreement and lack of consultation.[70] Northrop felt that it retained rights to all export sales.

McDonnell Douglas countersued, claiming that the Canadian sale involved the naval version on which it was prime contractor and not the F-18L. Litigation extended over more than five years. McDonnell Douglas later gained orders for the F/A-18 from Australia and Spain, but no F-18Ls were ever sold. The two parties settled out of court on April 8, 1985, after the navy refused to allow either to bill legal expenses as contract overhead, with McDonnell Douglas paying Northrop $50 million.[71] A new agreement reaffirmed McDonnell Douglas as prime contractor for all F/A-18 production.[72] Differences over the F-18, incidentally, did not preclude their later teaming on the ATF program.

The F/A-18 provided major factory work, but Northrop placed production hopes on the company-funded F-5G, powered by a GE F404 engine, which first flew on August 30, 1982. But marketing of the F-5G (later redesignated F-20 and named Tigershark to further differentiate it from earlier F-5 models) to smaller air forces failed due to what Northrop perceived to be lack of Reagan administration support.[73]

While a worthy design, and significantly less costly than the F-16, the F-20 also was handicapped by the air force NIH (not invented here) syndrome. An order from Taiwan was blocked, ostensibly due to opposition by China, and the government continued to promote the F-16 in other markets, squeezing out the F-20.[74] Tom Jones suspected the F-16 was favored due to its Texas factory and the influence of such politicians as Senator John Tower and House Speaker Jim Wright but was nonetheless reluctant to protest too strongly, fearful of losing support for the B-2. Northrop ended the F-20 on August 31, 1986, writing off $1.2 billion, after failing to sell the air force an air defense version. One result was that Jones was left disenchanted with a free-enterprise approach to military aircraft.[75]

Thomas V. Jones finally retired as chairman of Northrop on September 19, 1990, after disclosures of questionable payments to South Korean officials in connection with F-20 sales attempts. He and the company were again investigated for FCPA violations.[76] Kent Kresa, who joined Northrop in 1975, became president in December 1987 and chief executive in January 1989, then succeeded Jones as chairman. Kresa had gained a reputation for greater openness than the more secretive and autocratic Jones.[77] But Kresa also had been implicated in the Korean F-20 campaign and questionable billings on the B-2 program. Further, a special air force investigation criticized the company for systemic management weaknesses with regard to contracts and sales.[78] Numerous senior Northrop executives departed.

Northrop's loss with McDonnell Douglas of the ATF program was another major setback, especially since its YF-23 probably was more advanced aerodynamically than the YF-22 and possessed better stealth characteristics. Lockheed was regarded as having better cost controls, but Northrop felt that its well-publicized contracting and ethical problems had weighed heavily against it.[79]

All aircraft firms remained largely dependent on the next contract for survival in the field, but rarely had a single contract represented so heavy a stake to a firm as did the B-2 for Northrop.[80] The B-2 was a pure flying wing with no vertical surfaces, but control technology advances had overcome stability problems of earlier flying wings. The B-2 was increasingly pressured on cost, performance, and strategic rationale, and when it flew in 1988, it was behind schedule and the most expensive aircraft ever built.

LTV's Paul Thayer left in January 1983 to become deputy secretary of defense but eventually was forced from that office after being implicated in illegal stock trading. In an attempt to strengthen aerospace operations, Vought Corporation was renamed LTV Aerospace and Defense Company on April 4, 1983, divided into a Missiles and Advanced Programs Division and an Aero Products Division. LTV's losses in other fields eventually led to bankruptcy in July 1986, and litigation became particularly prolonged and expensive. On September 29, 1986, aerospace operations became the Aircraft Products Group and the Missiles and Electronics Group.

LTV developed the advanced YA-7F both as a conversion and upgrade of existing A-7s as well as for new production, for the Close Air Support (CAS) mission. The YA-7F first flew on November 29, 1989, but failed to attract interest, as the air force looked to a version of the F-16 to succeed the A-7 and A-10, while the navy began the A-12 program. The YA-7F thus marked the exit of LTV as a prime contractor.

Bell reached the production milestone of twenty-five thousand helicopters on January 18, 1981. The basic UH-1 series finally ended in the early 1980s, and having lost major new programs to Sikorsky and Hughes, Bell was increasingly relegated to R&D contracts. Previously a division of Textron, it became the wholly owned Bell Helicopter Textron, Inc., subsidiary on January 3, 1982. Bell teamed with Boeing Vertol in June 1982 to develop the Joint Services Vertical Lift Aircraft (JVX), designated XV-22, the contract for which was awarded in April 1983. Bell Helicopter Canada was formed in October 1983 to free U.S. facilities for the XV-22 Osprey and other military programs. Primarily commercial, BH Canada began production of the JetRanger and Long Ranger in 1987. The XV-22 completed its first vertical-horizontal transition on September 14, 1989, but generated further controversy as to cost and mission.

Textron acquired Avco Corporation, descended from the original Aviation Corporation of 1928, in December 1984. In so doing Textron incurred a heavy debt burden but also gained defenses against any possible takeover attempts. Textron in fact explored the sale of Bell early in 1985 to gain some relief. The Avco acquisition added the Lycoming engine division to Textron's existing aviation interests.

Sikorsky sales reached $1 billion in 1982, and with fourteen thousand employees it was the largest helicopter manufacturer in the Western world. The Stratford main factory also was the largest. Low-volume commercial S-61 production ended in 1980, but the S-70 and S-76 series continued for both military and commercial roles, along with the CH-53E. Robert F. Daniell, who headed Sikorsky from 1980 to 1983, became CEO of United Technologies in January 1987, determined to further reduce dependence on military contracts. Pratt and Whitney accounted for some 34 percent of UT total sales at the time, and Sikorsky sales were 85 percent military. Sikorsky definitely faced leaner times with defense reductions and lagging commercial markets. Army Black Hawk orders declined, and export competition was heated.[81] Internationalization expanded as UT acquired a significant equity interest in Westland, the financially troubled British helicopter firm and a longtime Sikorsky licensee, on February 12, 1986. The agreement included rights for the entire S-70 series. But Sikorsky felt that its long-term future lay with the LHX.

Boeing Vertol in suburban Philadelphia continued production of the CH-47 series and developed the Advanced Chinook program for the Army Advanced Cargo Aircraft requirement. Sikorsky offered the CH-53E as an alternative to the Advanced Chinook. Boeing Vertol also delivered components for Boeing airliners, and major future efforts were the XV-22 with Bell and the LHX with Sikorsky. The operation was renamed Boeing Helicopter Company on September 1, 1987, and the Vertol name disappeared.

The major rotary-wing prospect was indeed the army's LHX advanced battlefield helicopter, for light scout/utility/attack missions. To replace such long-serving types as the UH-1, AH-1, OH-58, and OH-6A, it was to perform with greater cost-effectiveness than the older types. Two teams, Bell–McDonnell Douglas and Boeing Sikorsky, won development contracts in April 1986. MBB of Germany joined Boeing Sikorsky, which won the production contract in 1991 and began to prepare the RAH-66 Cherokee for production. But the changing political and budgetary environments threatened delays in that program, among others, and a further concern was that massive development costs might never be recouped if the RAH-66 were delayed or orders reduced.[82]

Kaman continued stability with diversification, both in aerospace and other fields. A new order in 1981 for further SH-2F helicopters for the Mark 1 LAMPS requirement enabled Kaman again to restart production, albeit at low volume.

Before its sale to McDonnell Douglas, Hughes Helicopters had enjoyed strong growth and reached the production milestone of six thousand helicopters on December 16, 1981. Model 300 production was suspended in 1981, however, due to civil-market weakness, and the program was sold to the sailplane manufacturer Schweizer on July 13, 1983. Hughes received the army/DARPA no-tail-rotor (NOTAR) development contract in September 1980. NOTAR, involving air thrust through slots in a hollow tailboom, pulling rotor wash downward and to the crew's left, provided stability while making the helicopter quieter and safer. Along with AH-64 production, NOTAR continued under McDonnell Douglas with the MD520N, developed from the Hughes Model 500, and civil certification came on September 17, 1991.

The aerospace industry had managed a broad recovery during the 1980s, principally due to the Reagan defense buildup. But that expansion leveled off, and the industry again began to contract. Further, the commercial-aircraft sector suffered declining orders, and the space program became a prime target for budget cutting. Then the end of the Cold War made a permanent downsizing of the industry inevitable. Sheer survival in aircraft became the priority for most firms.

Chapter 16 ✈ Uncertainty after the Cold War

Whatever the doubts over the production future of aircraft, the frontiers of aeronautical knowledge continued to be pushed back in the 1990s. NASA, while stressing such ambitious programs as space station Freedom, also pursued a wide range of aeronautical research interacting with the industry. Efforts, pursuant to goals established in 1985, were centered in seven major areas.

1. Subsonic aeronautics: advanced propeller blades, drag reduction technology, and takeoff and landing safety.
2. Rotorcraft: higher speed capability, tilt-rotor design, and vibration control.
3. Supersonic aeronautics: High Speed Civil Transport (HSCT) project, focused on transpacific route growth.
4. Hypersonic technology: National Aero-Space Plane (NASP), jointly with the Department of Defense, with X-30 prototype development. But the X-30, a single-stage-to-orbit (SSTO) vehicle, using air-breathing scramjet engines for takeoff, was projected to cost in excess of $10 billion to develop, and thus failed to gain funding from Congress.
5. High-performance maneuverability (High-Alpha), using thrust vectoring and propulsive control. The Rockwell X-31A, a joint DARPA/NASA/USN/USAF/MBB project to test high-speed combat maneuverability, began with an MoU between the United States and Germany in May 1986.
6. Short Takeoff, Vertical Landing (STOVL) advances, jointly with the United Kingdom and based on the Harrier.
7. Forward Swept Wing research, a joint USAF/DARPA project, employing the Grumman X-29 prototype.[1] The X-29A contract was awarded on December 22, 1981.

The industry and the military pursued further advanced programs, some in cooperation with NASA, such as Multiaxis Thrust Vectoring (MATV) for combat aircraft, hypersonic Trans-Atmospheric Vehicles (TAVs), and unmanned aerial vehicles (UAVs) for reconnaissance. The division between aeronautics and astronautics was further blurred with TAVs, including the NASP. Additionally, there were persistent reports of more advanced stealth and hypersonic aircraft, developed under highly secret black programs.

Airliner development was spurred by growth prospects and international competition. Manufacturers strove to keep pace in technology and market coverage. The average age of the U.S. airliner fleet was on the order of twenty years and increasing, with replacement also spurred by safety and noise concerns. With the decline in military procurement, commercial aircraft bid to become the largest sector of the industry permanently. Civil aircraft represented 37 percent of the industry backlog for 1985 and grew to 63 percent by 1990.[2] But after a record sales year in 1989, the airliner market again experienced a broad decline, adding to the cost, risk, and competitive challenges of the sector.

U.S. airlines suffered aggregate financial losses on the order of $10 billion in the three calendar years 1990 through 1992, exceeding cumulative profits earned by the industry in its entire history and eroding their equity bases. A recession denting travel demand and vicious price-cutting were largely responsible. A ripple effect followed through the industry as orders were canceled and deliveries delayed. McDonnell Douglas was forced to slow MD-11 production at Long Beach, with substantial layoffs. In turn, Convair, manufacturing MD-11 fuselage sections, was forced into further layoffs, Pratt and Whitney reduced engine production and laid off still more workers, and numerous smaller suppliers also were affected. The impact of the airline recession was felt immediately.

The project definition phase for airliners, always extending over several years, lengthened with each new generation. The most important new program, announced in October 1990 after intensive study and consultations with airlines, was the Boeing 777, a long-range model intended to fill the gap between the 767 and 747. To some extent a response to the McDonnell Douglas MD-11, the 777 also faced strong competition from the new Airbus A330 and A340 series. Although the largest twin-engined aircraft ever built and a major risk for Boeing, proceeding during the recession, the 777 still was not the "bet the company" program the 707 and 747 had been. Powerplant competition again was heated, and the 777 could be powered by any of the new-generation large turbofans, the GE90, PW4000, and RR Trent, each offering eighty to ninety thousand pounds of thrust.

The Model 777 represented a major advance in being designed almost entirely by computer, as Boeing looked to minimize engineering man-hours.[3] Total development costs still were projected at $4 billion, and some 20 percent of the 777 was to be built by a Japanese consortium. Boeing maintained, however, that the core technology would remain in the

United States. The market breakthrough came with a major order from United, against spirited MD-11 and A330 competition, and orders mounted steadily. The first flight on June 15, 1994, was on schedule, and the 777 also gained ETOPS certification before service entry, a major boost.

Having decided against a new design in the 150-seat class, Boeing instead announced next-generation 737-X developments. Launched officially in September 1994, the new versions added -600, -700 and -800 to the 737 prefix, depending on passenger capacity, and orders flowed as the airliner market recovered in 1995 and 1996. Total 737 orders approached three thousand, and success of the new-generation developments vindicated Boeing's decision not to undertake a new design.

McDonnell Douglas, locked in third place in global airliner production, struggled to remain competitive. After considering larger MD-12X developments of the MD-11 for several years, the company finally announced a new MD-12 in 1992. A four-engined double-deck design larger than the 747–400, with a capacity of five hundred passengers, the MD-12 required a foreign partner, given the likely $4–$5 billion development cost. But the financial condition of the airlines at the time was such that no orders were placed, and the project ended.

Despite its commercial problems, McDonnell Douglas reached the production milestone of two thousand twin-jet airliners on June 11, 1992. But even combined with the 556 DC-8s and 389 DC-10s built, the jet total lagged far behind Boeing, and sales were slowing. McDonnell Douglas was further handicapped by being unable to match the family of airliners offered by Boeing and AI. But the MD-95 advanced-technology twin, after struggling to gain orders against new Airbus A320 and Boeing 737 developments, was finally launched in 1995. Assembly, originally planned for a factory in Texas, would remain at Long Beach, but with subassemblies coming from Canada, Japan, Korea, and Italy.

Both Boeing and AI studied six-hundred- to eight-hundred-passenger designs for service beyond the year 2000, under the Very Large Commercial Transport (VLCT) category. VLCTs would be four-engined ocean-spanning models weighing more than 1 million pounds. Boeing considered an original design, the New Large Aircraft (NLA), for this role. Then, in a surprising move at the end of 1992, Boeing announced it would explore the VLCT teamed with two AI consortium members, Deutsche Aerospace (DASA) and British Aerospace (BAe), but not formally with the AI entity. While viewed as a Boeing attempt to head off a European VLCT, the potential market in any event was so thin as to preclude two direct competitors. Boeing was also among those studying the HSCT, a Mach 2.4, three-hundred-passenger airliner for the twenty-first century, but it became apparent that the global market could not support both a VLCT and a new supersonic airliner. Boeing ended its VLCT and NLA studies on July 10, 1995, and turned to larger 747 developments for service after the year 2000.

Post–Cold War Strategy and the Industry

The drop in defense procurement growth from 1987 meant that the level of aircraft business was unlikely to be sustained. By 1991 production of older types such as the F-14 and F-15 was running down, and only five major aircraft programs remained, the Northrop B-2, the Lockheed-led F-22, the McDonnell Douglas C-17, the XV-22 tilt-rotor transport, and the LHX/RAH-66 advanced helicopter. All faced the dangers of delay, reduced orders, or full cancellation.[4] The unprecedentedly costly B-2 program was restricted to twenty units. The original LHX requirement of 4,595 units already was sharply reduced, and cancellation was distinctly possible, although that event would likely extend production of the McDonnell Douglas AH-64 and Sikorsky CH/UH-60 series.

The navy sought an interim replacement for the veteran A-6 while developing a successor program for the A-12. Grumman lobbied aggressively for a version of the F-14 for the mission, but the Pentagon instead chose a development of the F/A-18, the clumsily designated F/A-18E/F. The prototype F/A-18E flew in December 1995, and the navy planned to order some one thousand aircraft for service from 1999. McDonnell Douglas thus gained something of a reprieve, while Grumman's efforts were regarded as politically damaging to the firm. A successor A/F-X program launched in 1992 drew bids from three teams, led by Lockheed, General Dynamics, and Grumman, but was canceled by the Clinton administration. Then a joint navy/air force Multi-Role Fighter (MRF), to replace the F-16 and F/A-18, appeared but likewise did not proceed.

With the paucity of new military programs, the Joint Primary Aircraft Training System (JPATS) attracted intense interest. The potentially $6 billion program, for up to 711 aircraft plus related ground training equipment, appeared to be the last major production program of the twentieth century. The JPATS succeeded to some extent the canceled Fairchild T-46 and was intended to replace the air force T-37B, the navy T-34C, and possibly the navy T-2. It was further intended that existing designs be adapted for the role so that no development funding would be required. Illustrative of an increasingly global industry, not one of the original JPATS contenders was of American origin. Contenders were Vought teamed with FMA of Argentina, Beech with Pilatus of Switzerland,

Northrop with Embraer of Brazil, Grumman with Agusta of Italy, Rockwell with MBB of Germany, and Lockheed with Aermacchi of Italy. Short Brothers of Great Britain, producing a version of the Embraer Tucano for the RAF, explored teaming with Learjet, also owned by Bombardier, for the competition. Incidentally, the Rockwell/MBB (later DASA) teaming on the X-31 led to their JPATS association. Then Cessna made a late entry with a heavily modified development of its new CitationJet business aircraft, giving the competition one all-American entrant. After lengthy delays, the Beech-Pilatus design, known as the Mark II, was declared the winner in June 1995. Production was to begin in 1998, at low volume, but export prospects appeared promising.

The aerospace industry's technological capabilities for serving national security, commercial transportation, and space exploration were unquestioned. Doubts over its future related instead to rapid changes in both the strategic and economic environments. Among other uncertainties brought about by the end of the Cold War was the production future of aircraft in Southern California, its center for more than fifty years. The small city of Palmdale, for example, had grown rapidly in the 1980s due to expanding Lockheed and Rockwell operations but began to decline just as rapidly in the 1990s. Environmental restrictions and changing corporate ownerships were also factors, but the greatest factor simply was the decline in production and new contracts. Many aircraft workers faced permanent job loss.[5]

Rockwell's North American aircraft operations lacked a production successor to the B-1B, and Rockwell had strongly diversified away from aerospace. It left El Segundo in 1992, moving its headquarters to Seal Beach, where Rocketdyne launch engine development had begun. Northrop's aircraft future appeared linked primarily to the B-2, already sharply curtailed, and General Dynamics had not produced complete aircraft in San Diego since the 1960s. Lockheed's employment had dropped steadily from 1987 and accelerated with the close of Burbank and the transfers of LASC and the F-22 program to Marietta in 1991. Remaining P-3C production also shifted from Palmdale to Marietta. The famed Advanced Development Company (Skunk Works) was removed from LASC in May 1990 but moved from Burbank to Palmdale.

Douglas Aircraft Company experienced major employment cutbacks, and more operations shifted from Long Beach. Major components of the MD-11 and MD-80 already were produced in Salt Lake City, while portions of C-17 and T-45 work were transferred from California to Missouri to lessen the impact of the A-12 cancellation. Some C-17 work also was undertaken at a Georgia facility opened in response to financial incentives offered by the state, but which observers also viewed as an attempt to gain favor with Senator Sam Nunn, chairman of the Armed Services Committee.

The acquisition of Hughes Aircraft by General Motors also held unfavorable implications for Southern California, with engineering work transferred from Canoga Park to Mesa, Arizona. TRW Space and Defense operations were downsized, and Teledyne completed subcontract work on the Rockwell B-1B. Southern California's overall aerospace employment, unstable since the 1960s, continued to decline with defense reductions. Another factor, affecting many military as well as commercial aircraft, was that increasing proportions of their value were produced outside the United States.

Southern California was far from the only aerospace center affected by corporate restructuring and defense cuts. Grumman reduced employment steadily on Long Island but remained that region's major industrial employer. Even Boeing announced major employment cuts both in Seattle and Wichita due to the airline recession. Production, especially of the 747 and 767 models, remained profitable, but a slowdown was unavoidable.

The state of Connecticut, home to Pratt and Whitney, Sikorsky, and Kaman, provided financial support to those operations in an attempt to maintain employment, including $3 million to Kaman to develop the new K-Max helicopter for the commercial market.[6] Major U.S. manufacturing locations became more dispersed geographically than at any time since the Second World War. Texas loomed as an even more important center for the future, with Bell, General Dynamics (later Lockheed Martin), Loral, and Vought.

The vaunted Reagan defense buildup, even before the Soviet collapse and subsequent decline of the nuclear threat, had been blunted by unprecedented budget deficits.[7] Further, it had been established earlier that obsolescence of an operational aircraft no longer assured its replacement by a new design. Major upgrades and reconsiderations of operating needs resulted in certain older aircraft being maintained in service indefinitely. Such important aircraft as the B-52, C-130, C-141, A-6, KC-135, and T-37 served years, even decades, longer than originally anticipated. Accordingly, Service Life Extension Program (SLEP) contracts increased in importance. SLEP involved both structural strengthening and addition of upgraded electronics and other equipment, and so-called midlife updates (MLU) became routine for most aircraft. Contracts, which could amount to billions, were eagerly sought by the industry.

In addition, there were periodic proposals both from the industry and the military, given the severe difficulty of funding new-generation aircraft, to reinstate into production upgrades of such older designs as the C-141, C-5, A-7, F-14, and F-15. The F/A-18E/F was one upgrade that moved ahead, and Lockheed developed the privately funded C-130J, with new-technology Allison AE2100 engines and advanced propeller blades, to succeed earlier C-130 models. The F-22 thus competed for production funding not with another new design but with upgrades of the F-15 and F-16, while the costly C-17 faced competition from military versions of the Boeing 747 and 767 as well as from new production of the C-5. Such measures, while extending current production, ironically threatened new technology. Upgrade programs also were increasingly attractive to foreign air arms, which could undertake them locally, threatening exports. Israeli Aircraft Industries actively offered upgrades of U.S. aircraft for foreign air arms.

Military procurement behavior remained resistant to reform. But with the fundamental change in the strategic environment military spending, always cyclical, began to be cut to a degree not experienced since the end of the Second World War. Aircraft programs were among the most prominent cuts. The defense retrenchment and the severe contraction of the industry served to refute dramatically the long-held view in certain circles that the Cold War and heavy defense spending were primarily driven by military-industrial collusion or conspiracy. The military services remained parochial, however, reluctant to consolidate roles or eliminate duplications. The traditional preference for complex, costly weapon systems continued. The air force emphasized control of the "high battle" and still lacked enthusiasm for Close Air Support. But the Marine Corps air arm, largely duplicative of others, finally began to eliminate some aircraft categories and missions.

Yet while accepting the inevitability of downsizing, the military fought to preserve new and existing programs, even at inefficient production rates. The major change required by the budget crunch was a cultural change. The government was forced to become a smarter shopper, controlling byzantine specifications and inspection procedures that ran up costs. But deficit reduction through defense downsizing, while positive news for many taxpayers and for other public-policy priorities, translated for the industry into delays, reductions, cancellations, consolidations, plant closings, and massive layoffs.

The aerospace industry regarded itself as an irreplaceable national resource, not only for its production capability but also as a storehouse of technological and scientific skills. The industry argued for a coherent national policy directed toward its preservation, even with the changed strategic environment.[8] The Department of Defense likewise wished to maintain an industrial base of financially healthy, technically capable firms. Further, it wished to retain production capacity and the necessary mix of skilled personnel, factors also of concern for the broad economy. Debate was with regard to specific measures, and whether those measures should be determined by the government, raising the issue of industrial policy.

The defense/aerospace industry still consisted of privately owned firms. The United States had never adopted a formal industrial policy, and such a policy was strongly opposed by doctrinaire free-enterprise advocates. Yet defense policy, for which technology and hardware were provided by the private sector, constituted a de facto industrial policy. Additionally, NACA and NASA represented successful applications of government-led industrial policy. U.S. policy differed from that of Japan and its Ministry of International Trade and Industry (MITI) in that it was not necessarily by design, yet the impact was similar. Whatever the future, it was clear that a free-enterprise laissez-faire doctrine was unrealistic, and that the government would continue as the dominant influence over the structure and future of the aircraft/aerospace industry.

With the lag in developing a cohesive post–Cold War national-security strategy, the industry operated in what it perceived to be a strategic vacuum. Threats to the national security still existed, with the potential for deployment of combat forces, but precise military requirements and the timing and quantity of future weapon systems were even more uncertain. Major cuts in procurement and R&D programs were being determined on an individual basis, with no apparent assessment of their collective impact on national security.

Recognizing this deficiency, the Clinton administration in 1993 undertook a so-called Bottom-Up Review, redefining missions, force levels, and future requirements. The review still sought to cushion the impact on the defense industry, ensuring that declines would be gradual and that key programs would continue. One major change was that air force and navy combat aircraft requirements would be combined into a more affordable basic design. The Joint Advanced Strike Technology (JAST) program, succeeding the A/F-X and MRF and projected to replace the F/A-18E/F and the F-16, resulted. Then during 1995 the JAST was combined with the DARPA-directed Advanced STOVL (ASTOVL) program for a STOVL Strike Fighter (SSF), for which Boeing, Lockheed, and McDonnell Douglas held prime contracts. The combined program for a Joint Strike Fighter (JSF) then bid to replace the AV-8B, A-6, and effectively the canceled A-12 as well. Contenders were Lockheed Martin, McDonnell Douglas/Northrop Grumman/BAe, and Boeing, with two to be selected to build prototypes. The British Aerospace collaboration, significantly, marked the first foreign participation in a new American military aircraft program.

Major military reorganization proceeded during downsizing, including restructuring of operational commands. Of immediate concern to the industry, the Air Force Logistics and Systems Commands, dating from 1961, were effectively combined into a new Air Force Materiel Command on July 10, 1992. AFMC continued as the major point of interaction between the air force and the industry and was intended to improve that relationship.[9]

Defense downsizing also affected military personnel strength and the operation and maintenance budget, bringing the military increasingly into conflict with the industry in many modification, overhaul, and MLU programs. With force contraction, the military incentive was to expand SLEP and other work at military logistics centers rather than contract to the industry. Other interests, including some within the Department of Defense, felt that most military maintenance and modification work should be taken over by the industry, to maintain its structure and capabilities. The industry effectively won in 1994 as Defense moved, over air force opposition, to close more Air Logistics Centers, transfer work to the industry, and end the requirement for the industry to bid against the military for future modification and maintenance contracts.[10]

Despite the imperative for industry conversion and restructuring, its military capabilities simply were not readily transformable to other fields. The record of commercial diversification was not encouraging.[11] Observations along the lines of "you cannot go from submarines to washing machines" were voiced frequently. Related to the Bottom-Up Review, there was also considerable opposition to the conversion concept, with many feeling that the priority should be on restructuring that would preserve military capabilities rather than on commercial conversion.

Exports of aircraft and missiles could not provide a strong offset to domestic procurement declines as they had in the 1970s, since other nations were reducing imports as well as developing their own defense industries. Further, without the cushion of large domestic production programs as in the past, export prices would inevitably rise. Another complexity was the worldwide overcapacity in combat aircraft production. Former Soviet bloc aircraft producers were competing aggressively with Western companies on an economic basis for orders, particularly in developing countries. The commercial sector also suffered from overcapacity, and Russian airliners, many with Western electronics and engines, loomed as a future, although minor, competitive factor. But the political changes also brought increased collaborative opportunities with Russian manufacturers, as well as for upgrades of existing aircraft. Such developments did not fully offset the fall in demand, however; aerospace sales experienced an absolute decline in 1991 and 1992, and employment fell almost three hundred thousand from 1989 to 1992.[12] Aerospace employment dipped below one million in 1993, the lowest level since 1978, and continued to fall.

The aerospace industry still consisted of six sectors.

1. Military aircraft
2. Commercial aircraft
3. General-aviation aircraft
4. Helicopters
5. Missiles
6. Space

Electronics and engines/propulsion systems were closely related fields, of course. Increasingly critical to the industry, missiles and NASA space projects as well as aircraft programs were budget targets, so that declines in one sector were not offset by gains in another.

The Aerospace Industries Association attempted to increase political support by moving beyond narrow industry concerns, identifying eight key technologies for future support, including those not directly aerospace. The program called for an academic-industry-government partnership to expand and exploit those key technologies to combat the erosion of American technical and competitive leadership.[13] It represented in one sense advocacy of a more formal industrial policy. There was little question that the industry was required to shift a major portion of its resources and production capability to other endeavors, but the management of such a conversion remained unresolved. A major question remained of where the aerospace industry's capabilities could be best applied, given the fierce competition and overcapacity in such potential alternatives as surface mass transportation, electronics, and computers.

Rising concerns over American global competitiveness as military requirements decreased led to a role for DARPA in civil advanced technology. While a separate civil agency was considered, DARPA's defense prefix instead was dropped in 1993 as the agency became ARPA, its original name when established in 1959. It was anticipated that DARPA's military success would extend U.S. technological and competitive leadership across many fields. ARPA also was to implement the Pentagon goal of more effective integration of defense and nondefense sectors of industry.[14] Space activities were left almost entirely to NASA.

Regardless of its capabilities, the industry's problems were broadly representative of overall economic concerns. Declining demand and output, job losses, and stronger import competition affected many industries. Aerospace firms, in response, followed often politically controversial policies, including foreign sourcing and international production sharing, particularly in commercial aircraft, which many feared would lead to further erosion of the production base and intensified international competition. Labor relations became more tense over job shifting and outsourcing. Boeing suffered another strike in 1995 over the issue of foreign component sourcing. But industrial offsets were a fact of life in many industries, and aerospace felt that it had no choice but to pursue such policies.

The aerospace industry, unlike certain other industries, had never lost technological leadership or markets because of declining global competitiveness. Rather, it suffered from an absolute decline in demand, did not enjoy the broad customer base or product diversity of other major industries, and faced unique problems in nonaerospace diversification. The few remaining prime contractors continued to seek broader markets and to design aircraft to fill market gaps, including export markets, while striving to remain financially viable. But the greatest problem was the dearth of new programs and the low volume of current programs. Research-and-development contracts, long increasing in importance, became even more prominent with the production decline. Thus the aerospace industry could not anticipate absolute sales growth but would be forced to maintain or increase profitability in other ways.

In attempting to plan for future markets and demand, however, the industry again ran afoul of procurement rules and committed ethical violations. In the aftermath of the "Ill Wind" investigations, for example, industry spokesmen maintained that overall environmental pressures, including an impractical or unrealistic procurement system, rather than any inherent ethical deficiencies in the industry, carried substantial responsibility for recurring scandals. The AIA pointed out that in other industries and other markets, firms can work with their customers to ascertain likely future markets and needs, but that in dealing with the government, "Individuals now face criminal penalties for trying to find out what the customer may want five years hence."[15] Wolfgang Demisch, a prominent aerospace/defense security analyst, stated, "Well-publicized scandals do not reflect widespread dishonesty in the defense industry, but the government's impossible procurement systems."[16]

Procurement scandals remain a troubling aspect of the industry's record. Scandals within the industry and investigations thereof, dating from the First World War, led many to believe that the aircraft/aerospace/defense industry leadership was inherently exploitative or inclined toward unscrupulous business practices. Respect for the normal competitive interplay of the market and generally accepted ethics appeared almost nonexistent. Some continued to attribute problems to a military-industrial complex, which still existed despite defense downsizing. But there remained the questions of whether it constituted a conspiracy, and why unethical practices have continued.

The answer again hinges on the uniqueness of the industry. Unlike other corporations that may be large government contractors, such as General Motors and General Electric, aircraft/aerospace firms remain abjectly dependent upon continued government business for survival, which has accounted for numerous attempts to diversify. Other major corporations generally are free from the vulnerability to cancellation of or failure to win a single contract that would effectively end their production futures. Given such pressures, it should not be surprising that management might resort to unscrupulous practices in an attempt to gain an advantage. Yet such practices still would have been impossible if government control of the process had been effective. Critics of the industry-government relationship alleged that the industry exerted undue influence on, or even drove, government policy in exports, specifications, and operational requirements. If true, such influence necessarily requires the active cooperation or at least acquiescence of senior defense and military officials.

In other political areas, when the influence of certain advisors or groups with, for example, the president, has been questioned, it has been noted that no one can possess influence that the president does not wish the person to have. The same must be said of any contractor influence with the Pentagon or with Congress. Further, if officials act in the interests of particular firms and against the interests of the public and national security as determined by the president, then those officials, both military and civilian, can be removed, if not prosecuted. Yet before "Ill Wind" it was difficult to document a single case of such an action in consequence of improper relationships between officials and the industry. Senior officers, including chiefs of staff, have been fired for differences on national-security policy or for making impolitic public statements, but almost never for contract misconduct or questionable relationships with the industry. To the extent such improper relationships continued, it has been due to the failure of presidents and secretaries of defense to exercise their authority.

Despite reform efforts extending back to the 1950s, the acquisition process remains protracted and inefficient. A probable explanation for the failure of reform lies with broad societal choice. Society professes in the abstract to demand efficient, cost-effective military procurement, and nonduplicative, rational military missions and service roles. The reality, however, is that implementation results in lost jobs, closed factories and military installations, contract reductions or cancellations, and further shifting of risks to contractors. National-security reassessments result in similar measures with similar economic impacts. Society rather consistently has opted for maintaining unneeded installations, continuing unnecessary or troubled aircraft and weapons programs, and maintaining excess capacity in the industry. Elected representatives fought to preserve the status quo since their constituents rewarded them for doing so. Even in the 1990s, with huge and growing budget deficits and a consensus for military downsizing, individual congressmen still sought to alleviate high unemployment and bail out declining companies in their districts with defense dollars. Fitzgerald was uncomfortably close to a major truth when he alleged that inefficient defense contracting represented de facto national policy.

The same societal choice was manifested in military exports. An aircraft worker in Fort Worth was unlikely to be concerned with any negative impact on American relations with China or other destabilizing implications of an F-16 sale to Taiwan, but only with the fact that his job would be preserved for another few years. Stockholders and executives of affected firms tended to react similarly.

The industry's financial record argued against any antipublic conspiracy and also clashed with economic theory. Competitive theory held that profitability was strongly correlated to industry concentration and barriers to entry. Industries possessing those characteristics influence prices to the extent that above-average returns could be realized.[17] The aerospace industry was increasingly concentrated, yet financial crises appeared with distressing regularity, and long-term profitability remained elusive. The record reinforced the view that the aerospace/defense industry simply could not be encompassed within free-enterprise theory. The commercial sector also was so heavily politicized that pure market competition did not exist.

A consensus on the correct measurement of financial returns and risks for the industry remained elusive. Defense firms tended to be downgraded as investments by the capital markets, yet their stocks could rise rapidly on favorable earnings news. Complicating risk assessment was that a firm still could not retain proprietary rights to the technology it had developed. The Defense Department could and did transfer that technology to other firms.[18] One favorable ruling by President Bush in 1992 was that the policy of recoupment, in which firms were required to pay a levy to the Department of Defense on exports of aircraft for which development had been government funded, would end. The policy had been regarded by the industry as a deterrent to exports. Another promising benefit in the era of intensified competition was that the reluctance of the president and other high officials to promote aerospace products abroad had largely disappeared. President Clinton publicly and unabashedly stated that he would do so.

Defense contracting declined sharply as an attractive business to pursue. AIA president Don Fuqua contended that the government, in its contract reform attempts, had thrown the basic business equation out of balance. He called for both Congress and the Department of Defense to reexamine reforms, and to recognize the aerospace/defense industry as a free-enterprise industry to the extent that investors must realize a fair return. Otherwise, capital sources would dry up and the industry could not meet its national-security responsibilities.[19] Preserving the aerospace/defense industrial base remained the paramount issue.

Restructuring and Consolidation

As the aircraft/aerospace industry entered the 1990s, its long-standing problem, too many producers for the level of business, reached its most critical stage. Industry overcapacity had forced the exits of Fairchild, LTV and Rockwell as aircraft producers, Fairchild by a strategic decision. Rockwell, while having no production programs, remained active until the end of the X-31 program and its loss in the JPATS competition.

The aerospace industry thus faced a long-term secular decline in the new decade. By 1992 only seven firms, Boeing, General Dynamics, Grumman, Lockheed, McDonnell Douglas, Northrop, and the Rockwell North American Division, remained capable of developing advanced aircraft. However, certain firms were so structured that subordinate organizations possessed entirely separate prime contract capability, such as McDonnell and Douglas under McDonnell Douglas, and LASC and LADC (Skunk Works) under Lockheed.[20] But most forecasts were that no more than three or four prime aircraft contractors would be needed under any circumstances.[21] The number of major subcontractors also will decline sharply.[22] Further, Secretary of Defense William Perry emphasized that nothing in policy reviews would constitute an industry bailout or job preservation program, that the industry must continue to downsize and restructure, and that many smaller contractors inevitably would go out of business.

The chief business characteristic of the aerospace and defense industry remained risk. Diversification remained a priority. Increasingly, major firms were viewed as managers of product franchises that could be sold to others for financing or restructuring purposes. Fairchild's attempt to sell its T-46, and General Dynamics' F-16 sale, were examples.

Three major transactions implementing industry restructuring and consolidation occurred during 1992. LTV, attempting to work itself out of bankruptcy, placed its Aerospace and Defense operations up for sale on May 20, 1991. After a $350 million joint Lockheed–Martin Marietta bid was bested by a $400 million French bid, the resulting political controversy led to companion transactions. The LTV Aircraft Division was sold to Northrop, 49 percent, and the Carlyle Group, 51 percent, as Vought Aircraft Company, with Northrop having the option to buy the entire company later. Vought Aircraft continued with subassemblies for the Boeing 747, 757, and 767, promising synergism with Northrop's existing 747 work. Vought also subcontracted on the C-17 and B-2 but was eliminated as a JPATS contender. LTV Missiles and Space Division went to Loral Corporation as Loral Vought Systems. Both sales were effective on August 31, 1992. Although having exited complete aircraft, the Vought name, one of the oldest in the industry, survived.

Martin Marietta, already the dominant firm in defense/aerospace electronics and planning to remain so, bought General Electric Aerospace, primarily involved in defense electronics, but not including the aircraft engine operations, for $3 billion in November 1992. Then General Dynamics sold its Fort Worth Division to Lockheed, which outbid Boeing and Northrop, for $1.5 billion in March 1993. The division became the Lockheed Fort Worth Company (LFWC). Lockheed thus gained the highly successful F-16 program as well as the GD share of the Lockheed-led F-22 program, promising greater efficiency. With the sale General Dynamics exited aircraft after seventy years, dating from formation of Consolidated Aircraft in 1923. GD also sold its missile operations to GM-Hughes in May 1992 for $450 million but remained important in structures and electronics.

Major aerospace firms operated in a manner far different from that which typified their histories, with definite improvements in governance and accountability. The addition of more outside directors, greater director activism, and the decline of one-man or family control resulted in a governance profile more comparable to other major industries. One salient difference, however, was that aerospace firms cooperated and competed simultaneously, both domestically and internationally, in industry teaming for major defense contracts and for airliner development programs. The fraternal manner in which the industry had operated continued but became strained in the 1990s with merger pressures and intensified competition.

While vulnerable to a variety of adverse developments, aerospace firms increasingly returned to what they regarded as their core technologies, frequently closing or selling peripheral lines of business. Designs or new technology would not remain permanently proprietary, yet such firms as Northrop and General Dynamics continued their military emphasis, since they had no realistic alternatives. General Dynamics undertook a restructuring more fundamental than any firm in the industry, reducing the size and scope of its activities while remaining focused on defense. The sale in January 1992 of Cessna to Textron, after less than seven years under GD, reflected that strategy. But the sale of the Fort Worth Division represented a permanent downsizing, or even a partial liquidation, as proceeds were returned to stockholders as special dividends. Annual sales, approaching $10 billion in 1992, declined to some $3.5 billion, but the change was well received by the stock market.

William A. Anders, who became CEO on January 1, 1991, upon retirement of Stanley Pace, stated, "General Dynamics has a history of periodic unpleasant surprises."[23] He eliminated operations even within its four core technology areas, military aircraft, tanks, submarines, and space launch systems. A major problem was the apparent lack of a market for the submarine-producing Electric Boat Division, hence the sale of the Fort Worth Division. The Convair Division was closed in 1994, and MD-11 component production returned to McDonnell Douglas. What remained of General Dynamics could be merged with another firm, either in or out of aerospace.[24] Henry Crown died in 1990, but the Crown family still held more than 20 percent of GD stock, and a decision to sell that interest could lead to an immediate merger. GD and McDonnell Douglas eventually won a court reversal of the A-12 cancellation charges but once again faced charges of a taxpayer bailout.

McDonnell Douglas, the largest defense contractor at the beginning of the 1990s, faced major restructuring as well. It sold the information systems subsidiary, but the company's involvement in the commercial sector, which represented about one-third of total business, was a major complication. It was faced with securing major international financial and production support, and doubts persisted about its long-term future in commercial aircraft.[25] In yet another reflection of the complex contractor-government relationship, McDonnell Douglas faced accusations that it was being "bailed out" by favorable Pentagon financial treatment after a series of setbacks. The company had suffered the loss of five thousand jobs with the A-12 and encountered reliability problems with the AH-64 and major development problems with the C-17 and T-45. The results were severe cash flow pressures and a downgraded debt rating.[26] The helicopter operation, with declining prospects, was in fact offered for sale, although later withdrawn.

By 1992 Northrop faced a restructuring with a reduced B-2 program and loss of the ATF.[27] Major subcontracting on the F/A-18 series continued, but Northrop's only significant commercial activity remained 747 fuselage sections until the Vought acquisition. Even with the B-2 curtailment, Northrop was unwilling to accept a future as a subcontractor, leading to its purchase of Vought and its failed bid for GD–Fort Worth. A Boeing-Northrop merger, building on their subcontracting history, seemed logical but was not pursued.

Grumman, with its long-running F-14, A-6, and E-2 programs ending, faced an inevitable exit in aircraft, although it maintained major engineering and technical capabilities. The Calverton factory was closed, and remaining E-2C production was transferred to Melbourne. Overall employment, as high as thirty-four thousand in 1987, declined to twenty-three thousand by 1992 and continued to fall. In October 1993 Grumman announced its exit as an aircraft prime contractor, while retaining production capability for JPATS and other programs, including J-STARS at Melbourne. Its future was focused in systems integration, which already accounted for some $2 billion of its $3.5 billion annual sales.

More quietly, chairman Renso Caporali also placed the company on the market. On March 7, 1994, Martin Marietta announced a fifty-five-dollar-per-share, $1.9 billion offer, which Grumman's management accepted. Martin Marietta thus expected to add Grumman's electronics capabilities to its already impressive list but was stunned when Northrop quickly countered with a sixty-dollar-per-share, $2.1 billion offer. It was also revealed that Northrop had conducted negotiations for Grumman earlier on a friendly basis, but that Grumman eventually withdrew.[28] A Northrop-Grumman merger was regarded as a strong strategic fit, with few overlapping product lines, although both JPATS entries were retained.

Martin Marietta was widely expected to top Northrop's bid, presenting the prospect of an extremely rare takeover battle in the industry. Financial markets regarded the move as riskier for Northrop than for Martin Marietta, since Northrop lacked the latter's financial strength and planned to borrow the necessary funds. Ultimately Martin Marietta, true to its cautious approach in the past, declined to engage in a bidding war, and Grumman's board, sensitive to its 40 percent employee ownership and to its fiduciary responsibilities, accepted a higher Northrop offer of sixty-two dollars per share, or some $2.17 billion.[29] Further layoffs appeared inevitable at both operations postmerger, and speculation emerged that the Northrop Grumman Corporation, with a heavy debt burden, might become an acquisition target itself. Northrop appointed veteran executive Oliver C. Boileau to run the Grumman operations during the transition.

Martin Marietta announced that it would continue to explore acquisitions consistent with its business strategy. Northrop Grumman followed through with plans to acquire the remaining 51 percent of Vought Aircraft from the Carlyle Group for $131 million in cash. It offered to build an additional 20 B-2s over several years at a fixed price, although the proposal became politically controversial. The first twenty B-2s carried a program cost of some $2 billion each, comparable to a ballistic missile submarine. Northrop was widely felt to be involved in black programs as well.

Although many firms had been concerned that mergers, however necessary and inevitable, might encounter antitrust opposition, understandings were reached between Defense and Justice officials that recognized the special circumstances of the defense industry, and which eased the transition.[30] Few concerns were expressed about the Martin Marietta–GE or Lockheed-GD deals, and virtually none about the Northrop-Grumman merger. Perhaps most important, the accord recognized that consumer protection was not the issue it might be in other industries, since the defense industry had essentially a single buyer in the government.[31]

Boeing had no active military aircraft programs other than military versions of its commercial designs. Its role as a partner on the F-22 was of high importance, and if successful in the new Joint Strike Fighter program Boeing could return strongly to combat aircraft. Boeing's commercial priorities were efficiency and cost control, and market pressures had forced a sharp drop in employment, from a 1989 peak of 165,000 to below 120,000 by 1993. But an airliner market turnaround and success of the 777 portended a bright future, even leading to a capacity shortage.

McDonnell Douglas's effort to sell a 40 percent interest in its commercial operations to the government-controlled Taiwan Aerospace Corporation, although unsuccessful, presaged its commercial future.[32] Asia, with low-cost production and a rapidly growing commercial market, was the most logical locus of the necessary partnerships.[33] Such alliances also promised to improve the company's reputation on Wall Street, where heavy debt, cost overruns, and military contracting problems had taken a toll. John F. McDonnell was still criticized for the failure to fully integrate the St. Louis and California operations.[34] He shared authority with three other senior executives during the 1991–92 financial crisis and afterward focused increasingly on strategic planning. The company managed an impressive financial turnaround by 1993, with a rising stock price, and despite commercial difficulties was committed to reducing government business to less than 50 percent of the total by 1995.[35] At the end of 1994 Harry Stonecipher, who had turned around the defense contractor Sunstrand, was appointed CEO of McDonnell Douglas. With a reputation as a tough, hands-on manager, he dealt immediately with the commercial problems, as military business appeared strong. A major victory was that the C-17 eventually won support over the Boeing C-33 variant of the 747, and the full requirement of 120 was likely to be built.

Lockheed reached a major milestone with delivery of the two thousandth C-130 on May 15, 1992, and production bid to continue into the next century with the C-130J Hercules II. In purchasing GD–Fort Worth, Lockheed also departed from the prevailing view that defense contractors should lessen their dependence on the field.[36] Its earlier strategy in which aircraft would constitute only 20 percent of future sales was again recast. Even as the F-16 program ran down, Lockheed felt that it could prosper with a major military stake, but the rapid decline in that market led many to believe that it had seriously overpaid for GD–Fort Worth.

Regardless of doubts over its strategy, Lockheed remained probably the most technically capable of all aerospace firms. Then in a major development that surprised most observers, Lockheed and Martin Marietta made a joint announcement on August 29, 1994, of their plans to merge. A merger of equals, the transaction carried a value of $10 billion and would result in combined annual sales of $23 billion, second only to Boeing in aerospace. Lockheed Martin Corporation headquarters was to be in Bethesda, Maryland, with Daniel Tellep as chief executive, to be succeeded upon his retirement by Norman Augustine of Martin Marietta.[37] While a merger of that scale had been regarded as almost inevitable, Lockheed's diminished production prospects for the F-22, service entry having been postponed to 2003, and Martin Marietta's losing bid for Grumman helped draw the firms together. Each stated that they had considered no other partners. The Department of Defense, consistent with its policy of industry consolidation, approved the proposal after a competitive review, as did the Justice Department. The merger became effective on March 15, 1995, and promised defense cost savings, but also major job losses. The deal heightened further merger speculation.[38] The trend also applied to engines, as Rolls-Royce, with a collaborative history dating from the TF41 turbofan, purchased Allison from GM in 1995.

Another surprise announcement occurred in October 1995, of a possible affiliation between Boeing and McDonnell Douglas. Complicated by their competition in the commercial sector, talks soon ended with no basis for agreement. Lockheed Martin acquired Loral for $9 billion in April 1996, becoming the largest defense contractor in the process. Then in 1996, with continuing problems in the commercial sector, and deciding against a major stretch of the MD-11, McDonnell Douglas announced that it would cooperate with Boeing on future airliner development.

Only two months later, on December 15, 1996, came the announcement of the largest aerospace consolidation of all. Boeing announced a bid for outright acquisition of McDonnell Douglas, for $13 billion in stock. Boeing would dominate management and the board. The imperative for McDonnell Douglas remained its troubled airliner operation. The MD-95 is likely to be the only long-term Douglas airliner survivor, cast as a smaller alternative to the 737 series. The firm agreed to merger as it appeared to be simply losing out to Boeing and Airbus in the market. In addition, it had been eliminated in the JSF competition, leaving its BAe partner embittered. The result was a cloudy military future after completion of the C-17, F/A-18E/F, and T-45 programs. Boeing, while negotiating from a position of strength, felt that it could benefit in the JSF competition from McDonnell Douglas military experience.

In the rotary-wing sector, production remained predominantly military, although civil uses continued to grow in variety and number. Sikorsky and Bell continued their leadership, and production of the primarily military Sikorsky S-70 series reached two thousand by summer 1994. Expense of operation will remain a limiting factor in civil market growth, but Sikorsky announced the advanced S-92 Helibus transport for 1998, a strongly multinational project involving participation from Japan, China, Taiwan, Brazil, and Spain. With downsizing, Boeing Helicopters became the Helicopter Division of the Space and Defense Group in 1992. Consolidation among the four major helicopter producers, all subsidiaries of larger firms, appeared inevitable. The Boeing–McDonnell Douglas merger meant that the helicopter operations of both firms would become rationalized in some manner.

Total military aircraft requirements are unlikely to exceed several hundred per year in the future, compared with the thousands in Cold War years, with manufacturers finding only low-volume production possibilities. All sought areas of potential profitability in a reduced market. It appeared highly doubtful that the United States required or could afford an F-22, an F/A-18E/F, and a Joint Strike Fighter in succession. The JSF proposal is single-engined, as a cost-saving measure. Further, export prospects for such aircraft, given a potential unit cost approaching $100 million, were dim. The industry had achieved its success with technology, but it was clear that technological leadership would no longer be sufficient to assure survival.[39]

Possibly in recognition of the reduced prospects, in July 1997 Northrop Grumman Corporation, only three years after its formation, agreed to be acquired by Lockheed Martin. The coming end of the B-2 program and of remaining Grumman programs spurred the merger, which still appeared to raise more national security and antitrust concerns than other recent combinations. The merger would leave the industry with only two large aircraft producers, although Raytheon had gained stature in missiles and electronics.

A Global Future

Global competition will increasingly determine the industry's future. Examples of an aircraft completely designed, built, and equipped in a single country will become rare. Teaming on military programs and global collaborations will characterize the operations of the industry. In this respect the aerospace industry is becoming similar to the already global automobile industry.

The commercial sector already was strongly globalized. Not only was production of components and subassemblies internationalized, but design, development, and even management of airliner programs became globalized. Airliner components were produced in as many as ten countries for final assembly at a central location.

There will be increasing links between U.S. and major European aerospace firms. Future strategic alliances are a matter of simple economics, with the world market simply unable to support three or four competitors for one class of aircraft.

Little public pressure was evident for restrictions on international commercial joint ventures, either for aircraft or engines. While some concerns were voiced that joint ventures would serve only to create new international competitors in the future, it seemed likely that competition would emerge more in components and subassemblies, which in turn would benefit the American prime contractor.[40] The Department of Justice in practice took a permissive view of such arrangements as GE-RR and P & W–RR, though they appeared anticompetitive on the surface. The General Dynamics (now Lockheed Martin) joint venture with Japan to develop a next-generation FS-X fighter based on the F-16 raised fears of creating future military competition. But the industry was irreversibly global, with a global market.

The global industry also became recognized as critical to U.S. strategic interests. But with deepening involvement in international collaboration or strategic alliances, the question remained as to whether such ventures were really attractive for the firms or were a necessary evil into which they were forced for survival.[41] On that question the researcher Materna, using McDonnell Douglas as a primary case study, began by identifying seven characteristics of the contemporary aerospace industry.

1. Export leadership
2. Capital intensity
3. Technology
4. Civil-military relationship
5. Diversification
6. Historical global dominance
7. Internationalization trend

McDonnell Douglas proved a particularly interesting choice, given its history as a military company founded and led by an individual not known for his international orientation. Yet Materna observed that 24 percent of McDonnell Douglas total revenues in 1988 came from international sales.[42] Such variables as international business experience of key executives, and a sense of the importance of, and commitment to, internationalization were the most important determinants in the transformation. Key managers were required to accept the need to become more geocentrically oriented to survive.[43] The study concluded that the firm was indeed emerging from ethnocentrism into a global corporation staking its future on Global Strategic Partnerships.[44] There were already, for example, McDonnell Douglas Canada, a major subcontractor to the parent, and McDonnell Douglas China, for the license-produced MD-82 airliner. All aerospace firms would become geocentric.[45]

Despite increasing internationalization, U.S.-based aerospace firms remain unlikely to develop into true multinational enterprises. In fact they cannot, due to national-security considerations, including restrictions on foreign ownership. The controversy over government subsidies to commercial aircraft, especially to Airbus, remained. Restrictions appeared to gain broader political support, especially with recently approved GATT reforms. U.S. protests diminished, however, as Boeing continued to dominate the market, gaining almost 70 percent of new airliner orders worldwide. The political sensitivity of the aerospace industry was indeed global, and there was strong pressure for more subdued rhetoric from all parties.

Aerospace may largely disappear as a distinct industry. It appears increasingly likely to evolve into a more diversified high-technology industry along the Lockheed Martin model, remaining committed to defense, but with aircraft a minority of total business. Boeing ultimately may become the only firm with aircraft constituting a majority of sales. Remaining firms thus will be less vulnerable to loss or cancellation of a single major contract, but a consequent danger is that the strong impetus toward aeronautical progress and development may also decline.

The aircraft/aerospace industry has rarely enjoyed stability, but the end of the Cold War and the rise of globalization have forced a permanent change in its size and structure. The industry has been reduced to a small core of active aircraft firms, albeit more technically sophisticated than at any time before, and that consolidation seems a good point to bring our story of the industry to an end.

It would perhaps be satisfying to conclude that the industry had gained a stable position in the American economy and society, and that its structure and its relationship with the government were also clearly defined and stable. Yet we cannot. Perhaps it would be rewarding also to state that the industry was fully understood and well regarded by the public and taxpayers, but this is far from the case. Unquestionably the industry has contributed greatly to American economic superiority and national security and has demonstrated astonishing technological achievements. It will remain the preeminent technological industry. Further, aircraft production, and other sectors of the broad aerospace industry, will have a future in the United States, but the future size, structure, and societal role of the industry are unresolved. The moral dilemma of development of ever more potent destructive weapons by an industry that provides numerous civil benefits, in technology, taxes, and employment, persists. The industry argues for its continued support more on those grounds than on any strategic or national-security needs. Further, as with most government programs and government-supported industries, the fact of its existence is widely argued as the prime rationale for its preservation.

Epilogue: What Manner of Man?

In concluding the ninety-year history of the American aircraft/aerospace industry, it seems fitting to address a final time the nature or qualities of the men who were its major builders. What set them apart from others, and what enabled them to succeed for so long in a business with a notoriously high failure rate? It would appear prima facie that at least three characteristics were necessary.

1. An obsession with aviation, quite apart from any business or profit opportunities, and a dedication to the industry regardless of economic conditions, market structure, or the political environment at any given time
2. Managerial skills, including the ability to attract capable designers, engineers, and production managers. Also included were the ability to attract financing, organize production, and market to both military and commercial customers, given frequent market shifts and strong competition. Such industry leaders as Robert Gross, William Allen, and Reuben Fleet were known primarily as master salesmen. Another essential skill, and in this the aircraft industry is set apart from most others, at least in degree, is the ability to maintain strong relationships with and confidence of the major customer, the government.
3. An ability to keep abreast of or adapt to both changing technology and operating environments. In this many early industry leaders fell short. They persisted too long with established designs, with established ways of doing business, and with maintaining personal control as the industry became more complex and diversified and faced new environmental and competitive challenges. Contemporary managers must be more cost-conscious than ever before, while the procurement system enabled most early leaders to avoid rigid cost controls, although there were exceptions.

Conspicuous in their absence are flying skills and design skills. As important and as characteristic as they were of the industry pioneers, these skills simply were not essential for long-term success. Such individuals as Allen of Boeing, Gross of Lockheed, and Kindelberger of North American owed their success to vision and leadership rather than to any capability as designers, engineers, or aviators. The principle of universality of management, that its functions are the same regardless of industry or external environment, applies to the aircraft/aerospace industry, but management skills applied in other industries do not appear to be readily transferrable to aerospace. There are few instances of successful executives in other industries being recruited to run an aerospace firm and continuing their success. Bunker of Martin was a significant exception.

A further difference in personal qualities may be observed among those primarily associated with helicopters. That quality was inventiveness. Igor Sikorsky, Frank Piasecki, Stanley Hiller, and Charles Kaman were also inventors, gaining recognition for accomplishments in areas other than rotary-winged aircraft.

Now, what further personal characteristics may be discerned from study of those who contributed most prominently to the development of the industry? It is already established that most who started their own companies did so when young. Many had experience working for other early companies and industry pioneers. Others, particularly after the First World War, also had military experience, either as aviators or engineers. All shared an obsession with aviation, with the importance of building an aircraft industry, and with its potential.

A conventional home and family life did not appear to be the norm for industry leaders. The Wright brothers, as is well known, were lifelong bachelors, and socially somewhat reclusive. Glenn Martin also was a lifelong bachelor, and may well have been homosexual. His mother, who lived with him until her death, was the strongest influence on his life. Sherman Fairchild also was a bachelor, but differed in being independently wealthy and socially prominent.

Larry Bell, Dutch Kindelberger, Jack Northrop, and Donald Douglas were divorced after long marriages. Bell did not remarry. Reuben Fleet was married three times, and Allan Lockheed also was divorced. Howard Hughes, although not meriting inclusion among the leaders, was twice divorced, a playboy in his younger days, and of doubtful sanity during the final two decades of his life. While it is not the purpose of this book to examine and discuss the personal lives of the builders of the industry, it nonetheless warrants comment in that most who built the industry did not appear to conform in lifestyle to those who built other major industries, or to contemporary business leaders.

Many leaders were of the American Midwest, or were at least born there. This included the Wright brothers, Glenn Martin, Claude Ryan, Larry Bell, Bill Boeing, Frederick Rentschler. Others were from the two coasts. Few were from the mountain states, northern plains, or the South.

As noted earlier, many early industry leaders were immigrants. To a major extent the locational pattern of the industry reflected the geographical locations of the founders when they began their careers. Early industrial location was more a matter of happenstance than of any plan or the natural advantages of any region, although Southern California certainly possessed distinct advantages in manufacturing climate and flying weather. Education does not appear to be an explanatory factor in industry success. Many pioneers and leaders lacked formal education, although others were degreed in nonaeronautical fields. The great designer Ed Heinemann possessed only a high school education.

Another pattern that emerges is that of strong family influences, not only in initiating manufacturing ventures but also across generations. To an extent this has been characteristic of American business in general, in that small businesses frequently encounter difficulty in attracting financing and managerial talent, given their inability to offer rewards and security equal to those of larger, better established firms. Thus both financing and management are more likely to come from within the family. In addition to the famous Wright brothers partnership, there were also the Thomas brothers, the Loughead brothers, the Loening brothers, and later the Gross brothers of Lockheed. Larry Bell was first drawn into the industry by his brother Grover, an early exhibition pilot. Chance Vought relied on his family to a major extent in establishing his company, and Frederick Rentschler received major assistance from his brothers in organizing and financing United Aircraft and Transport. James S. McDonnell relied on his banker brother William for financial management for many years. The Kellett brothers worked together in early autogiro and helicopter development. Donald Douglas Jr. succeeded his father as president of his company, and successive chief executives of McDonnell Douglas were Sanford McDonnell and John F. McDonnell, nephew and son of the founder. While such family influences are hardly unusual in American business, they remain especially characteristic of the aircraft industry.

Ranking the greatest figures of the American aircraft industry is inherently subjective and perhaps even unfair, in that individuals are recognized for achievements to which thousands of people made major contributions. The following alphabetical list nonetheless represents the author's determination of the most significant and successful leaders of the American aircraft industry.

- William M. Allen, Boeing
- Donald W. Douglas, Douglas
- Robert E. Gross, Lockheed

- James H. Kindelberger, North American
- Glenn L. Martin, Martin
- James S. McDonnell, McDonnell

Of the six, two require clarifying comments perhaps more than others. William Allen did not build Boeing but took over leadership of a long-established and widely respected company. He merits inclusion because of the direction in which he led Boeing during his tenure, making it the largest aerospace firm and also the leading commercial-aircraft manufacturer in the world. Boeing became perhaps the most representative firm in the industry with regard to such vital measures as productivity, export success, sound management, and design leadership.

Glenn Martin in some respects is the most difficult individual to include in the list, since his firm, during his life, became the least successful of those represented. It also had perhaps the most troubled overall history, with Martin himself being forced out in his final years. Martin seldom was regarded as a strong manager but merits inclusion because he alone of the group began his career in the pioneer days, and because his firm alone survived from that period into the jet age while still under the direction of the founder. He also enabled more future industry leaders to gain the experience and skills to succeed on their own than any other industry pioneer. Douglas, Bell, Kindelberger, and McDonnell worked for Martin at some point before leading their own firms, and Boeing and Fleet learned to fly in Martin trainers. The industry might not have developed as it did without Martin.

Now we may turn to those who, while also major leaders, are not quite in the first rank.

- Lawrence D. Bell, Bell
- William E. Boeing, Boeing and United Aircraft
- Reuben H. Fleet, Consolidated
- Leroy R. Grumman, Grumman
- Clement M. Keys, Curtiss and North American
- John K. Northrop, Northrop
- Frederick B. Rentschler, United Aircraft

Of these, the choice of Rentschler requires some clarification, since his primary orientation was toward engines rather than aircraft. His United Aircraft was of course a significant aircraft manufacturer with Chance Vought and Sikorsky, but never of the scale of the largest firms. Yet Rentschler was a major force in developing the structure of the modern aircraft, later aerospace, industry, making the industry stronger than it would have been had a UATC never been organized.

Ranking below these figures are several who either were not active long enough or who did not make a sufficient long-term impact to merit inclusion in the short lists, but who still were significant figures in the development of the industry. Included are Grover Loening, Allan Lockheed, Claude Ryan, and Sherman Fairchild. Possibly Glenn Curtiss should be included as well. Igor Sikorsky certainly must be mentioned for his work in helicopters. Chance Vought also remains an important figure who would have attained greater prominence if not for his early death. The same might be said of Jerry Vultee.

There have been others who have made a major impact on their companies and on the industry, but who for various reasons may not be included. Lee Atwood, who followed Kindelberger at North American, and T. A. Wilson of Boeing are examples. Tom Jones of Northrop, Dan Haughton of Lockheed, and David Lewis of both McDonnell Douglas and General Dynamics also were extraordinary leaders in many respects, but their careers ended in ethical controversy.

It is to the questions of political influence and ethics that we turn for a final look. Were industry leaders, as frequently alleged, unscrupulous businessmen bent on exploiting the military procurement system for private gain, oblivious to the public interest, national-security needs, or the international repercussions of their actions? Such allegations have appeared throughout the industry's history, but with increasing frequency from the 1960s. Indeed, there have long been instances of improper industry conduct, most recently in the "Ill Wind" scandals of the 1980s. A casual observer might well conclude that industry leaders have been ethically impaired, but a comprehensive answer is elusive.

Much of the industry's poor reputation for ethics and integrity unquestionably has been self-inflicted, and always such lapses must be laid at the leadership, not at lower-level employees. It must be restated, however, that the nature of the defense procurement system provided strong incentives toward such actions as bribery and exertion of improper influence, however unethical or illegal. Further, the economic vulnerability of the aerospace industry as compared with other major industries is well understood and documented. Questions of ethics, the public interest, or international stability are unlikely to take precedence over the issue of sheer survival. Recent stresses have served only to heighten the pressures.

Still, if industry leaders have been guilty as charged, the question remains as to what critics would have had those executives do. During the 1930s, for example, the firms that were to grow into the wartime industrial giants were struggling simply to survive yet were frequently investigated over such matters as contract overcharges, competitive practices, and exports to areas experiencing international tensions. Yet a government that did not then support a critical industry later faulted that industry for its sometimes extreme survival tactics. Yet cautiously ethical conduct would almost inevitably have meant losing out to those less ethical. The cost of highly ethical conduct is immediate and certain. A look at contemporary political campaigns, in which candidates increasingly "do whatever is necessary" to win, provides an insight. It has been observed that negative campaign tactics are pursued because they tend to be effective with voters. Taking the high road may well lead to defeat. Many of America's most successful political leaders have not been the most ethical or scrupulous of men.

Whatever the ethical lapses or failings evidenced by leaders of the aircraft industry, those lapses or failings have been influenced by the lack of diversification of the industry, by the demands of advanced technology, and by the politicization of the industry, which in turn has been driven by the societal choice against efficiency and performance accountability. But the government has always been the controlling factor, and the failure has been most acute in the political leadership.

Appendix ✈ Chronology of the Aircraft Industry

```
1910                1920                1930                1940
```

Herring-Curtiss (1909)
(bankrupt, 1910) ||

Curtiss Aeroplane (1910) → Curtiss A&M (1916) ——————————————→ Curtiss-Wright (1929)

Curtiss Motor (1910) → (to Curtiss A&M, 1916; absorbed, 1919)

Burgess Co. & Curtis (1910)—-to Burgess Co. (1914)

Wright Co. (1909); sold, 1915 → Wright Aeronautical (1919) ————————→ (to Wright Aeronautical, 1928)

Huff Daland (1920) ---- name changed to Keystone (1927)

Wright-Martin (1916)

Glenn L. Martin Co. (1912)
(new) Martin Co. (1917) ————————————-to Baltimore (1929); bankrupt, 1934——

Loening (1917) ————————————————to Keystone (1928)

Gallaudet Engineering (1910)-----(reorg.) Gallaudet Aircraft (1917)

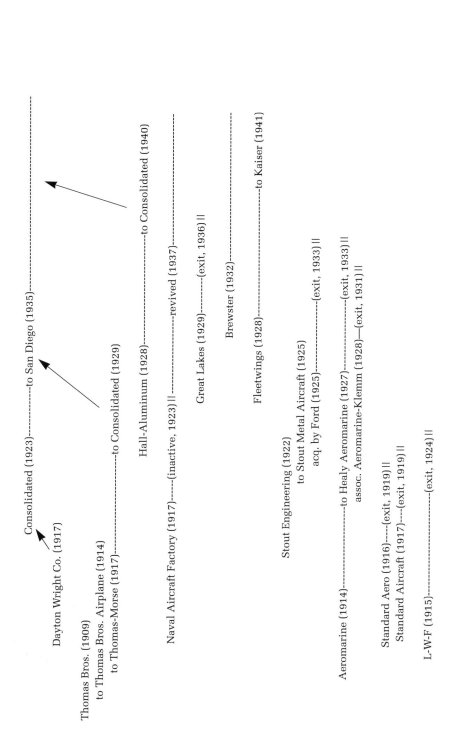

| 1910 | 1920 | 1930 | 1940 |

Loughead Bros. (1916)------[bankrupt, 1921] ||
  (new) Lockheed (1926)----to Detroit (1929); bankrupt (1931)
    (reorg.) Lockheed, 1932--------------------
      Vega Div. (1937)--------------

Lewis & Vought (1917)-----(reorg.) Chance Vought (1922) ⟶
  Sikorsky Aero (1923) ⟶
  Sikorsky Mfg. (1925)
    Sikorsky Aviation (1928) ⟶
  Pratt & Whitney (1925) ⟶
    United Aircraft and Transport (1929)-----United Aircraft (1935)--------------
      Vought-Sikorsky Div. (1939)---
    Stearman (1927) ↗
    Avion (1928) ↗

Pacific Aero (1916)
  Boeing Airplane (1917)-----to Boeing Airplane and Transport (1928) ↗
    UATC Northrop-Stearman Div. (1931)--(dissolved, 1932)
    Northrop Aircraft Sub. (1930) ↙

      (new) Boeing (1934)--------------
      Stearman Div. (1934)-------------

Davis-Douglas (1920)
Douglas Co. (1921)--------Douglas Aircraft (1928)------------------------(new) Northrop Div. (1932) ---(dissolved, 1937)

McDonnell (1939)--------

(new) Northrop (1939)--------

Grumman (1929)--------

Lawrence Sperry (1917)------(exit, 1923)||

Hughes (1936)--------

Ryan (1922)
   to Ryan Airlines (1925)
   to B. F. Mahoney, as subsid. (1927)
   to Mahoney Aircraft (1927)
   to Mahoney-Ryan (1928)
      to Detroit (1929); bankrupt (1931)

(new) Ryan Aeronautical (1934)--------

1910	1920	1930	1940

Sturtevant (1915)----(exit, 1919) ||

Wittemann Bros. (1912?)
 to Wittemann Aircraft (1915)
  to Wittemann-Lewis (1917)
   to Wittemann Aircraft (1918)----(exit, 1924) ||

North American Aviation (1928)----autonomous (1934); to Calif. (1935)

(to North American, 1930)

Berliner (1926)----to Berliner-Joyce (1929)

Atlantic Aircraft (1923)
 to Fokker (1925)----to General Motors (1929)

General Motors--
 acq. Dayton Wright (1919); sold rights (1923)
  acq. Fokker (1929), formed GAMC subsid.
   formed General Aviation Co. (1930); to NAA (dissolved, 1933)

Seversky (1931)--------(reorg.) Republic (1939)--------

Fairchild Aerial Survey (1920)
   Fairchild Aviation (1925)
      Fairchild Aviation Corp. (1927)
         → to AVCO (1929)

Kreider-Reisner (1925)
   to Fairchild (1928)

Aviation Corp. (AVCO, 1929)
forms American Airplane and Engine from Fairchild (1931)----(exit, 1935) ‖

   AVCO (subsid.) Airplane Dev. Corp. (1932)
      to Vultee Div., AMC (1934)
         to Vultee Airplane Div. (1936)
            to Vultee Aircraft (1939)-----------

(new) Fairchild Aviation (1931)
   to Hagerstown (1932)
      to Fairchild E & A (1936)-----------

```
1940        1950        1960        1970        1980        1990        2000

Fairchild E & A ──────────Fairchild Stratos (1961)
                                    Fairchild Hiller (1963)
                                         ↗
       Hiller (1942)─────────to Hiller Aircraft (1963)
                                              ─────(exit aircraft, 1987)‖
                                         ↗
                                    to Fairchild Corp. (1971)
Republic Aviation  ─────────────────to Fairchild Hiller (1965)

Consolidated
    ↘
Consolidated Vultee (1942)──────to GD Corp. (1953)────────────────┐
    ↗                                                             │
Vultee Aircraft                                                   │
                                              GD Military Aircraft Div. to Lockheed (1993)
Lockheed ─────────────────────────────────────────────────────┐   ↓
  (Vega absorbed, 1943)                                       Lockheed Martin Corp. (1994)
                                                                  ↑
Martin Co. ───────────────to Martin Marietta Corp. (1961)─────────┘
            ↗
 American-Marietta Corp.

GM–Fisher Body (1942–44)‖
```

Curtiss-Wright (exit aircraft, 1950) ||

------Brewster (susp., 1944; exit, 1946) ||

Sikorsky Div. (1943)------

Vought-Sikorsky Div.

Chance Vought Div. (1943)------

Chance Vought Aircraft (1954)

to LTV Corp. (1961)

Vought aircraft operations to Northrop (1992)

Northrop Grumman Corp. (1994)

Northrop --

Grumman --

Bell ------to Textron (1961)------

(Helicopter Div., 1941)

Kaman (1945) ---

Chase (1944)------(to Kaiser, 1953) ||

	1940	1950	1960	1970	1980	1990	2000

Ryan ----------------(exit aircraft, 1960)----to Teledyne (1968)

Hughes (exit aircraft, 1947) ||

Fleetwings Div., Kaiser (exit, 1945) ||

Naval Aircraft Factory (exit, 1945) ||

McDonnell -----------------------------------↗

                              McDonnell Douglas Corp. (1967)-----------------to Boeing (1997)

Douglas ----------------------------------↗

                  (new) Hughes Helicopters (1955)-----------------to MDCC (1984)↗

P-V Engr. Forum (1943)

    Piasecki (1946)--------Vertol (1956)---→ (to Boeing, 1960)

Boeing --↖

(Stearman Div. reorg., 1941)   aerospace operations to Boeing (1996)

North American ----------------------North American Rockwell (1967)----to RI Corp. (1973)---------------↗

                           Rockwell-Standard Corp.

Notes

Chapter 1

1. Tom Crouch, *The Bishop's Boys* (New York: W. W. Norton, 1989), 169–70.
2. Ibid., 243–44.
3. Grover C. Loening, *Takeoff into Greatness* (New York: Putnam, 1968), 21.
4. Crouch, *The Bishop's Boys*, 300.
5. Crouch, 301–5, contains a general discussion of the Wrights' view of the world.
6. A. G. Renstrom, *Wilbur and Orville Wright: A Chronology* (Washington, D.C.: Library of Congress, 1975), 27.
7. Crouch, *The Bishop's Boys*, 348.
8. Ibid., 378.
9. Fred Kelly, *The Wright Brothers* (New York; Harcourt Brace, 1943), 166; and Elsbeth Freudenthal, *Flight into History* (Norman: University of Oklahoma Press, 1949), 214.
10. Freudenthal, *Flight into History*, 219.
11. *Jane's All the World's Airships, 1909* (New York: Arco, 1969), 277.
12. Crouch, *The Bishop's Boys*, 312.
13. Ibid., 359.
14. Curtis Prendergast, *The First Aviators*, Epic of Flight Series (New York: Time, Inc., 1980), 33–34.
15. Crouch, *The Bishop's Boys*, 390; and Freudenthal, *Flight into History*, 205.
16. *Aircraft Year Book, 1919* (New York: Manufacturers Aircraft Association, 1920), 141–45.
17. Prendergast, *The First Aviators*, 30.
18. William K. Kaiser, ed., *The Development of the Aerospace Industry on Long Island, 1904–1964*, vol. 1 (Hempstead, NY: Hofstra University, 1968), 2; *Aircraft Year Book, 1951* (Washington, DC: Aircraft Industries Association, 1951), 30; and John Rae, *Climb to Greatness* (Cambridge, MA: MIT Press, 1968), 1.
19. C. R. Roseberry, *Glenn Curtiss: Pioneer of Flight* (Garden City, NY: Doubleday, 1972), 81.
20. Ibid., 101.
21. Prendergast, *The First Aviators*, 54.
22. Roseberry, *Glenn Curtiss*, 151; and Fred Howard, *Wilbur and Orville: A Biography of the Wright Brothers* (New York: Knopf, 1987), 309.
23. Roseberry, *Glenn Curtiss*, 159–60.
24. Ibid.

25. *Jane's, 1909,* 262.
26. Roseberry, *Glenn Curtiss,* 180.
27. Ibid., 251.
28. "Internal Fight in Herring-Curtiss Company," *Aeronautics,* February 1910, 69.
29. Bartlett Gould, "The Story of Burgess Aircraft," part 1, *American Aviation Historical Society Journal* 10 (summer 1965): 82.
30. Roseberry, *Glenn Curtiss,* 254.
31. Ibid., 259.
32. Ibid., 366.
33. Ibid., 397.
34. Ibid., 398.
35. Ibid., 396.
36. "Burgess Company and Curtis Withdraw from Aviation Field," *Aeronautics,* February 28, 1914, 57.
37. Bartlett Gould, "The Story of Burgess Aircraft," part 3, *American Aviation Historical Society Journal* 11 (winter 1966): 272.
38. "The Burgess Company Formed," *Aeronautics,* February 28, 1914, 58.
39. "The Curtiss Aeroplane and Motor Company Acquires Services of W. Starling Burgess and Burgess Company," *Aerial Age Weekly,* February 14, 1916, 519.
40. *Jane's All the World's Aircraft, 1927* (London: Sampson Low, Marston and Co., 1927), 271.
41. Jacob Vander Meulen, "The American Aircraft Industry to 1936: A Political History of a Military Industry," Ph.D. diss., University of Toronto, 1989, 19–20.
42. E. L. Jones, in *Jane's All the World's Aircraft, 1913* (New York: Arco, 1969), 201.
43. *History of the Bureau of Aircraft Production,* vol. 1 (Wright-Patterson Air Force Base, OH: Air Materiel Command, 1951), 104.
44. *Jane's, 1913,* 204–5.
45. Ibid.
46. Ibid.
47. Loening, *Takeoff into Greatness,* 51–56.
48. Renstrom, *Wilbur and Orville,* 41.
49. Crouch, *The Bishop's Boys,* 404.
50. Ibid., 402.
51. Loening, *Takeoff into Greatness,* 32.
52. Crouch, *The Bishop's Boys;* and Loening, *Takeoff into Greatness,* 46.
53. Crouch, *The Bishop's Boys,* 465; and Loening, *Takeoff into Greatness,* 45.
54. Crouch, *The Bishop's Boys,* 466.
55. Loening, *Takeoff into Greatness,* 67.

56. Crouch, *The Bishop's Boys*, 464.
57. Ibid., 471–72.
58. Loening, *Takeoff into Greatness*, 83–86.
59. Ibid., 114–15.
60. *Dynamic America: A History of General Dynamics Corporation and Its Predecessor Companies* (New York: General Dynamics Corporation, 1960), 6.
61. William Wagner, *Reuben Fleet and the Story of Consolidated Aircraft* (Fallbrook, CA: Aero Publishers, 1975), 4.
62. Ibid.
63. Henry Still, *To Ride the Wind: A Biography of Glenn L. Martin* (New York: Julian Messner, 1964), 103.
64. Ibid., 117.
65. Loening, *Takeoff into Greatness*, 89.
66. Still, *To Ride the Wind*, 119. Other sources state total capital to be only $5 million.
67. *Aircraft Year Book, 1919*, 287.
68. Still, *To Ride the Wind*, 120.
69. "New Interests Paid Curtiss $5,000,000 in Cash," *Aerial Age Weekly*, January 24, 1916, 447. Roseberry reports that Curtiss received almost $7 million in cash and stock in the deal (*Glenn Curtiss*, 407).
70. *Aircraft Year Book, 1919*, 113.
71. Bartlett Gould, "The Story of Burgess Aircraft," part 5, *American Aviation Historical Society Journal* 13 (spring 1968): 39–40.
72. Loening, *Takeoff into Greatness*, 98–100.
73. Ibid., 101.
74. Grover C. Loening, *Our Wings Grow Faster* (Garden City, NY: Doubleday, Doran, 1935), 21–23.
75. Ibid., 69.
76. Ibid.
77. Ibid., 78.
78. Gerard P. Moran, *Aeroplanes Vought, 1917–77* (Temple City, CA: Historical Aviation Album, 1978), 10.
79. Ibid., 13–14.
80. Arthur L. Schoen, *Vought: Six Decades of Aviation History* (Plano, TX: Aviation Quarterly Publishers, 1978), 44.
81. *Aircraft Year Book, 1919*, 225.
82. *Aircraft Year Book, 1919*, 175.
83. William M. Leary, Jr., "At the Dawn of Commercial Aviation: Inglis M. Uppercu and Aeromarine Airways," *Business History Review* 53 (summer 1979): 182.
84. Engel Papers, Army Air Corps Records, Box 15, Record Group 18, National Archives.
85. *Aircraft Year Book, 1919*, 219.

Chapter 2

1. *Bureau of Aircraft Production*, 1:104.
2. *Jane's All the World's Aircraft, 1919* (London: Sampson Low, Marston and Co., 1919), 433.
3. Loening, *Takeoff into Greatness*, 179.
4. James C. Fahey, *U.S. Army Aircraft, 1908–1946* (Washington, D.C.: Ships and Aircraft, 1946), 6.
5. Loening, *Takeoff into Greatness*, 72–73.
6. Fahey, *U.S. Army Aircraft*, 6.
7. Ibid.
8. Adrian O. Van Wyen, ed., *Naval Aviation in World War I* (Washington, D.C.: Chief of Naval Operations, 1969), 6.
9. Ibid., 8.
10. Benjamin D. Foulois, *From the Wright Brothers to the Astronauts* (New York: McGraw-Hill, 1968), 142.
11. Ibid., 143.
12. Vander Meulen, "American Aircraft Industry," 7.
13. Loening, *Takeoff into Greatness*, 106.
14. I. B. Holley Jr., *Ideas and Weapons* (New Haven: Yale University Press, 1953), 52–54, 60. Holley provides a well-researched account of wartime production problems.
15. *Bureau of Aircraft Production*, 1:104; and *Jane's, 1919*, 429.
16. *Jane's, 1919*, 429.
17. Edward O. Purtee, *History of the Army Air Service, 1907–1926* (Wright-Patterson Air Force Base, OH: Air Materiel Command, 1948), 62.
18. See "Evolution of the U.S. Air Force," exhibit 12, in *Air Force Magazine*, May 1975, 128.
19. Purtee, *Army Air Service*, 67.
20. "Aircraft Production Cost, 1917–1918," *Aviation and Aircraft Journal*, April 18–25.
21. *Jane's, 1919*, 429.
22. Fahey, *U.S. Army Aircraft*, 8. There is a small discrepancy in accounts here. Loening, *Takeoff into Greatness* reports a figure of 13,894 aircraft produced for the war, against Fahey's 14,071. William G. Cunningham, in *The Aircraft Industry: A Study of Industrial Location* (Los Angeles: Lorrin L. Morrison, 1951), reports a total of 13,111 from April 6, 1917, to November 1, 1919, a figure also in Purtee, *Army Air Service*, 104.
23. Loening, *Our Wings Grow Faster*, 76.
24. John B. Rae, "Financial Problems of the American Aircraft Industry, 1906–1940," *Business History Review* 39 (1965): 102.
25. Loening, *Takeoff into Greatness*, 107.
26. Loening, *Our Wings Grow Faster*, 69.
27. Loening, *Takeoff into Greatness*, 104–9.
28. Ibid., 112.
29. Purtee, *Army Air Service*, 103.

30. Izetta Winter Robb, "The Navy Builds an Aircraft Factory," in Van Wyen, *Naval Aviation,* 35.

31. William F. Trimble, *High Frontier: A History of Aeronautics in Pennsylvania* (Pittsburgh: University of Pittsburgh Press, 1982), 100.

32. Ibid.

33. Van Wyen, *Naval Aviation,* 89.

34. *Jane's, 1919,* 428–90.

35. Rae, *Climb to Greatness,* 2.

36. *Aircraft Year Book,* various editions.

37. *Aircraft Year Book, 1920* (New York: Manufacturers Aircraft Association, 1920), 101–22.

38. Ibid., 101.

39. Vander Meulen, "American Aircraft Industry," 46; and Purtee, *Army Air Service,* 40.

40. Howard Mingos, "The Birth of an Industry," in G. R. Simonson, ed., *The History of the American Aircraft Industry: An Anthology* (Cambridge, MA: MIT Press, 1968), 31.

41. Jacob Vander Meulen, *The Politics of Aircraft: Building an American Military Industry* (Lawrence: University Press of Kansas, 1991), 64.

42. Mingos, "Birth of an Industry," 39.

43. *Aircraft Year Book, 1924* (New York: Manufacturers Aircraft Association, 1924), appendix, 289.

44. Purtee, *Army Air Service,* 108.

Chapter 3

1. *Aircraft Year Book, 1920,* 101.

2. Mingos, "Birth of an Industry," 50–51.

3. Loening, *Takeoff into Greatness,* 129–31.

4. *Aircraft Year Book, 1925* (New York: Manufacturers Aircraft Association, 1925), 2.

5. John W. Underwood, *The Stinsons* (Glendale, CA: Heritage Press, 1969), 26.

6. Army Air Corps Records, Box 15, Record Group 18, National Archives.

7. Loening, *Takeoff into Greatness,* 124.

8. Eugene E. Bauer, *Boeing in Peace and War* (Enumclaw, WA: TABA Publishing, 1990), 30.

9. Fahey, *U.S. Army Aircraft,* 9.

10. Loening, *Takeoff into Greatness,* 125; and Loening, *Our Wings Grow Faster,* 105.

11. Roseberry, *Glenn Curtiss,* 429.

12. Ibid.

13. Gerard P. Moran, "Rex B. Beisel," in *Historical Aviation Album,* vol. 15 (Temple City, CA: Historical Aviation Album, 1984), 348.

14. Ibid., 350.
15. Rae, *Climb to Greatness,* 12.
16. Ed Cray, *Chrome Colossus: General Motors and Its Times* (New York: McGraw-Hill, 1980), 263.
17. Donald J. Norton, *Larry: A Biography of Lawrence D. Bell* (New York: Nelson-Hall, 1981), 40.
18. Loening, *Our Wings Grow Faster,* 125.
19. Grover Loening, letter to Clement Keys, March 10, 1924, Army Air Corps Records, Box 15, Record Group 18, National Archives.
20. William Wyatt Davenport, *Gyro! The Life and Times of Lawrence Sperry* (New York: Scribners, 1978), 112.
21. Ibid., 220.
22. Frank Cunningham, *Sky Master: The Story of Donald Douglas* (Philadelphia: Dorrance and Company, 1943), 43.
23. Ibid., 54–55.
24. Ibid., 55.
25. Ibid., 66; and Still, *To Ride the Wind,* 117.
26. Cunningham, *Sky Master,* 88–90.
27. Ibid., 105.
28. Ibid., 126–30.
29. Ibid., 137–38.
30. Rae, *Climb to Greatness,* 10.
31. Ibid., 142–43.
32. René Francillon, *McDonnell Douglas Aircraft since 1920* (London: Putnam, 1979), 53.
33. Trimble, *High Frontier,* 134.
34. Ibid., 144.
35. William Wagner, *Ryan, the Aviator* (New York: McGraw-Hill, 1971), 35.
36. Ibid., 59.
37. Ibid., 85.
38. Ibid., 113.
39. Ibid., 145.
40. Ibid., 166.
41. Wagner, *Reuben Fleet,* 2.
42. Ibid., 12.
43. Ibid., 18.
44. Memo, Rear Admiral Moffett to Major General Patrick, June 17, 1927, and memo, Patrick to Moffett, August 3, 1927, Army Air Corps Records, Box 14, Record Group 18, National Archives; and John Wegg, *General Dynamics Aircraft and Their Predecessors* (Annapolis, MD: Naval Institute Press, 1990), 43.
45. Igor I. Sikorsky, *The Story of the Winged-S* (New York: Dodd, Mead, 1938), 143–46.
46. Ibid., 147.
47. Ibid., 153–57.

48. Ibid., 156–59.
49. Ibid., 170–78.
50. Wagner, *Reuben Fleet,* 110.
51. Sikorsky, *Story of the Winged-S,* 183.
52. *Aircraft Year Book, 1919,* 74.
53. Ibid., 74–75.
54. *Jane's All the World's Aircraft, 1929* (London: Sampson Low, Marston and Co., 1929), 269; and *Aircraft Year Book, 1929* (New York: Manufacturers Aircraft Association, 1929), 74.
55. Richard P. Hallion, *Designers and Test Pilots,* Epic of Flight Series (New York: Time, Inc., 1983), 44.
56. *Of Men and Stars: A History of Lockheed Aircraft Corporation, 1913–1957* (Burbank, CA: Lockheed Aircraft Corporation, 1957), chap. 2, pp. 10–12.
57. Ibid.
58. Douglas J. Ingells, *L-1011 TriStar and the Lockheed Story* (Fallbrook, CA: Aero Publishers, 1973), 16.
59. Ibid., 20.
60. Ibid., 22.
61. George Vecsey and George C. Dade, *Getting Off the Ground* (New York: E. P. Dutton, 1979), 235–36.
62. Ingells, *L-1011 TriStar,* 22–23.
63. *Of Men and Stars,* chap. 3, p. 2.
64. Ingells, *L-1011 TriStar,* 26–30.
65. *Aircraft Year Book, 1921,* 229.
66. Ibid.
67. *Aircraft Year Book, 1923* (New York: Manufacturers Aircraft Association, 1923), 303–6.
68. *Aviation Facts and Figures, 1953* (Baltimore: Lincoln Press, 1953), 24.
69. *Aircraft Year Book, 1924* (New York: Manufacturers Aircraft Association, 1924), 298–300.
70. Ibid.
71. *Aircraft Year Book, 1926* (New York: Manufacturers Aircraft Association, 1926), 8–11.
72. Foulois, *Wright Brothers to Astronauts,* 199.
73. *Aircraft Year Book, 1926,* 8–15.
74. Loening, *Takeoff into Greatness,* 177.
75. Foulois, *Wright Brothers to Astronauts,* 202.
76. *Aircraft Year Book, 1928* (New York: Manufacturers Aircraft Association, 1928), 6.
77. Loening, *Takeoff into Greatness,* 143.
78. Ibid., 161.
79. S. A. Zimmerman, *Procurement in the United States Air Force, 1938–1948* (Wright-Patterson Air Force Base, OH: Air Materiel Command, 1950), 1:66–67.
80. Loening, *Takeoff into Greatness,* 172.

81. Ibid.
82. Quoted in Vander Meulen, *The Politics of Aircraft,* 46.
83. Robert W. Fausel, *Whatever Happened to Curtiss-Wright?* (Manhattan, KS: Sunflower University Press, 1990), 77–78.

Chapter 4

1. Vander Meulen, *The Politics of Aircraft,* 88–89.
2. *Aircraft Facts and Figures, 1953,* 24.
3. John W. R. Taylor, ed., *The Lore of Flight* (New York: Crescent Books, 1987), 685–87.
4. "Success in Santa Monica," *Fortune,* May 1935, 178.
5. I. B. Holley, *Buying Aircraft: Materiel Procurement for the AAF,* U.S. Army in World War II, Special Studies (Washington, D.C.: Office of Military History, 1964), 37.
6. Ibid.
7. Wagner, *Reuben Fleet,* 119.
8. Ibid., 135.
9. Norton, *Larry,* 45; and Wegg, *General Dynamics Aircraft,* 45.
10. Wagner, *Reuben Fleet,* 142–43.
11. *Aircraft Year Book, 1930* (New York: Manufacturers Aircraft Association, 1930), 7.
12. Loening, *Our Wings Grow Faster,* 171–72; and Richard Thruelson, *The Grumman Story* (New York: Praeger Publishers, 1976), 24.
13. Loening, *Takeoff into Greatness,* 206.
14. Ibid.; and Bill Gunston, *One of a Kind: The Story of Grumman* (Bethpage, NY: Grumman Corporation, 1988), 8.
15. Gunston, *One of a Kind,* 8.
16. Thruelson, *The Grumman Story,* 17.
17. Gunston, *One of a Kind,* 8.
18. Ibid., 19.
19. Ibid., 22.
20. Trimble, *High Frontier,* 208.
21. Edward T. Maloney, *Sever the Sky: Evolution of Seversky Aircraft* (Corona del Mar, CA: Planes of Fame Publishers, 1979), 10.
22. Ibid., 15.
23. Walter Boyne, "The Hall-Aluminum Story," *Flying Review International,* January 1969, 61–62.
24. Bill Yenne, *Rockwell: The Heritage of North American* (New York: Crescent Books, 1989), 12.
25. Alfred P. Sloan, *My Years with General Motors* (Garden City, NY: Doubleday, 1963), 364.
26. Ibid.
27. Rae, *Climb to Greatness,* 14.
28. Trimble, *High Frontier,* 151–52.
29. Loening, *Our Wings Grow Faster,* 193–94.

30. Loening, *Takeoff into Greatness*, 171–72.
31. *A History of Lockheed* (Burbank, CA: Lockheed Horizons, 1983), 15.
32. Ingells, *L-1011 TriStar*, 22–23.
33. Ibid., 34–35.
34. [Paul Fisher], *The Pratt and Whitney Aircraft Story* (East Hartford, CT: United Aircraft Corp., 1950), 22.
35. Frederick B. Rentschler, *An Account of the Pratt and Whitney Aircraft Company, 1925–1950* (East Hartford, CT: United Aircraft Corp., 1950), 12; and Eugene E. Wilson, *Slipstream: Memoirs of an Air Craftsman* (New York: McGraw-Hill, 1950), 50.
36. Wilson, *Slipstream*, 51.
37. Fisher, *Pratt and Whitney*, 45–47.
38. Ibid., 67.
39. Rentschler, *Account*, 28.
40. Ibid., 73.
41. Moran, "Rex B. Beisel," 351–52.
42. Rentschler, *Account*, 28.
43. Ted Coleman, *Jack Northrop and the Flying Wing* (New York: Paragon House, 1988), 47.
44. Richard S. Allen, *The Northrop Story, 1929–1939* (New York: Orion, 1990), 156.
45. Elsbeth E. Freudenthal, *The Aviation Business: From Kitty Hawk to Wall Street* (New York: Vanguard Press, 1940), 248.
46. Ibid., 81.
47. Harold Mansfield, *Vision: The Story of Boeing* (New York: Duell, Sloane, and Pearce, 1966), 46.
48. Bauer, *Boeing*, 78.
49. *AVCO Corporation: The First Fifty Years* (Greenwich, CT: Avco Corporation, 1979), 8.
50. Fairchild Papers, Box 1, Garber Facility, National Air and Space Museum.
51. *AVCO Corporation*, 11.
52. Freudenthal, *The Aviation Business*, 82.
53. *AVCO Corporation*, 16.
54. *Yesterday, Today, and Tomorrow: Fifty Years of Fairchild Aviation* (Germantown, MD: Fairchild Hiller, 1970), 22.
55. Ibid., 24.
56. Genevieve Brown, *Development of Transport Airplanes and Air Transport Equipment* (Wright Field, OH: Air Technical Service Command, April 1946), 67.
57. Peter C. Smith, *Vengeance! The Vultee Vengeance Dive Bomber* (Washington, D.C.: Smithsonian Institution Press, 1986), 4.
58. *AVCO Corporation*, 19.
59. Roseberry, *Glenn Curtiss*, 438; and Fausel, *Whatever Happened?* 3.
60. Frederick W. Roos, "Curtiss-Wright St. Louis," *American Aviation Historical Society Journal* 35 (winter 1990): 298.

61. Loening, *Takeoff into Greatness,* 206; and Loening, *Our Wings Grow Faster,* 193–94.

62. Francis Walton, *Miracle of World War II: How American Production Made Victory Possible* (New York: Macmillan, 1956), 446.

63. Howard Mingos, "Birth of an Industry," in Freudenthal, *The Aviation Business,* 78.

64. Frank Kingston Smith, *Legacy of Wings: The Story of Harold Pitcairn* (New York: J. Aronson, 1981), 160.

65. Yenne, *Rockwell,* 16.

66. Ralph B. Oakley, "North American, Its Products and People," part 1, *American Aviation Historical Society Journal* 19 (summer 1974): 135.

67. Yenne, *Rockwell,* 13.

68. Oakley, "North American," 138; and Paul R. Matt, "Formation of North American Aviation, Inc.: A Brief History," *Historical Aviation Album* 13 (1974): 183.

69. Matt, "Formation of North American," 183.

70. Yenne, *Rockwell,* 16.

71. *A History of Lockheed,* 21.

72. Ingells, *L-1011 TriStar,* 36.

73. *Of Men and Stars,* chap. 1, pp. 8–9.

74. *Of Men and Stars,* chap. 4, p. 6.

75. Wagner, *Ryan, the Aviator,* 166.

76. Ibid.

77. "Salesman at Work," *Time,* January 14, 1946, 82.

78. *A History of Lockheed,* 15; and *Of Men and Stars,* chap. 3, p. 7.

Chapter 5

1. Freudenthal, *The Aviation Business,* 83.

2. Douglas J. Ingells, *The Plane That Changed the World* (Fallbrook, CA: Aero Publishers, 1966), 81.

3. Ibid.

4. Foulois, *Wright Brothers to Astronauts,* 234, 261–72.

5. *Pedigree of Champions: Boeing since 1916,* 6th ed. (Seattle: Boeing Company, 1985), 12.

6. Foulois, *Wright Brothers to Astronauts,* 228.

7. Robin Higham, *Air Power: A Concise History* (New York: St. Martin's, 1972), 72.

8. Foulois, *Wright Brothers to Astronauts,* 226.

9. Ibid., 229.

10. Ibid., 265.

11. Ibid., 274.

12. Vander Meulen, "American Aircraft Industry," 245–49.

13. Ibid.

14. Holley, *Buying Aircraft,* 122.

15. Ibid., 131.

16. Ibid., 56–58.
17. *Aircraft Year Book, 1931* (New York: Manufacturers Aircraft Association, 1931), 9–10.
18. Cunningham, *Sky Master,* 216.
19. Ibid., 214; and Ingells, *Plane That Changed World,* 13.
20. "Success in Santa Monica," 81.
21. Cunningham, *Sky Master,* 218; and Ingells, *Plane That Changed World,* 18.
22. Ingells, *Plane That Changed World,* 36.
23. "Success in Santa Monica," 83.
24. Ibid., 213.
25. Ingells, *Plane That Changed World,* 29.
26. "Success in Santa Monica," 79.
27. Ibid., 225; and Ingells, *Plane That Changed World,* 57.
28. Ingells, *Plane That Changed World,* 73.
29. Ibid., 38.
30. "Success in Santa Monica," 176. *McDonnell Douglas Aircraft since 1920,* 53, reports a figure of $7,526,705 for that year.
31. Ingells, *Plane That Changed World,* 85.
32. Cunningham, *Sky Master,* 253.
33. Bauer, *Boeing,* 79.
34. W. F. Craven and J. L. Cate, eds., *The U.S. Army Air Forces in World War II,* vol. 6, *Men and Planes* (Chicago: University of Chicago Press, 1955), 300; and Holley, *Buying Aircraft,* 159.
35. W. A. M. Burden, "Postwar Status of the Aircraft Industry," *Harvard Business Review,* 22 winter 1944: 212.
36. Rae, *Climb to Greatness,* 97.
37. *Aerospace Facts and Figures,* (Washington, D.C.: Aerospace Industries Association, 1959), 6–7.
38. Rae, *Climb to Greatness,* 84.
39. Cunningham, *The Aircraft Industry,* 8.
40. Vander Meulen, *The Politics of Aircraft,* 109. Also see 197.
41. Robert E. Gross, letter to Federal Aviation Commission, November 19, 1934, Box 13, Gross Papers, Library of Congress.
42. Benjamin S. Kelsey, *The Dragon's Teeth? The Creation of United States Air Power for World War II* (Washington, D.C.: Smithsonian Institution Press, 1982), 92.
43. Richard Barclay Harding, *The Aviation Industry* (New York: Charles D. Barney, 1937), vii–viii.
44. Grover Loening, *Aviation and Banking* (New York: Chase National Bank, October, 1938), 1.
45. Ibid., 5.
46. Trimble, *High Frontier,* 172.
47. Ibid., 173.
48. Ibid.

49. Elsbeth E. Freudenthal, "The Aviation Business in the 1930s," in Simonson, *American Aircraft Industry,* 101.

50. Ibid.

51. Tom Lilley et al., *Problems of Accelerating Aircraft Production during World War II* (Boston: Division of Research, Harvard Business School, January 30, 1946), 5.

52. Army Air Corps, *Aircraft Production and Export Policies, 1931–1939* (n.p: n.p., n.d.).

53. Tom Lilley et al., "Problems of Accelerating Aircraft Production during World War II," in Simonson, *American Aircraft Industry,* 171.

54. In Great Britain, by contrast, the Royal Air Force controlled naval aircraft procurement, flying training, and specifications until the Royal Navy finally gained control of the Fleet Air Arm (FAA) in 1937. One unhappy result was that the FAA did not possess a competitive shipboard fighter for wartime and was forced to rely on American Grumman and Chance Vought fighters and naval conversions of the RAF Hurricane and Spitfire. Higham, *Air Power,* 150.

55. Foulois, *Wright Brothers to Astronauts,* 227–28.

56. Ibid., 232; and Mansfield, *Vision,* 59.

57. Wagner, *Reuben Fleet,* 210; and Mansfield, *Vision,* 59.

58. Gene Gurney, *The War in the Air* (New York: Bonanza Books, 1962), 27.

59. Rae, *Climb to Greatness,* 60.

60. John B. Rae, "Wartime Technology and Commercial Aviation," *Aerospace Historian,* September 1973, 132.

61. Zimmerman, *Procurement,* 1:6.

62. Army Air Corps Records, Box 12, Record Group 18, National Archives; and Henry H. Arnold, letter to various contractors, October 14, 1938, Henry H. Arnold Papers, Box 222, Library of Congress.

63. Holley, *Buying Aircraft,* 181–83.

64. Craven and Cate, *Men and Planes,* 187.

65. Ibid., 189.

66. Donald C. Barlett and James B. Steele, *Empire: The Life, Legend, and Madness of Howard Hughes* (New York: W. W. Norton, 1979), 69.

67. Ingells, *L-1011 TriStar,* 46.

68. Rae, *Climb to Greatness,* 95.

69. Mansfield, *Vision,* 75.

70. Rae, *Climb to Greatness,* 89, 92.

71. Rae, "Wartime Technology," 132.

72. "Commercial Airlines in the United States," in *The Aviation Annual of 1944,* ed. Reginald N. Cleveland and Frederick P. Graham (Garden City, NY: Doubleday, Doran, 1945), 145.

73. Douglas Ingells, *747: Story of the Boeing Super Jet* (Fallbrook, CA: Aero Publishers, 1970), 79.

74. Peter M. Bowers, *The DC-3: 50 Years of Legendary Flight* (Blue Ridge Summit, PA: Tab Books, 1986), 46.

75. Douglas Ingells, *The McDonnell Douglas Story* (Fallbrook, CA: Aero Publishers, 1979), 117–18.
76. Ingells, *L-1011 TriStar*, 72.
77. Ibid., 73.
78. Craven and Cate, *Men and Planes,* 174–76.
79. *Of Men and Stars,* chap. 5, p. 2.
80. Craven and Cate, *Men and Planes,* 174–76.
81. Rae, *Climb to Greatness,* 105.

Chapter 6
1. Wagner, *Reuben Fleet,* 177.
2. Ibid.
3. Ibid., 178.
4. Trimble, *High Frontier,* 207.
5. Wagner, *Reuben Fleet,* 203–4.
6. Rae, *Climb to Greatness,* 127.
7. Ibid.
8. *Avco Corporation,* 20.
9. Rae, *Climb to Greatness,* 126.
10. Ibid., 127.
11. Roos, "Curtiss-Wright St. Louis," 299.
12. Ibid.
13. *Jane's All the World's Aircraft, 1945–6* (New York: Arco Publishing Co., 1970), 251.
14. *Yesterday, Today, and Tomorrow,* 32.
15. "Fairchild's Blitz," *Newsweek,* July 18, 1949, 54; and "Multifarious Sherman Fairchild," *Fortune,* May 1960, 220.
16. *Of Men and Stars,* chap. 5, p. 3.
17. Ibid., 10.
18. Ibid., 6.
19. Ibid.
20. Ingells, *L-1011 TriStar,* 62–63.
21. *Of Men and Stars,* chap. 5, p. 9.
22. James R. Wilburn, "Social and Economic Aspects of the Aircraft Industry in Metropolitan Los Angeles during World War II," Ph.D. diss., UCLA, 1971, 11–12; and Robert E. Gross, letter to Courtlandt S. Gross, March 31, 1937, Box 13, Gross Papers, Library of Congress.
23. Wilburn, "Social and Economic Aspects"; and Robert E. Gross, letter to Courtlandt S. Gross, September 16, 1937, Box 13, Gross Papers, Library of Congress.
24. George Bryant Woods, *The Aircraft Manufacturing Industry: Present and Future Prospects* (New York: White, Weld, 1946), 59.
25. *Jane's, 1945–6,* 210.

26. Ralph B. Oakley, "North American Goes to War," part 2, *American Aviation Historical Society Journal* 20 (spring 1975): 46.
27. Sloan, *My Years*, 367.
28. Joshua Stoff, *The Thunder Factory: An Illustrated History of the Republic Aviation Corporation* (London: Arms and Armour Press, 1990), 25.
29. Ibid., 32.
30. Army Air Corps Records, Box 12, Record Group 18, National Archives.
31. Ibid.
32. Maloney, *Sever the Sky*, 7.
33. "The Founder Complains," *Time*, September 2, 1940, 56.
34. Stoff, *The Thunder Factory*, 19.
35. Woods, *The Aircraft Manufacturing Industry*, 86.
36. "The Founder Complains," 57.
37. Thruelson, *The Grumman Story*, 100.
38. Gunston, *One of a Kind*, 26.
39. Trimble, *High Frontier*, 173.
40. Barlett and Steele, *Empire*, 126–27; and Noah Dietrich, *Howard: The Amazing Mr. Hughes* (New York: Fawcett, 1973), 169.
41. Barlett and Steele, *Empire*, 124; and Dietrich, *Howard*, 135.
42. Barlett and Steele, *Empire*, 129; and Dietrich, *Howard*, 162.
43. Wagner, *Reuben Fleet*, 180.
44. Ibid.; and Norton, *Larry*, 49, 54.
45. Norton, *Larry*, 56.
46. Ibid., 74.
47. Ibid., 92.
48. Maynard, *Flight Plan for Tomorrow*, 41.
49. Allen, *The Northrop Story*, 157.
50. Cunningham, *The Aircraft Industry*, 44.
51. Frank J. Taylor and Lawton Wright, *Democracy's Air Arsenal* (New York: Duell, Sloane, and Pearce, 1947) 168.
52. Coleman, *Jack Northrop*, 54, 57.
53. Allen, *The Northrop Story*, 115.
54. Coleman, *Jack Northrop*, 64.
55. Fred Anderson, *Northrop: An Aeronautical History* (Los Angeles: Northrop Corporation, 1976), 16.
56. Ibid., 31.
57. Sanford N. McDonnell, "Much May Be Made of a Scotsman: The Story of McDonnell Douglas," Address to the Wings Club, New York, May 16, 1990, 3.
58. Ingells, *The McDonnell Douglas Story*, 31.
59. McDonnell, "Much May Be Made," 4.
60. Ibid.; and James J. Horgan, *City of Flight: A History of Aviation in St. Louis* (Gerald, MO: Patrice Press, 1984), 338–42.
61. Ingells, *The McDonnell Douglas Story*, 42.

Chapter 7

1. H. H. Arnold Papers, Box 18, Library of Congress.
2. Rae, *Climb to Greatness,* 108.
3. Ibid., 151–52.
4. Richard G. O'Lone, "Ames Enjoys 50 Years of Aerospace Leadership," *Aviation Week and Space Technology,* December 24, 1990, 65.
5. Mary L. McMurtrie, *History of the Army Air Forces Materiel Command, 1926 through 1941* (Wright Field, OH: Historical Office, Air Materiel Command, 1943), appendix E.
6. T. P. Wright, "50,000 Planes a Year," *Aviation,* July 1940, 100; and Zimmerman, *Procurement,* 1:9.
7. Rae, *Climb to Greatness,* 114.
8. *Aviation Facts and Figures, 1945* (Washington, D.C.: Aircraft Industries Association, 1945), 54.
9. Craven and Cate, *Men and Planes,* xxvii.
10. Cunningham, *Sky Master,* 293.
11. Donald M. Nelson, *Arsenal of Democracy: The Story of American War Production* (New York: Harcourt, Brace, and Co., 1946), 122.
12. Ibid.
13. Ibid.
14. Gerald T. White, *Billions for Defense* (University: University of Alabama Press, 1980), 21.
15. Holley, *Buying Aircraft,* 284.
16. Rae, *Climb to Greatness,* 121.
17. Holley, *Buying Aircraft,* 289.
18. Craven and Cate, *Men and Planes,* 15.
19. McMurtrie, *Air Forces Materiel Command,* appendix E.
20. Nelson, *Arsenal of Democracy,* 117.
21. Zimmerman, *Procurement,* 1:9.
22. Nelson, *Arsenal of Democracy,* 232.
23. Wilburn, "Social and Economic Aspects," 27.
24. Lilley et al., "Accelerating Aircraft Production," 166–67.
25. Taylor and Wright, *Democracy's Air Arsenal,* 46.
26. Nelson, *Arsenal of Democracy,* 234.
27. Walton, *Miracle,* 367.
28. Craven and Cate, *Men and Planes,* xvii; and Thomas M. Coffey, *Hap* (New York: Viking Press, 1982), 112.
29. Cunningham, *The Aircraft Industry,* 77.
30. Nelson, *Arsenal of Democracy,* 120.
31. Cunningham, *The Aircraft Industry,* 77.
32. R. Elberton Smith, *The Army and Economic Mobilization,* The U.S. Army in World War II series (Washington, D.C.: Department of the Army, 1959), 482.
33. Cunningham, *The Aircraft Industry,* 77.
34. Zimmerman, *Procurement,* 1:2.
35. Ibid., 3.

36. Roos, "Curtiss-Wright St. Louis," 301.
37. Cunningham, *The Aircraft Industry,* 124–26.
38. Zimmerman, *Procurement,* Boeing Case History file, unpaginated.
39. White, *Billions for Defense,* 22.
40. Ibid.
41. Craven and Cate, *Men and Planes,* 317.
42. Rae, *Climb to Greatness,* 129.
43. Cunningham, *The Aircraft Industry,* 93.
44. *Aviation Facts and Figures, 1945,* 17.
45. Smith, *Army and Economic Mobilization,* 291.
46. Ibid., 293.
47. S. J. Zeigler and John W. Meador, *An Investigation of the Financial Condition and Recent Earnings of Aircraft Manufacturers* (Washington, D.C.: Department of the Navy, May 14, 1942, 3.
48. Ibid., 1, 37–38.
49. Ibid., 32.
50. Ibid., 24.
51. Ibid., 14.

Chapter 8

1. Virginia N. Anderson, "GALCIT's Aero Library: Fruit of a Timely Investment," in Catherine D. Scott, Ed. *Aeronautical and Space Flight Collections* (New York: The Haworth Press, 1985), 112, 130.
2. Taylor and Wright, *Democracy's Air Arsenal,* 203.
3. Holley, *Buying Aircraft,* 471.
4. Yenne, *Rockwell,* 48–49.
5. Ibid., 50.
6. Walton, *Miracle,* 89.
7. John K. Northrop, letter to Henry H. Arnold, July 10, 1941, Box 18, Arnold Papers, Library of Congress.
8. Lilley et al., "Accelerating Aircraft Production," 130–31.
9. Ibid., 137.
10. Ibid., 139.
11. Nelson, *Arsenal of Democracy,* 238.
12. Ibid.
13. Ibid., 163.
14. Lilley et al., "Accelerating Aircraft Production," 162.
15. Ibid.
16. President's Air Policy Commission, *Survival in the Air Age* (Washington, D.C.: GPO, 1948), 27.
17. Trimble, *High Frontier,* 218–19.
18. William A. Trimble, *Wings for the Navy: A History of the Naval Aircraft Factory* (Annapolis, MD: Naval Institute Press, 1990), 224.
19. Trimble, *High Frontier,* 218–19.
20. McDonnell, "Much May Be Made," 6.

21. Ingells, *The McDonnell Douglas Story*, 42.
22. Ibid.
23. Richard J. Whalen, "Banshee, Demon, Voodoo, Phantom—and Bingo!" *Fortune,* November 1964, 258.
24. Ingells, *The McDonnell Douglas Story*, 70.
25. Anderson, *Northrop*, 55.
26. Ibid., 47.
27. Schoen, *Vought*, 20.
28. Hugh Allen, *Goodyear Aircraft* (Cleveland: Corday and Gross, 1947), 36.
29. Ibid., 136.
30. Cunningham, *Sky Master*, 298.
31. "The Passionate Engineer," *Time,* November 22, 1943, 77.
32. McMurtrie, *Air Forces Materiel Command,* appendix E, 84.
33. Ibid.
34. Roy A. Anderson, "A Look at Lockheed," address to the Newcomen Society, November 1, 1982 (New York: Newcomen Society, 1983), 28.
35. Mansfield, *Vision,* 176; and Taylor and Wright, *Democracy's Air Arsenal*, 82.
36. *Pedigree of Champions,* 35.
37. Mansfield, *Vision,* 238.
38. Norton, *Larry,* 97.
39. Coffey, *Hap,* 258; and "Experiment at Republic," *Fortune,* February 1947, 124.
40. Woods, *The Aircraft Manufacturing Industry,* 86.
41. *Yesterday, Today, and Tomorrow,* 34.
42. Ibid., 36.
43. Ibid., 48.
44. Oakley, "Goes to War," 48.
45. Ibid., 49.
46. Ibid., 47.
47. Wagner, *Reuben Fleet,* 242.
48. Rae, *Climb to Greatness,* 128.
49. Ibid.
50. Wagner, *Reuben Fleet,* 278.
51. *Jane's, 1945–6,* 220.
52. Dietrich, *Howard,* 166.
53. Wagner, *Reuben Fleet,* 253.
54. *AVCO Corporation,* 27.
55. *A History of Eastern Aircraft Division, General Motors Corporation* (Linden, NJ: Eastern Aircraft Division, 1944), 19.
56. Ibid., 20–21.
57. *Jane's, 1945–6,* 250.
58. *Eastern Aircraft Division,* 64.
59. *Jane's, 1945–6,* 255–56.
60. Ibid., 256.

61. Ibid.
62. Rae, *Climb to Greatness*, 136.
63. Ibid., 159–60.
64. *New York Times*, May 20, 1942, 10.
65. Rae, *Climb to Greatness*. Also see Holley, *Buying Aircraft*, 518–29, for an extensive account of the Willow Run operation.
66. Brown, *Development of Transport Airplanes*, 76.
67. Roos, "Curtiss-Wright St. Louis," 304.
68. *Jane's, 1945–6*, 215; and Trimble, *High Frontier*, 224.
69. *Jane's, 1945–6*, 215.
70. Brown, *Development of Transport Airplanes*, 77.
71. Smith, *Legacy of Wings*, 283.
72. Trimble, *High Frontier*, 219.
73. Smith, *Legacy of Wings*, 271.
74. Ibid., 283.
75. Robert Williams, "The Greening of the Helicopter," *Aviation Heritage*, March 1991, 23.
76. Ibid.
77. Trimble, *High Frontier*, 227.
78. Ingells, *The McDonnell Douglas Story*, 251.
79. Trimble, *High Frontier*, 227–28.
80. W. J. Holt Jr., "He Likes to Fly Straight Up," *Saturday Evening Post*, March 11, 1951, 33.
81. Rae, *Climb to Greatness*, 164; and *A History of Lockheed*, 24.
82. *Of Men and Stars*, chap. 7, p. 6.
83. Rae, *Climb to Greatness*, 166.
84. Walton, *Miracle*, 409–10.
85. Rae, *Climb to Greatness*, 166.
86. Ibid.
87. Ibid.
88. Ibid., 167; and *A History of Lockheed*, 134.
89. *A History of Lockheed*, 134.
90. Jim Maas, "Fall from Grace: Brewster Aeronautical Corporation, 1931–1942," *American Aviation Historical Society Journal* 30 (summer 1985): 125.
91. Trimble, *High Frontier*, 221.
92. Walton, *Miracle*, 549.
93. Trimble, *High Frontier*, 223.
94. Barlett and Steele, *Empire*, 113–14.
95. Ibid., 117.
96. Ibid., 130.
97. Ibid., 130–31; and Dietrich, *Howard*, 166.
98. Barlett and Steele, *Empire*, 136–37; and Dietrich, *Howard*, 175.
99. Barlett and Steele, *Empire*, 115.
100. Ibid., 130.
101. Ibid., 131–37; and Dietrich, *Howard*, 166.

102. Barlett and Steele, *Empire,* 137; and Dietrich, *Howard,* 175.

103. Donald H. Riddle, *The Truman Committee: A Study in Congressional Responsibility* (New Brunswick, NJ: Rutgers University Press, 1964), 154.

104. Ibid., 136.

105. Lilley et al., *Accelerating Aircraft Production,* 1; and Craven and Cate, *Men and Planes,* 187.

Chapter 9

1. John K. Northrop, "The Air Industry and Air Power," lecture presented at Air War College, Maxwell Air Force Base, October 23, 1952 (Air Force Historical Research Center, Maxwell Air Force Base, AL), 8.

2. Lynn L. Bollinger and Tom Lilley, *Financial Position of the Aircraft Industry* (Boston: Harvard University, 1943).

3. Ibid., 9–10.

4. Ibid., 25.

5. Nelson, *Arsenal of Democracy,* 238.

6. Zimmerman, *Procurement,* 2:227.

7. Aeronautical Chamber of Commerce, *The Aircraft Industry Prepares for the Future* (Air Force Research Center, Maxwell Air Force Base, AL, 1944).

8. Report of the Air Coordinating Committee, 44–46; also in Royal D. Frey, *USAF Postwar Chronology* (Wright-Patterson Air Force Base, OH: Air Materiel Command, Historical Office, 1950).

9. "Planemakers Prospects," *Time,* September 17, 1945, 81.

10. Burden, "Postwar Status," 212.

11. Woods, *The Aircraft Manufacturing Industry,* 45.

12. Robert E. Gross, Statement to President's Air Policy Commission, October 14, 1947, Box 15, Gross Papers, Library of Congress.

13. President's Air Policy Commission, *Survival in Air Age,* 51.

14. Derek Wood, *Project Cancelled: The Disaster of Britain's Abandoned Aircraft Projects,* rev. ed. (London: Jane's Publishing, 1986), 16.

15. Robert E. Gross, letter to Lynn Bollinger, October 22, 1946, Gross Papers, Box 13, Library of Congress.

16. Allen, *The Northrop Story,* 130.

17. Statement of Floyd Odlum before House Armed Services Committee, August 25, 1949, Arnold Papers, Box 214, Library of Congress.

18. Coleman, *Jack Northrop,* 144.

19. Ibid., 195.

20. William G. Cunningham, "The Consequences of the Postwar Contraction," in Simonson, *American Aircraft Industry,* 181.

21. Ibid., 182.

22. Oliver P. Echols, "The Aircraft Industry, 1947–48," speech to Aircraft Industries Association, (Air Force Historical Research Center, Maxwell Air Force Base, AL).

23. Cunningham, "Consequences of Postwar Contraction," 181 and 206.

24. Cunningham, *The Aircraft Industry,* 186.

25. Ibid., 168–69.

26. Wood, *Project Cancelled,* 7–11.

27. Roger Bilstein, *Orders of Magnitude: A History of the NACA and NASA, 1915–1990* (Washington, D.C.: NASA, 1989), 37.

28. Letter and conference memorandum, July 6, 1945, Arnold Papers, Box 10, Library of Congress.

29. Northrop, "Air Industry," 14.

30. R. E. G. Davies, *Airlines of the U.S. since 1914,* rev. ed. (Washington, D.C.: Smithsonian Press, 1982), 389.

31. Richard K. Smith, "Fifty Years of Transatlantic Flight, Part II, 1945–1984," *American Aviation Historical Society Journal* 35 (fall 1990): 199.

32. I. H. Taylor, Eastern Representative, Douglas Aircraft Company, letter to Donald W. Douglas, April 11, 1944, Arnold Papers, Box 11, Library of Congress.

33. Ibid.

34. Smith, "Fifty Years," 199; and Ronald Miller and David Sawers, *The Technical Development of Modern Aviation* (New York: Praeger Publishers, 1970), 134.

35. Ingells, *The McDonnell Douglas Story,* 119.

36. Charles D. Bright, *The Jet Makers: The Aerospace Industry from 1945 to 1972* (Lawrenceville, KS: Regents Press, 1978), 85.

37. Almarin Phillips, *Technology and Market Structure* (Lexington, MA: Lexington Books, 1971), 88.

38. Ingells, *The McDonnell Douglas Story,* 124.

39. Maynard, *Flight Plan for Tomorrow,* 30.

Chapter 10

1. Cunningham, "Consequences of Postwar Contraction," 186–87.

2. *Jane's All the World's Aircraft, 1950–51* (London: Sampson Low, Marston and Co., 1951), 214.

3. *Jane's All the World's Aircraft, 1952–53* (London: Sampson Low, Marston and Co., 1953), 190.

4. "Kaiser-Frazer: Roughest We Ever Tackled," *Fortune,* July 1951, 161.

5. Richard C. Palmer, Fairchild vice president, interview by Anthony Brandt, June 11, 1969, p. 41, Fairchild Papers, Archives of the National Air and Space Museum.

6. Barlett and Steele, *Empire,* 149.

7. Ibid., 156.

8. Ibid., 157.

9. Ibid., 160.

10. Charles J. V. Murphy, "The Blowup at Hughes Aircraft," *Fortune,* February 1954, 118.

11. Barlett and Steele, *Empire,* 171–72.

12. Murphy, "Blowup at Hughes Aircraft," 188.
13. Barlett and Steele, *Empire,* 191.
14. Murphy, "Blowup at Hughes Aircraft," 193.
15. Barlett and Steele, *Empire,* 193.
16. Murphy, "Blowup at Hughes Aircraft," 117; and *Aviation Week,* June 28, 1954, 19.
17. Barlett and Steele, *Empire,* 199–200.
18. Ibid., 349.
19. "Mr. Odlum Gets the Business," *Fortune,* August 1949, 91.
20. Roger Franklin, *The Defender* (New York: Harper and Row, 1986), 133.
21. Ibid., 134.
22. Ibid.
23. Ibid.
24. "Mr. Odlum Gets Business," 134.
25. Bright, *The Jet Makers,* 191.
26. Frey, *USAF Postwar Chronology.*
27. Bright, *The Jet Makers,* 191.
28. Richard M. Langworth, *Kaiser-Frazer: The Last Onslaught on Detroit* (Kutztown, PA: Automobile Quarterly Publications, 1975), 158.
29. "So They Named It General Dynamics," *Fortune,* April 1953, 136.
30. Franklin, *The Defender,* 17–18.
31. Ibid., 119. "So They Named It" puts the figure as $12 million.
32. Ibid., 123–24.
33. "So They Named It," 136.
34. Franklin, *The Defender,* 144.
35. *Jane's All the World's Aircraft, 1956–57,* (London: Sampson Low, Marston and Co., 1957), 253.
36. Franklin, *The Defender,* 134.
37. John Mecklin, "The Rockwells Take Off for Outer Space," *Fortune,* June 1967, 166.
38. Yenne, *Rockwell,* 80.
39. Sloan, *My Years,* 374.
40. Fausel, *Whatever Happened?* 33–34.
41. Ibid., 30.
42. Roos, "Curtiss-Wright St. Louis," 304.
43. Fausel, *Whatever Happened?* 33.
44. Ibid.
45. Ibid., 60.
46. Ibid., 64.
47. Maynard, *Flight Plan for Tomorrow,* 49.
48. Ibid.
49. Nicholas M. Williams, "The X-Rays," *American Aviation Historical Society Journal* 22 (winter 1977): 243.
50. *Washington Post,* July 30, 1949.
51. Glenn L. Martin Papers, Box 73, Library of Congress.

52. Still, *To Ride the Wind,* 236.
53. Bright, *The Jet Makers,* 189.
54. Still, *To Ride the Wind,* 239.
55. *A History of Lockheed,* 120–21.
56. Ibid., 72–74.
57. Mansfield, *Vision,* 262.
58. Bauer, *Boeing,* 161–62.
59. Higham, *Air Power,* 202.
60. Norton, *Larry,* 198.
61. Ibid., 204.
62. Daniel Seligman, "Barrier-Breaking Bell Aircraft," *Fortune,* March 1956, 227.
63. Ibid.
64. Woods, *The Aircraft Manufacturing Industry,* 53–54.
65. Norton, *Larry,* 219–221.
66. Cunningham, *The Aircraft Industry,* 206.
67. "Happy Days at Grumman," *Fortune,* June 1948, 112.
68. Ibid., 115.
69. Thruelson, *The Grumman Story,* 245.
70. Moran, *Aeroplanes Vought,* 105; and Cunningham, *The Aircraft Industry,* 190.
71. Cunningham, *The Aircraft Industry,* 190.
72. Moran, "Rex B. Beisel," 114.
73. Moran, *Aeroplanes Vought,* 354.
74. *Aviation Week,* October 20, 1947, 20.
75. Sherman M. Fairchild, letter to J. Carlton Ward, October 22, 1946, Fairchild Papers, Archives of the National Air and Space Museum.
76. J. Carlton Ward, letter to Sherman M. Fairchild, October 23, 1946, Fairchild Papers.
77. "Fairchild's Blitz," 54–55.
78. Ibid.
79. *Jane's All the World's Aircraft, 1953–54* (London: Sampson Low, Marston and Co., 1954), 213.
80. Ibid., 86.
81. *Fortune,* July 1948, 108.
82. Coleman, *Jack Northrop,* 189.
83. Coleman, *Jack Northrop,* 167; and Vecsey and Dade, *Getting Off the Ground,* 240.
84. "Experiment at Republic," *Fortune,* February 1947, 167.
85. Stoff, *The Thunder Factory,* 81.
86. Ibid., 94.
87. Paul Lambermont, *Helicopters and Autogyros of the World,* rev. ed. (New York: A. S. Barnes, 1970), 381.
88. Trimble, *High Frontier,* 246.
89. Ibid., 247.
90. Ibid.

91. Ibid.
92. Ibid., 248.
93. Lambermont, *Helicopters and Autogyros,* 213.
94. Ibid.; and Leston Faneuf, "Lawrence D. Bell: A Man and His Company," address to Newcomen Society, May 15, 1958 (New York: Newcomen Society, 1958), 19.
95. Charles H. Kaman, *Kaman: Our Early Years* (Indianapolis: Curtis Publishing, 1985), 46.
96. Ibid., 52.
97. Ibid., 61.
98. Ibid., 75–81.
99. Ibid., 96.
100. Ibid.
101. *Yesterday, Today, and Tomorrow,* 71–73.
102. Ibid., 74.
103. Lambermont, *Helicopters and Autogyros,* 303–4.
104. Barlett and Steele, *Empire,* 350–151.
105. Smith, *Legacy of Wings,* 326.

Chapter 11

1. Donald J. Mrozek, "The Truman Administration and the Enlistment of the Aviation Industry in Postwar Defense," *Business History Review* 48 (spring 1974): 93.
2. Ibid., 73.
3. President's Air Policy Commission, *Survival in the Air Age,* 1948.
4. Ibid.
5. Ibid., 25.
6. Ibid., 25.
7. Ibid., 66, 99.
8. Frederic M. Scherer, *The Weapons Acquisition Process: Economic Incentives* (Boston, Harvard Business School, 1964), 343.
9. *Aviation Week,* August 26, 1957, 31.
10. Gordon Adams, *The Iron Triangle: The Politics of Military Procurement* (New York: Council on Economic Priorities, 1981).
11. Monro McCloskey, *The United States Air Force* (New York: Praeger Publishers, 1967), 63.
12. Ibid., 64.
13. Alfred Goldberg, ed., *History of the United States Air Force* (Princeton, NJ: Van Nostrand, 1974), 219.
14. Charles J. V. Murphy, "The Wild Blue Chip Yonder," *Fortune,* July 1955, 73.
15. Ibid., 118.
16. In fact, probably too rapidly, with the flight test phase still incomplete. Severe stability problems required substantial redesign to solve.
17. Norton, *Larry,* 225.

18. Bilstein, *Orders of Magnitude*, 40–41.

19. See "Weapons: Has Complexity Gone Too Far?" editorial, *Fortune*, April 1952, 91–93.

20. *Aircraft Year Book, 1954* (Washington, D.C.: Aircraft Industries Association, 1954), 79.

21. *Aviation Week*, December 20, 1954, 12.

22. Thomas G. Miller Jr., *Strategies for Survival in the Aerospace Industry* (Cambridge, MA: Arthur D. Little, 1964), table 2, 11.

23. Office of Management and Budget, *Budget of the United States Government*, annual.

24. Bright, *The Jet Makers*, 170; and Barry Bluestone, Peter Jordan, and Mark Sullivan, *Aircraft Industry Dynamics: An Analysis of Competition, Capital, and Labor* (Boston: Auburn House, 1981), 111–12.

25. Higham, *Air Power*, 204.

26. Scherer, *The Weapons Acquisition Process*, 389.

27. "Flying Low," *Time*, September 14, 1959, 56.

28. G. R. Simonson, "Missiles and Creative Destruction in the American Aircraft Industry, 1956–1961," *Business History Review* 38 (autumn 1964): 313–14.

29. Ibid., 306.

30. The Douglas 1959 loss was $33.8 million, the Lockheed 1960 loss $42.9 million, and the General Dynamics 1961 loss $143.2 million. *Fortune*, May 5, 1980, 93.

31. Miller, *Strategies for Survival*, 6.

32. See Herman O. Stekler, *The Structure and Performance of the Aerospace Industry* (Berkeley and Los Angeles: University of California Press, 1965), 31, for a definition of *aerospace*.

33. Aerospace Industries Association, *Aerospace Facts and Figures, 1968* (Washington, D.C.: Aerospace Industries Association, 1968), 8.

34. From a total of 943 for all three services in 1953 to only 431 in 1954. The 1953 figure would not be exceeded until 1964. *Aerospace Facts and Figures, 1968*, 17.

35. Trimble, *High Frontier*, 257.

36. Ibid.

37. Ibid., 258.

38. Ibid.; and *Pedigree of Champions*, 84.

39. Lambermont, *Helicopters and Autogyros*, 247.

40. Barlett and Steele, *Empire*, 354.

41. Kaman, *Kaman*, 133–39.

42. Ibid., 160.

43. Ibid., 140.

44. John W. Fozard, "Vertical Heroes," *Air Power History* 37 (fall 1990): 54–62.

45. See John McDonald, "Jet Airliners: Year of Decision," part 1, *Fortune*, April 1953, 125–28, 242–48.

46. John McDonald, "Jet Airliners: Year of Decision," part 2, *Fortune*, May 1953, 129.

47. "Offering the Airlines Their First U.S.-Built Jet," *Business Week*, March 6, 1954, 112.

48. Robert J. Serling, *The Jet Age*, Epic of Flight series (New York: Time, Inc., 1982), 59.

49. *Vision*, 199; and Ingells, *747*, 120.

50. Ingells, *747*, 119.

51. Ibid.

52. Richard G. O'Lone, "Pan Am Pushed Aircraft Builders to Advance Transport State of the Art," *Aviation Week and Space Technology*, December 16–23, 1991, 34.

53. Richard G. Hubler, *Big Eight: A Biography of an Airplane* (New York: Duell, Sloane, and Pearce, 1960), 87.

54. Arthur E. Raymond, "Reminiscences of Douglas: 1925–1960," *American Aviation Historical Society Journal* 32 (summer 1987): 122.

55. Serling, *The Jet Age*, 62.

56. Ibid., 69.

57. *Lockheed Horizons*, 123.

58. John Newhouse, *The Sporty Game* (New York: Alfred A. Knopf, 1982), 131.

59. Richard Austin Smith, *Corporations in Crisis* (New York: Doubleday, 1963), 64.

60. Ibid., 68.

61. Franklin, *The Defender*, 152–53; and Barlett and Steele, *Empire*, 218–22.

62. Smith, *Corporations in Crisis*, 80.

63. Ibid.

64. Franklin, *The Defender*, 150.

65. Franklin, *The Defender*, 149; and Barlett and Steele, *Empire*, 222.

66. Smith, *Corporations in Crisis*, 82.

67. Franklin, *The Defender*, 153.

68. Ibid.; and Jonathan Proctor, "The Convair 880," *American Aviation Historical Society Journal* 21 (fall 1976): 203.

69. Smith, *Corporations in Crisis*, 87; and Franklin, *The Defender*, 165–66. Howard Hughes soon was forced by his creditors to relinquish managerial control of TWA and to place his 78 percent stock holding in a trust. In a highly publicized transaction, Hughes sold his TWA stock on May 3, 1966, for $546 million, the largest sum ever paid to one individual to that time.

70. Franklin, *The Defender*, 163–64; and Proctor, "The Convair 880," 210.

71. Proctor, "The Convair 880," 210, 213.

72. Franklin, *The Defender*, 158.

73. Ibid., 159.

74. Ibid., 161.

75. Bright, *The Jet Makers,* 95.
76. Stekler, *Structure and Performance,* 160.

Chapter 12

1. Wagner, *Ryan, the Aviator,* 239.
2. Ibid.
3. Seligman, "Barrier-Breaking Bell Aircraft," 218.
4. Howard Wolko, National Air and Space Museum, former Bell engineer, interview with author, August 29, 1991.
5. Charles J. V. Murphy, "The Plane Makers under Stress," part 1, *Fortune,* July 1960, 234.
6. Franklin, *The Defender,* 168.
7. Ibid., 170.
8. Ibid.; and "Turmoil at General Dynamics," *Forbes,* September 1, 1961, 11–12.
9. Franklin, *The Defender,* 230.
10. *Of Men and Stars,* chap. 9, p. 13.
11. Herbert Solow, "North American: A Corporation Deeply Committed," *Fortune,* June 1962, 145.
12. Mecklin, "Rockwells Take Off," 166.
13. Solow, "North American," 175.
14. Ingells, *The McDonnell Douglas Story,* 156.
15. Ibid., 154–56.
16. As part of military reform and reorganization, aircraft designations were extensively revised on September 18, 1962, creating potential confusion in this narrative. Designations were standardized along the air force system, and ascending series numerical designations for each category were stopped with those then under development. For example, fighters, which had gone as high as F-111, were subsequently restarted with F-1, including those naval models already out of production but still in the inventory. Bombers, which had reached as high as B-70, were restarted with B-1, and transports and later trainers were accorded the same treatment. Navy and Marine aircraft, whose designations had contained letters assigned to the manufacturers, were standardized along the air force pattern. The North American FJ Fury became the F-1, the Douglas F4D became the F-6, the North American A3J the A-5, the Chance Vought F8U the F-8, and so on. Helicopter designations were amended with functional prefixes, CH for transport, SH for antisubmarine, UH for utility, OH for observation, and AH for attack. Also, the attack prefix, abandoned by the air force in 1948 but retained by the Navy, was reinstated. The Douglas AD became the A-1, and the air force later ordered the LTV A-7 and Fairchild Republic A-10. The army liaison prefix was abandoned in favor of utility and observation prefixes.

This narrative attempts to use the designation in effect at the time the particular aircraft is being discussed. For example, before 1962 reference is to the McDonnell F4H Phantom II, but after 1962 it is referred to as the F-4, and after 1967 as the McDonnell Douglas F-4. To add to the potential confusion, there were inconsistencies in treatment. For example, the Lockheed C-130 and C-141 transports were unchanged, but the next Lockheed transport was the C-5, while the C-140 acquired a utility prefix.

17. Whalen, "Banshee, Demon, Voodoo, Phantom," 137.
18. Bright, *The Jet Makers,* 193.
19. Edward H. Heinemann and Rosario Rausa, *Ed Heinemann: Combat Aircraft Designer* (Annapolis, MD: U.S. Naval Institute, 1980), 244.
20. Ibid., 195–96.
21. Newhouse, *The Sporty Game,* 135.
22. T. A. Wise, "How McDonnell Won Douglas," *Fortune,* March 1967, 156.
23. *Time,* December 16, 1966, 90.
24. Wise, "How McDonnell Won Douglas," 234.
25. Ingells, *The McDonnell Douglas Story,* 159.
26. "A Place in Space," *Time,* October 27, 1961, 89.
27. Stoff, *The Thunder Factory,* 122.
28. "When One Product Is Not Enough," *Business Week,* April 14, 1962, 86–90.
29. Raymond P. Kucera, *The Aerospace Industry and the Military* (Beverly Hills, CA: Sage Publications, 1973), 41.
30. Randolph P. Kucera, "Grumman Aircraft Engineering Corporation and Its 'Familial' Relationship with the U.S. Navy," Ph.D. diss., Syracuse University, 1973, 211.
31. Gunston, *One of a Kind,* 74.
32. "Hard-Headed Planemaker," *Forbes,* December 15, 1960, 13–16.
33. *Yesterday, Today, and Tomorrow,* 77–79.
34. *Aviation Week and Space Technology,* May 11, 1964, 28.
35. *Time,* September 18, 1964, 101.
36. Ibid.
37. Stanley H. Brown, *Ling: The Rise, Fall, and Return of a Texas Titan* (New York: Atheneum, 1972), 82–83.
38. Ibid., 87.
39. Higham, *Air Power,* 208.
40. Fozard, "Vertical Heroes," 60.
41. Bright, *The Jet Makers,* 148.
42. F. M. Scherer, "The Aerospace Industry," in *The Structure of American Industry,* ed. Walter Adams (New York: Macmillan, 1970), 358–59.
43. Stekler, *Structure and Performance,* 64, 77.
44. Ibid., 165.
45. J. Ronald Fox, *Arming America: How the U.S. Buys Weapons* (Cambridge, MA: Harvard University Press, 1974), 449.
46. "The Cats of MIG Alley," *Time,* June 29, 1953, 88.
47. Merton J. Peck and Frederic M. Scherer, *The Weapons Acquisition*

Process: An Economic Analysis (Boston: Division of Research, Harvard Business School, 1962), 33–34.

48. Fox, *Arming America,* 195–96.
49. Bright, *The Jet Makers,* 165; and William Green, *The World Guide to Combat Planes,* vol. 2 (Garden City, NY: Doubleday, 1967), 162.
50. Stekler, *Structure and Performance,* 155.
51. Bluestone, Jordan, and Sullivan, *Aircraft Industry Dynamics,* 44; and Kucera, *Aerospace Industry and Military,* 54.
52. Bluestone, Jordan, and Sullivan, *Aircraft Industry Dynamics,* 55; Kucera, *Aerospace Industry and Military,* 55.
53. Fox, *Arming America,* 21; and Stekler, *Structure and Performance,* 210.
54. Miller, *Strategies for Survival,* 24.
55. "Skybolt's American Shadow," *Economist,* January 12, 1963, 127–28.
56. Phillips, *Technology and Market Structure,* 130.
57. Higham, *Air Power,* 209.
58. *Time,* November 22, 1963, 88.
59. Bright, *The Jet Makers,* 188–89.
60. Solow, "North American," 145.
61. Jacques Gansler, *The Defense Industry* (Cambridge, MA: MIT Press, 1981), 28.
62. Bluestone, Jordan, and Sullivan, *Aircraft Industry Dynamics,* 27.
63. Robert J. Art, *The TFX Decision: McNamara and the Military* (Boston: Little, Brown, 1968), 3.
64. As reported in Kenneth S. Davis, ed., *Arms, Industry, and America* (New York: H. W. Wilson, 1971), 214–215.
65. Smith, *Corporations in Crisis,* 174.
66. Higham, *Air Power,* 199, 221.
67. Smith, *Corporations in Crisis;* and Franklin, *The Defender,* 182.
68. Art, *The TFX Decision,* 22.
69. Robert F. Coulam, *Illusions of Choice: The F-111 and the Problem of Weapons Acquisition Reform* (Princeton: Princeton University Press, 1977), 43, 100.
70. Ibid., 53.
71. Smith, *Corporations in Crisis,* 192.
72. Ibid., 178. As it developed, McNamara served more than seven years, a record.
73. Ibid.
74. Franklin, *The Defender,* 173–74.
75. Smith, *Corporations in Crisis,* 192.
76. Ibid., 182.
77. Art, *The TFX Decision,* 97–98.
78. Smith, *Corporations in Crisis,* 181.
79. Ibid., 185.
80. Art, *The TFX Decision,* 74.
81. Ibid., 187.

82. Ibid., 193.
83. Ibid., 199; and Franklin, *The Defender,* 194.
84. Franklin, *The Defender,* 196–97.
85. Bauer, *Boeing,* 234–35.
86. Franklin, *The Defender,* 201.
87. Tom Alexander, "McNamara's Expensive Economy Plane," *Fortune,* June 1967, 91.
88. Coulam, *Illusions of Choice,* 69.
89. Ibid., 78.
90. Ibid., 203.
91. Ibid., 308.
92. Alexander, "McNamara's Expensive Economy Plane," 89.
93. Coulam, *Illusions of Choice,* 5.
94. Ibid., 111.

Chapter 13

1. *Aerospace Facts and Figures, 1968,* 8; and *Aerospace Facts and Figures, 1976* (Washington, D.C.: Aerospace Industries Association, 1976), 119.
2. Tai Sheng Shin, "A Financial Analysis of the Airframe-Turned-Aerospace Industry," Ph.D. diss., University of Illinois, 1969, 134–36, 142–43, 146.
3. Ibid.
4. F. M. Scherer, *Industrial Market Structure and Economic Performance* (Chicago: Rand-McNally, 1980), 205.
5. Reported in Fox, *Arming America,* 48–49.
6. General Accounting Office, *Defense Industry Profit Study,* Report No. B-159896 (Washington, D.C.: General Accounting Office, 1971), 28.
7. Ibid.
8. Stekler, *Structure and Performance,* 51.
9. Newhouse, *The Sporty Game,* 142.
10. Ibid., 135–36.
11. Ibid., 149.
12. Mecklin, "Rockwells Take Off ," 101.
13. Thomas Derdak, ed., *International Directory of Company Histories,* vol. 1 (New York: St. James Press, 1988), 79.
14. Bright, *The Jet Makers,* 160–61.
15. Bilstein, *Orders of Magnitude,* 103–5.
16. The term *convertiplane,* which had fallen into disuse, was replaced by the acronyms VTOL and STOVL. Functionally, such aircraft were referred to by whatever means of transition from horizontal to vertical, such as tilt-rotor thrust or deflected thrust, was employed.
17. Scherer, "The Aerospace Industry," 351.
18. *Aerospace Facts and Figures, 1976,* 31.
19. Stoff, *The Thunder Factory,* 169.
20. Coulam, *Illusions of Choice,* 76.

21. Ibid., 251.
22. "Incredible Contract: General Dynamics Gain Is the Nation's Loss," *Barron's,* May 11, 1970, 1.
23. Ibid., 10.
24. *Aviation Week and Space Technology,* December 21, 1970, 19–20.
25. Ibid., 20.
26. Gunston, *One of a Kind,* 106.
27. Ibid., 108.
28. Ingells, *The McDonnell Douglas Story,* 263–64.
29. Gansler, *The Defense Industry,* 99.
30. John Mecklin, "The Ordeal of the Plane Makers," *Fortune,* December 1965, 284.
31. Ibid.
32. A. Ernest Fitzgerald, *The High Priests of Waste* (New York: W. W. Norton, 1972), 99.
33. Fox, *Arming America,* 86.
34. Fitzgerald, *High Priests of Waste,* 305–6.
35. Nick Kotz, *Wild Blue Yonder: Money, Politics, and the B-1 Bomber* (New York: Pantheon Books, 1988), 95–95.
36. Ibid., 108–10.
37. Ibid., 119.
38. Ibid., 128.
39. Ibid., 16.
40. Ibid., 130.
41. The revival of the B-1 in a considerably redeveloped version is treated in chapter 15.
42. "Spending More, Getting Less," *Forbes,* April 15, 1975, 28.
43. Lt. Gen. James T. Stewart, quoted in *A Man for the Time,* Oral History project (Wright-Patterson Air Force Base, OH: History Office, ASD, Air Force Systems Command, January 1990), 137–39.
44. Newhouse, *The Sporty Game,* 159–60.
45. Laurence S. Kuter, *The Great Gamble: The Boeing 747* (University: University of Alabama Press, 1973), 11–12.
46. Ibid., 3.
47. Ibid., 9.
48. O'Lone, "Pan Am Pushed Builders," 35.
49. Newhouse, *The Sporty Game,* 166–67.
50. Ibid., 163.
51. Ibid., 15, 22–23.
52. Ibid., 172.
53. Ibid., 22.
54. Kuter, *The Great Gamble,* 91.
55. Newhouse, *The Sporty Game,* 132.
56. Ibid., 173.
57. Ibid., 153–55.
58. Ibid., 135–36.

59. "Rumor of Nixon Pledge on Lockheed Heard as Firm Briefs Bankers on British Pact," *Wall Street Journal,* April 9, 1971, 2.
60. *Wall Street Journal,* July 2, 1971, 24.
61. A. E. Reinhardt, "Break-Even Analysis for Lockheed's Tri Star," *Journal of Finance* 28 (September 1973): 821.
62. Ibid.
63. Fitzgerald, *High Priests of Waste,* 291, 310.
64. *Wall Street Journal,* September 20, 1971, 31.
65. Dan Cordtz, "The Withering Aircraft Industry," *Fortune,* September 1970, 199.
66. Fitzgerald, *High Priests of Waste,* 203.
67. Ingells, *The McDonnell Douglas Story,* 167.
68. Ibid.; and Newhouse, *The Sporty Game,* 125.
69. Newhouse, *The Sporty Game,* 178.
70. John W. R. Taylor, *Jane's All the World's Aircraft, 1968–69* (London: Sampson Low, Marston and Co., 1968), i.
71. Newhouse, *The Sporty Game,* 125.
72. Department of Commerce, *A Competitive Assessment of the U.S. Civil Aircraft Industry* (Washington, D.C.: Department of Commerce, 1984), 128.

Chapter 14
1. Franklin, *The Defender,* 242–44.
2. Ibid., 245.
3. Quoted in *Man for the Time,* 107.
4. Franklin, *The Defender,* 245.
5. Herbert J. Coleman, "F-16 Team Forming in Europe," *Aviation Week and Space Technology,* June 16, 1975, 16–17; and Clyde H. Farnsworth, "The F-16 and How It Won Europe," *New York Times,* July 27, 1975, F1, 12.
6. Farnsworth, "F-16 and Europe," F1, 12.
7. Quoted in Fox, *Arming America,* 6.
8. Lt. Gen. J. T. Stewart, in *Man for the Time,* 189.
9. Ibid.
10. Ibid., 193.
11. Scherer, "The Aerospace Industry," 374.
12. Thomas V. Jones, "For a Sound Defense Industry," *New York Times,* November 23, 1976, 33.
13. Quoted in Eugene Kozicharow, "Disciplined Defense Procurement Urged," *Aviation Week and Space Technology,* October 3, 1977, 20.
14. Coulam, *Illusions of Choice,* 369.
15. Fitzgerald, *High Priests of Waste,* 321–22.
16. Cordtz, "The Withering Aircraft Industry," 114.
17. From 6.0 percent of GNP in FY 73 to 4.8 percent of GNP for FY 79. See Office of Management and Budget, *Budget of the United States Government,* annual.
18. Gansler, *The Defense Industry,* 9–11.

19. Ibid., 21.
20. Ibid., 26.
21. Jacques Gansler, "The U.S. Aircraft Industry: A Case for Sectoral Planning," *Challenge,* July–August, 1977, 13.
22. Gansler, *The Defense Industry,* 213.
23. "Flying into Trouble," *Time,* May 4, 1981, 14–16.
24. Gansler, *The Defense Industry,* 27.
25. Ibid., 170.
26. Ibid., 86.
27. Ibid., 179–82.
28. Fitzgerald, *High Priests of Waste,* 44.
29. Fox, *Arming America,* 81, 301.
30. Ibid., 19, 449.
31. Ibid., 174.
32. Adapted from Gansler, "The U.S. Aircraft Industry," 17–19.
33. Ibid., 14.
34. Gansler, *The Defense Industry,* 91.
35. Fox, *Arming America,* 148–50.
36. Gansler, *The Defense Industry,* 96.
37. Anthony Sampson, *The Arms Bazaar* (New York: Viking, 1977), 119.
38. Ibid., 125.
39. Ibid., 133.
40. Newhouse, *The Sporty Game,* 132.
41. Sampson, *The Arms Bazaar,* 216.
42. Ibid., 139.
43. Ibid., 147.
44. Ibid., 279.
45. Phillips, *Technology and Market Structure,* 130.
46. Suk H. Kim and Sam Barone, "Is the Foreign Corrupt Practices Act of 1977 a Success or Failure?" *Journal of International Business Studies* 12, no. 3 (winter 1981): 123–26.
47. Ibid., 124.
48. Stoff, *The Thunder Factory,* 167.
49. Ibid., 168.
50. Gunston, *One of a Kind,* 125.
51. Thruelson, *The Grumman Story,* 368–69.
52. Ibid., 370.
53. "Court Fight Seems Likely as Navy Orders 48 More Planes at Old Price," *Wall Street Journal,* December 12, 1972, 2.
54. Ibid.
55. Ibid.
56. Thruelson, *The Grumman Story,* 372–74.
57. "Takeover," *Forbes,* June 1, 1973, 36–37.
58. "Grumman Is Pressed by Pentagon to Find Commercial Credit," *Wall Street Journal,* August 16, 1974, 7; and "Grumman Nears Its Moment of Truth" *Business Week,* August 24, 1974, 15–16.

59. *Aviation Week and Space Technology,* September 22, 1975, 15.
60. "Takeover," 36–37.
61. "Boeing: Slimming Down an Aerospace Giant," *Business Week,* April 1, 1972, 41–46.
62. "Master of the Air," *Time,* April 7, 1980, 54.
63. "Boeing," 45.
64. Bauer, *Boeing,* 295–97.
65. Derdak, *International Directory,* 58; and Franklin, *The Defender,* 231–33.
66. Franklin, *The Defender,* 236.
67. Derdak, *International Directory,* 59.
68. Anderson, "A Look at Lockheed," 48.
69. Sampson, *The Arms Bazaar,* 216.
70. Louis Kraar, "How Lockheed Won Back Its Wings," *Fortune,* October 1977, 199.
71. Ibid.
72. Derdak, *International Directory,* 72.
73. Ingells, *The McDonnell Douglas Story,* 285.
74. Derdak, *International Directory,* 77.
75. "Happy Days for North American," *Business Week,* June 13, 1970, 28.
76. Ibid.
77. Bluestone, Jordan, and Sullivan, *Aircraft Industry Dynamics,* 110.
78. Kaman, *Kaman,* 162.
79. David L. Mowery, *Alliance Politics and Economics: Multinational Joint Ventures in Commercial Aircraft* (Cambridge, MA: Ballinger Publishing Company, 1987), 42–45.
80. Ibid., 47.
81. Keith Hayward, *International Collaboration in Civil Aerospace* (New York: St. Martin's Press, 1986), 19.
82. Mowery, *Alliance Politics and Economics,* 77.
83. Ibid., 95; and Newhouse, *The Sporty Game,* 148.
84. Mowery, *Alliance Politics and Economics,* 97.
85. Ingells, *The McDonnell Douglas Story,* 210–11; and Howard Banks, "The Aeroplane Makers," *Economist,* August 30, 1980, 8.
86. Hayward, *International Collaboration,* 132.
87. Ibid., vi.
88. Mowery, *Alliance Politics and Economics,* 82.
89. Richard Moxon and J. Michael Geringer, "Multinational Ventures in the Commercial Aircraft Industry," *Columbia Journal of World Business* 20 (summer 1985): 58.

Chapter 15

1. $3,743.9 million 1979 vs. $5,336.7 million in 1967 in current dollars. *Aerospace Facts and Figures, 88–89* (Washington, D.C.: Aerospace Industries Association, 1989), 64.

2. Allen, *The Northrop Story,* 132.

3. Virginia C. Lopez and Loren Yager, *The U.S. Aircraft Industry and the Trend toward Internationalization* (Washington, D.C.: AIA Research Center, March, 1988), 43.

4. Newhouse, *The Sporty Game,* 59.

5. Anderson, "A Look at Lockheed," 11.

6. Reinhardt, "Break-Even Analysis," 834. Reinhardt projected $1 billion, 10 percent, and 5 percent, respectively.

7. "The Higher the Flight, the Farther to Fall," *Economist,* May 3, 1986, 24.

8. "Master of the Air," 54.

9. Mowery, *Alliance Politics and Economics,* 164–65.

10. Newhouse, *The Sporty Game,* 20.

11. Alexander Stuart, "Boeing's New Beauties Are a Hard Sell," *Fortune,* October 18, 1982, 114.

12. Newhouse, *The Sporty Game,* 52.

13. Ibid., 148.

14. Lee Smith, "McDonnell Douglas Is Flying Scared," *Fortune,* August 25, 1980, 41.

15. Stuart, "Boeing's New Beauties," 114.

16. Mowery, *Alliance Politics and Economics,* 144.

17. Ibid., 148.

18. Moxon and Geringer, "Multinational Ventures," 55.

19. Mowery, *Alliance Politics and Economics,* 72.

20. Keith Hayward, *Government and British Civil Aerospace* (Manchester, UK: University of Manchester Press, 1983), 170.

21. Hayward, *International Collaboration,* 111; and "McDonnell Douglas/Fokker Cancel 150-Seat Aircraft," *Aviation Week and Space Technology,* February 15, 1982, 34.

22. Bruce A. Smith, "Douglas, China Sign MD-82 Coproduction Agreement," *Aviation Week and Space Technology,* April 22, 1985, 31.

23. "Lockheed, Airbus Consider U.S. A330/340 Production," *Aviation Week and Space Technology,* June 29, 1987, 31; and Eileen White Read, "Aircraft Boom May Be Only Beginning," *Wall Street Journal,* September 26, 1988, 6.

24. *Aerospace Facts and Figures, 88–89,* 15.

25. Banks, "The Aeroplane Makers," 5.

26. *Aerospace Facts and Figures, 88–89,* 131.

27. Ibid., 15.

28. Richard W. Stevenson, "New Jet Fighter: Risks Are High," *New York Times,* December 27, 1989, D1.

29. Mowery, *Alliance Politics and Economics,* 43.

30. Buck, John T., *Blueprint for Tomorrow* (Wright-Patterson Air Force Base, OH: Aeronautical Systems Division, Air Force Systems Command, 1984).

31. Quoted in *Man for Our Time,* 128.

32. Kotz, *Wild Blue Yonder,* 189.
33. Ibid., 215.
34. John Lancaster, "As World Changes, B-2 Backers Struggle to Keep Costly Jet 'Relevant,'" *Washington Post,* October 23, 1991, A4.
35. President's Blue Ribbon Commission on Defense Management, *Conduct and Accountability* (Washington, D.C.: Government Printing Office, 1986), 2–3.
36. Ibid.
37. Ibid., 17.
38. Mowery, *Alliance Politics and Economics,* 86–87.
39. Stoff, *The Thunder Factory,* 190.
40. Ibid., 182; and "Fairchild Withdrawing from 340 Aircraft Project," *Aviation Week and Space Technology,* October 21, 1985, 23.
41. *Aviation Week and Space Technology,* November 25, 1985, 17; and December 23, 1985, 16.
42. "Why Carlyle Group Is Taking a Flyer on Fairchild," *Business Week,* October 24, 1988, 30.
43. Derdak, *International Directory,* 62.
44. "Grumman Fighting Acquisition Bid by LTV," *Aviation Week and Space Technology,* October 5, 1981, 27.
45. Ibid., 63.
46. "General Dynamics, McDonnell Douglas Win Competition to Build Navy ATA," *Aviation Week and Space Technology,* January 4, 1988, 33; and William M. Carley, "Grumman's Future in Military Aircraft Is Called into Question by Recent Setback," *Wall Street Journal,* January 22, 1988, 6.
47. "Master of the Air," 52.
48. Rick Wartzman and David J. Jefferson, "Boeing Tries to Vector Its Defense Operations Out of Recent Clouds," *Wall Street Journal,* July 30, 1991, A10.
49. Andy Pasztor and Rick Wartzman, "How a Spy for Boeing and His Pals Gleaned Data on Defense Plans," *Wall Street Journal,* January 15, 1990, A1–A4.
50. *Aviation Week and Space Technology,* December 9, 1985, 28.
51. Boeing Company, *Annual Report,* 1986.
52. Jeff Rowe, "Boeing Co. Plans to Create Group on Defense, Space," *Wall Street Journal,* October 30, 1989, A12.
53. "General Dynamics: Striking It Rich in Defense," *Business Week,* May 3, 1982, 102.
54. Ibid.; and Franklin, *The Defender,* 237.
55. Derdak, *International Directory,* 59.
56. John D. Morrocco, "Navy Weighs Alternatives after Cheney Kills Avenger 2," *Aviation Week and Space Technology,* January 14, 1991, 19.
57. Editorial, "Protecting the Defense Industrial Base," *Aviation Week and Space Technology,* February 11, 1991, 7.
58. Rick Wartzman, "Lockheed May Abandon Major Facility," *Wall Street Journal,* May 9, 1990, A4.

59. Stevenson, "New Jet Fighter," D4.

60. Rick Wartzman, "Lockheed's Chief Faces His Biggest Test," *Wall Street Journal,* January 16, 1990, A6.

61. Lockheed Corporation, *1989 Annual Report,* 2.

62. "The Decision to Restructure the Lockheed Aeronautical Systems Group," LASC Information Brief, n.d.

63. Wartzman, "Lockheed's Chief Faces Test," A6.

64. "Harold Simmons Is Playing a Crafty War Game," *Business Week,* March 12, 1990, 40; and Rick Wartzman and Frederick Rose, "Lockheed Management Claims Victory, Simmons Insists Battle 'Too Close to Call,'" *Wall Street Journal,* March 30, 1990, 2.

65. Quoted in J. R. Wilson, "McDonnell Douglas: Recovering from Major Surgery," *Interavia,* November 1991, 29.

66. The autonomous Hughes Aircraft, with $5 billion in annual sales, was sold soon after. GM won the bidding over Boeing, paying Howard Hughes Medical Institute $2.7 billion cash and 50 million shares of new Class H stock (*Forbes,* October 22, 1984, 44–45). The sale, given problems with the Phoenix missile and GM's troubles in the auto market, generated some controversy. GM formed a new GM Hughes Electronics Corporation.

67. Paul Mann, "High Federal Deficits Blight U.S. Aerospace Industry," *Aviation Week and Space Technology,* July 23, 1990, 18.

68. Andy Pasztor and Rick Wartzman, "Lawmakers Question Pentagon on Plans to Extend Help to McDonnell Douglas," *Wall Street Journal,* July 25, 1991, B2.

69. Kotz, *Wild Blue Yonder,* 223.

70. Derdak, *International Directory,* 77; and Coleman, *Jack Northrop,* 212.

71. Coleman, *Jack Northrop;* and "McDonnell Douglas, Northrop Settle Lengthy F/A-18 Dispute," *Aviation Week and Space Technology,* April 15, 1985, 23.

72. "McDonnell Douglas, Northrop," 23.

73. Coleman, *Jack Northrop,* 215–17.

74. Ibid., 218.

75. Ibid., 223.

76. Rick Wartzman and Roy J. Harris Jr., "Northrop Draws Panel's Rebuke; Chairman Quits," *Wall Street Journal,* September 20, 1990, A3.

77. Richard W. Stevenson, "Can a Tarnished Northrop Regain Its Reputation?" *New York Times,* October 21, 1990, F11.

78. Ibid.

79. "Northrop's Biggest Foe May Have Been Its Past," *Business Week,* May 6, 1991, 30–31.

80. Richard W. Stevenson, "Northrop's 'Awesome' B-2 Gamble," *New York Times,* August 20, 1989, sec. 3, p. 6.

81. The French Aerospatiale subsidiary in Texas won a U.S. Coast Guard import order for ninety SA 365 Dauphin 2 helicopters in 1985, sparking protests from Bell, Sikorsky, and others.

Notes to Pages 343–57 • 419

82. "It'll Take More Than Saddam for Sikorsky to Soar Again," *Business Week,* October 1, 1990, 37.

Chapter 16

1. *NASA Aeronautics* (Washington, D.C.: NASA, Office of Aeronautical and Space Technology, 1991), 12–27.
2. *Aviation Week and Space Technology,* January 6, 1992, 19.
3. John Holusha, "Pushing the Envelope at Boeing," *New York Times,* November 10, 1991, 1–3.
4. Andy Pasztor, "Cheney Is Expected to Seek the Survival of Four Costly, Hi-Tech Plane Programs," *Wall Street Journal,* March 30, 1990, 2.
5. Bruce A. Smith, "California Aerospace Firms Fighting on Three Fronts," *Aviation Week and Space Technology,* November 9, 1992, 47–51.
6. Fred R. Bleakley, "New York and Its Neighbors, Still Hurt by the Recession, Scramble to Save Jobs," *Wall Street Journal,* September 16, 1993, A2.
7. Mann, "High Federal Deficits," 18.
8. Aerospace Industries Association, *Leadership through Technology, Annual Report,* 1991.
9. David A. Fulgham, "Materiel Command Open to Ease USAF Relations with Contractors," *Aviation Week and Space Technology,* May 6, 1991, 64.
10. Thomas E. Ricks, "Shift of Military Maintenance Work to Private Sector Is Urged by Pentagon," *Wall Street Journal,* March 24, 1994, A4.
11. Jerrold T. Lundquist, "The False Promise of Defense Conversion," *Wall Street Journal,* March 18, 1993, A12; and "Restructure Defense, Don't Convert It," editorial, *Aviation Week and Space Technology,* January 11, 1993, 70.
12. "Aerospace Sales and Jobs," *Air Force Magazine,* February 1992, 12.
13. Aerospace Industries Association, *Annual Report,* 1991, 64.
14. DARPA, incidentally, played a role in development of the Internet, as it sought a direct link between the Pentagon and the industry on development activities.
15. Aerospace Industries Association, *Maintaining a Strong Aerospace Industry* (Washington, D.C.: Aerospace Industries Association, 1991), 4.
16. Quoted in John Stodden, "Harsher Contract Terms Erode U.S. Defense Industrial Base," *Aviation Week and Space Technology,* January 4, 1988, 80.
17. Scherer, *Industrial Market Structure,* 267.
18. Stodden, "Harsher Contract Terms," 81.
19. Don Fuqua, untitled, *Air Power History* 36 (summer 1989): 49.
20. John D. Morrocco, "U.S. Military Aircraft Design Base Eroding," *Aviation Week and Space Technology,* October 19, 1992, 22.
21. Howard Banks, "Skakeout," *Forbes,* May 1, 1989, 114.

22. Anthony Velocci, "Declining Business Base Spells Industry Consolidation," *Aviation Week and Space Technology,* March 18, 1991, 63.

23. General Dynamics Corporation, *Annual Report,* 1991, 6.

24. Velocci, "Declining Business Base," 65.

25. Thomas G. Donlon, "Riding for a Fall?" *Barron's,* October 21, 1991, 9.

26. "McDonnell Douglas: Disarmed and Dangerous," *Economist,* January 19, 1991, 66.

27. Velocci, "Declining Business Base," 64.

28. Jeff Cole, "Northrop Seeks Grumman in Hostile $2.04 Billion Bid," *Wall Street Journal,* March 11, 1994, A3; and Anthony Velocci, "Northrop Braces for Martin Response," *Aviation Week and Space Technology,* March 28, 1994, 24.

29. Roy J. Harris Jr., "Northrop Offer of $2.17 Billion Wins Grumman," *Wall Street Journal,* April 5, 1994, A3.

30. Thomas E. Ricks, "Antitrust Pact Aims to Support Defense Mergers," *Wall Street Journal,* April 12, 1994, A4.

31. Ibid.

32. *Business Week,* December 2, 1991, 55; and *Business Week,* March 9, 1992, 49.

33. Anthony Velocci, "Douglas Eyes MD-12X Alliance as Financial Health Improves," *Aviation Week and Space Technology,* September 15, 1991, 42.

34. Ibid., 43.

35. Ibid., 45.

36. "Lockheed Sticks to Its Guns," *Business Week,* April 26, 1993, 100.

37. Jeff Cole, "Merger of Lockheed and Martin Marietta Pushes Industry Trend," *Wall Street Journal,* August 30, 1994, A1, A5; Anthony Velocci, "Megamerger Points to Industry's Future," *Aviation Week and Space Technology,* September 5, 1994, 36–38; and Robert F. Dorr, "The Urge to Merge Reemerges," *Aerospace America,* November 1994, 10–11.

38. Jeff Cole and Andy Pasztor, "Lockheed-Marietta Merger Sends Industry Scrambling," *Wall Street Journal,* August 31, 1994, A3.

39. Artemis March, "The Future of the U.S. Aircraft Industry," *Technology Review,* January 1990, 28.

40. Mowery, *Alliance Politics and Economics,* 157.

41. A finding of Moxon and Geringer, "Multinational Ventures," 62.

42. Robert D. Materna, "An Examination of Internationalization in the U.S. Aircraft Industry: A Case Study of McDonnell Douglas Corporation," Ph.D. diss., Georgia State University, 1988, 6, from McDonnell Douglas Corporation, 1987 *Annual Report.*

43. Ibid., 26–28.

44. Ibid., 207; and Lopez and Yager, *Trend toward Internationalization,* 9.

45. Ibid., 216.

Select Bibliography

Articles, Monographs, Reports, and Speeches

Aeronautical Chamber of Commerce. *The Aircraft Industry Prepares for the Future.* Prepared for Senate Military Affairs Committee. N.p., July 10, 1944.

Aerospace Industries Association. *Maintaining a Strong U.S. Aerospace Industry.* Washington, D.C.: Aerospace Industries Association, 1991.

"The Aircraft Boom." *Fortune,* March 1940, 68–71, 104–12.

"Aircraft Industry: Tomorrow's Pterodactyls?" *Economist,* May 30, 1981, 3–12.

Alexander, Tom. "McNamara's Expensive Economy Plane." *Fortune,* June 1967, 89–91, 184–87.

Anderson, Roy A. "A Look at Lockheed." Address to the Newcomen Society, November 1, 1982. New York: Newcomen Society, 1983.

Army Air Corps. *Aircraft Production and Export Policies, 1931–1939.* N.p., n.d.

Banks, Howard. "The Aeroplane-Makers." *Economist.* August 30, 1980, 5–24.

———. "Running Ahead, but Running Scared." *Forbes,* May 13, 1991, 34–38.

———. "Shakeout." *Forbes,* May 1, 1989, 114–19.

"The Big Deal McDonnell Douglas Turned Down." *Business Week,* December 1, 1980, 81–82.

"The Big Payoff." *Time,* February 23, 1976, 28–36.

Bleakley, Fred R. "New York and Its Neighbors, Still Hurt by the Recession, Scramble to Save Jobs." *Wall Street Journal,* September 16, 1993, A2.

"Boeing: Slimming Down an Aerospace Giant." *Business Week,* April 1, 1972, 41–46.

"Boeing: The Higher the Flight, the Farther to Fall." *Economist,* May 3, 1986, 23–26.

Bollinger, Lynn L., and Tom Lilley. *Financial Position of the Aircraft Industry.* Boston: Harvard University, 1943.

Bower, Peter T., et al. *The Impact on Defense Industrial Capability of Changes in Procurement and Tax Policy, 1984–1987.* Washington, D.C.: MAC Group, 1988.

Boyne, Walter. "The Hall-Aluminum Story." *Flying Review International,* January 1969, 58–62.

Brown, David A. "Competing Bid Expected for LTV Aerospace Divisions." *Aviation Week and Space Technology,* March 2, 1992, 31.

———. "Defense Base Erosion Could Hurt Military Aircraft Exports." *Aviation Week and Space Technology,* March 2, 1992, 66.

Buck, John T. *Blueprint for Tomorrow.* Wright-Patterson Air Force Base, OH: Aeronautical Systems Division, Air Force Systems Command, 1984.
Burden, William A. M. "Postwar Status of the Aircraft Industry." *Harvard Business Review* 22 (winter 1944): 211–16.
Carley, William M. "Grumman's Future in Military Aircraft Is Called into Question by Recent Setback." *Wall Street Journal,* January 22, 1988, 6.
"The Cats of MIG Alley." *Time,* June 29, 1953, 82–89.
Cole, Jeff. "Merger of Lockheed and Martin Marietta Pushes Industry Trend." *Wall Street Journal,* August 30, 1994, A1, A5.
———. "Northrop Seeks Grumman in Hostile $2.04 Billion Bid." *Wall Street Journal,* March 11, 1994, A3–A4.
Cole, Jeff, and Andy Pasztor. "Lockheed-Marietta [sic] Merger Sends Industry Scrambling." *Wall Street Journal,* August 31, 1993, A3.
Coleman, Herbert J. "F-16 Team Forming in Europe." *Aviation Week and Space Technology,* June 16, 1975, 16–17.
Cordtz, Dan. "The Withering Aircraft Industry." *Fortune,* September 1970, 114–17, 199–201.
Department of Commerce. *A Competitive Assessment of the U.S. Civil Aircraft Industry.* Washington, D.C., 1984.
Donlon, Thomas G. "Riding for a Fall?" *Barron's,* October 21, 1991, 8–9.
Dorr, Robert F. "The Urge to Merge Reemerges." *Aerospace America,* November 1994, 1–11.
Driscoll, Lisa. "It'll Take More Than Saddam for Sikorsky to Soar Again." *Business Week,* October 1, 1990, 37.
Echols, Oliver P. "The Aircraft Industry, 1947–48." Speech to the Aircraft Industries Association, n.d. Air Force Historical Research Center, Maxwell Air Force Base, AL.
"The Embattled Farmers." *Time,* September 11, 1944, 79–88.
"Experiment at Republic." *Fortune,* February 1947, 123–25, 164–72.
"Fairchild Withdrawing from 340 Aircraft Project." *Aviation Week and Space Technology,* October 21, 1985, 23.
Faneuf, Leston. "Lawrence D. Bell: A Man and His Company." Address to the Newcomen Society, May 15, 1958. New York: Newcomen Society, 1958.
Farnsworth, Clyde H. "The F-16 and How It Won Europe." *New York Times,* July 27, 1975, F1, 12.
"Flying Low." *Time,* September 14, 1959, 92–93.
"The Founder Complains." *Time,* September 2, 1940, 56–57.
Fozard, John W. "Vertical Heroes." *Air Power History* 37 (fall 1990): 54–62.
Frank, Allan Dodds. "Killing the Goose?" *Forbes,* October 22, 1984, 44–45.
Fulghum, David A. "Materiel Command Headquarters Opens to Ease USAF Relations with Contractors." *Aviation Week and Space Technology,* May 6, 1991, 64.
Gansler, Jacques. "The U.S. Aircraft Industry: A Case for Sectoral Planning." *Challenge,* July–August 1977, 13–20.

General Accounting Office. *Defense Industry Profit Study.* Washington, D.C.: GAO, March 17, 1971.

"General Dynamics: Striking It Rich on Defense." *Business Week,* May 3, 1982, 102–6.

"General Dynamics, McDonnell Douglas Win Competition to Build Navy ATA." *Aviation Week and Space Technology,* January 4, 1988, 33.

Gould, Bartlett. "The Story of Burgess Aircraft." *American Aviation Historical Society Journal* 10 (summer and winter 1965), 79; 11 (winter 1966), 230; 12 (winter 1967), 273; 13 (spring 1968), 37.

"Happy Days at Grumman." *Fortune,* June 1948, 112–15, 174–86.

"Hard-Headed Planemaker." *Forbes,* December 15, 1960, 13–16.

Harris, Roy J., Jr. "Northrop Offer of $2.17 Billion Wins Grumman." *Wall Street Journal,* April 5, 1994, A3, A5.

Holt, W. T., Jr. "He Likes to Fly Straight Up." *Saturday Evening Post,* March 11, 1951, 32–33, 61–65.

Holusha, John. "Pushing the Envelope at Boeing." *New York Times,* November 10, 1991, sec. 3, 1–2.

Horner, H. M. "The United Aircraft Story." Address to the Newcomen Society, New York, November 20, 1958. New York: Newcomen Society, 1958.

"Incredible Contract: General Dynamics Gain Is the Nation's Loss." *Barron's,* May 11, 1970, 1, 8, 10.

Jones, Thomas V. "For a Sound Defense Industry." *New York Times,* November 23, 1976, 33.

———. "Kaiser-Frazer: Roughest We Ever Tackled." *Fortune,* July 1951, 74–77, 154–62.

Kim, Suk H., and Sam Barone. "Is the Foreign Corrupt Practices Act of 1977 a Success or Failure?" *Journal of International Business Studies* 12 (winter 1981): 123–26.

Klemin, Alexander. "Investing in Aviation." *Scientific American,* February 1929, 148–53.

Kozicharow, Eugene. "Disciplined Defense Procurement Urged." *Aviation Week and Space Technology,* October 3, 1976, 20.

Kraar, Louis. "How Lockheed Got Back Its Wings." *Fortune,* October 1977, 198–210.

Lancaster, John. "As World Changes, B-2 Backers Struggle to Keep Costly Jet 'Relevant.'" *Washington Post,* October 23, 1991, A4.

Leary, William M., Jr. "At the Dawn of Commercial Aviation: Inglis M. Uppercu and Aeromarine Airways." *Business History Review* 53 (summer 1979): 180–93.

Lilley, Tom, and LaVerne Horton. *Future Financial Problems of Conversion in the Aircraft Industry.* Cambridge, MA: Division of Research, Harvard Business School, 1944.

Loening, Grover C. *Aviation and Banking.* Prepared for Chase National Bank, New York, October 1938.

Lopez, Virginia C., and Loren Yager. *The U.S. Aerospace Industry and the*

Trend toward Internationalization. Washington, D.C.: Aerospace Research Center, March 1988.

Lundquist, Jerrold T. "The False Promise of Defense Conversion." *Wall Street Journal,* March 18, 1993, A12.

Maas, Jim. "Fall from Grace: The Brewster Aeronautical Corporation, 1932–1942." *American Aviation Historical Society Journal* 30 (summer 1985): 118–35.

"Masters of the Air." *Time,* April 7, 1980, 52–59.

Mann, Paul. "High Federal Deficits Blunt U.S. Aerospace Industry." *Aviation Week and Space Technology,* July 23, 1990, 18–19.

March, Artemis. "The Future of the U.S. Aircraft Industry." *Technology Review,* January 1990, 26–36.

McDonald, John. "Jet Airliners: Year of Decision." *Fortune,* April 1953, 125–28, 242–48, May 1953, 128–31, 216–18.

McDonnell, John F. "Between the Devil and the Deep Blue Sea," Speech, Council on Foreign Relations, New York, May 19, 1992. Reprinted by Aerospace Industries Association.

McDonnell, Sanford N. "A Look at the Aerospace Industry: The Last 20 Years and the Next 20 years." *Columbia Journal of World Business* 20 (1985): 79–81.

"McDonnell Douglas: Disarmed and Dangerous." *Economist,* January 19, 1991, 66.

Mecklin, John. "The Biggest, Cheapest Lift Ever." *Fortune,* November 1965, 179–80, 268–84.

———. "The C-5: The Ordeal of the Plane Makers." *Fortune,* December 1965, 158–59, 280–92.

———. "The Rockwells Take Off for Outer Space." *Fortune,* June 1967, 100–104, 165–71.

Moran, Gerard P. "Rex B. Beisel." *The Historical Aviation Album,* vol. 15, 345–55. Temple City, CA: Historical Aviation Album, 1984.

Morrocco, John D. "Navy Weighs Alternatives after Cheney Kills Avenger 2." *Aviation Week and Space Technology,* January 14, 1991, 18–20.

———. "U.S. Military Aircraft Design Base Eroding." *Aviation Week and Space Technology,* October 19, 1992, 22–24.

Moxon, Richard W., and J. Michael Geringer. "Multinational Ventures in the Commercial Aircraft Industry." *Columbia Journal of World Business* 20 (summer 1985): 55–62.

"Mr. Odlum Gets the Business." *Fortune,* August 1949, 90–94, 134–45.

Mrozek, Donald J. "The Truman Administration and the Enlistment of the Aviation Industry in Postwar Defense." *Business History Review* 48 (spring 1974): 73–94.

"Multifarious Sherman Fairchild." *Fortune,* May 1960, 170–74, 220–25.

Murphy, Charles J. V. "The Anxious Aircraft 'Primes.'" *Fortune,* September 1957, 148–49, 276–78.

———. "The Blowup at Hughes Aircraft." *Fortune,* February 1954, 116–18, 188–200.

———. "The Plane Makers under Stress." *Fortune,* June 1960, 134–37, 292–304, July 1960, 111–13, 227–44.
———. "The Wild Blue Chip Wonder." *Fortune,* July 1955, 72–77, 170–72.
"New Interests Paid Curtiss $5,000,000 in Cash." *Aerial Age Weekly,* January 24, 1917, 447.
Northrop, John K. "The Air Industry and Air Power." Lecture presented at Air War College, Maxwell Air Force Base, October 23, 1952. Air Force Historical Research Center, Maxwell Air Force Base, AL.
Oakley, Ralph B. "North American: Its Products and People, Part I." *American Aviation Historical Society Journal* 19 (summer 1974): 133–44.
———. "North American Goes to War, Part II." *American Aviation Historical Society Journal* 20 (spring 1975): 46–57.
"Offering the Airlines Their First U.S.-Built Jet." *Business Week,* March 6, 1954, 112–16.
O'Lone, Richard G. "Pan Am Pushed Aircraft Builders to Advance Transport State of the Art." *Aviation Week and Space Technology,* December 16–23, 1991, 34–35.
———. "Strong Commuter Market Leads Boeing to de Havilland." *Aviation Week and Space Technology,* December 9, 1985, 28–29.
"The Passionate Engineer." *Time,* November 22, 1943, 77–84.
Pasztor, Andy. "Cheney Is Expected to Seek the Survival of Four Costly, Hi-Tech Plane Programs." *Wall Street Journal,* March 30, 1990, 2.
Pasztor, Andy, and Rick Wortzman. "How a Spy for Boeing and His Defense Pals Gleaned Data on Defense Plans." *Wall Street Journal,* January 15, 1990, A1, A4.
Peppers, Jerome, G. *A Brief Historical Review of U.S. Military Logistics, 1939–1985.* Typescript.
"A Place in Space." *Time,* October 27, 1961, 89–94.
"Planemakers Prospects." *Time,* September 29, 1945, 81–82.
President's Blue Ribbon Commission on Defense Management. *Conduct and Accountability.* Washington, D.C.: GPO, 1986.
Proctor, Jonathan, "The Convair 880." *American Aviation Historical Society Journal* 21 (fall 1976): 202–13.
Rae, John B. "Financial Problems of the American Aircraft Industry, 1906–1940." *Business History Review* 39 (spring 1965): 99–114.
———. "Wartime Technology and Commercial Aviation." *Aerospace Historian* 20, no. 3 (1973): 131–36.
Ramirez, Anthony. "Boeing's Happy, Harrowing Times." *Fortune,* July 17, 1989, 40–48.
Raymond, Arthur E. "Reflections of Douglas, 1925–1960." *American Aviation Historical Society Journal* 32 (summer 1987): 110–25.
Read, Ellen White. "Aircraft Boom May Be Only Beginning." *Wall Street Journal,* September 26, 1988, 6.
Reinhardt, A. E. "Break-Even Analysis for Lockheed's Tri Star [*sic*]: An Application of Financial Theory." *Journal of Finance* 28 (September 1973): 821–38.

Report to the Air Coordinating Committee of the Subcommittee on Demobilization of the Aircraft Industry. N.p., October 11, 1945.

Ricks, Thomas E. "Antitrust Pact Aims to Support Defense Mergers." *Wall Street Journal,* April 12, 1994, A4.

———. "Shift of Military Maintenance Work to Private Sector Is Urged by Pentagon." *Wall Street Journal,* March 24, 1994, A4.

Ross, Donald. *An Appraisal of Prospects for the Aircraft Manufacturing Industry.* New York: White, Weld, 1940.

Rowe, Jeff. "Boeing Co. Plans to Create Group on Defense, Space." *Wall Street Journal,* October 30, 1989, A12.

"Rumor of Nixon Pledge on Lockheed Heard as Firm Briefs Bankers on British Pact." *Wall Street Journal,* April 9, 1971, 2.

"Salesman at Work." *Time,* January 14, 1946, 77–86.

Scherer, F. M. "The Aerospace Industry." In *The Structure of American Industry,* ed. Walter Adams, 335–79. 4th ed. New York: Macmillan, 1970.

Schine, Eric. "Lockheed Sticks to Its Guns." *Business Week,* April 26, 1993, 100–101.

Seligman, Daniel. "Barrier-Breaking Bell Aircraft." *Fortune,* March 1956, 125–27, 218–29.

Simonson, G. R. "The Demand for Aircraft and the Aircraft Industry, 1907–1958." *Journal of Economic History* 20 (September 1960): 367–82.

Smith, Bruce A. "California Aerospace Firms Fighting on Three Fronts." *Aviation Week and Space Technology,* November 9, 1992, 47–51.

———. "Douglas, China Sign MD-82 Coproduction Agreement. *Aviation Week and Space Technology,* April 22, 1985, 31.

Smith, Lee. "McDonnell Douglas Is Flying Scared." *Fortune,* August 25, 1980, 41–42.

Smith, Richard K. "Fifty Years of Transatlantic Flight, Part II." *American Aviation Historical Society Journal* 35 (fall 1990), 197–208.

Solow, Herbert. "North American: A Corporation Deeply Committed." *Fortune,* June 1962, 145–49, 164–82.

"So They Named It General Dynamics." *Fortune,* April 1953, 134–37, 228–41.

"Spending More, Getting Less." *Forbes,* April 15, 1975, 22–32.

Stevenson, Richard W. "Can a Tarnished Northrop Regain Its Momentum?" *New York Times,* October 21, 1990, F11.

———. "New Jet Fighter: Risks Are High." *New York Times,* December 27, 1989, D1–4.

———. "Northrop's 'Awesome' B-2 Gamble." *New York Times,* August 20, 1989, sec. 3, 1–3.

Stodden, John. "Harsher Contract Terms Erode U.S. Defense Industrial Base." *Aviation Week and Space Technology,* January 4, 1988, 80–81.

Stuart, Alexander. "Boeing's New Beauties Are a Tough Sell." *Fortune,* October 18, 1982, 114–20.

"Success in Santa Monica." *Fortune,* May 1935, 79–84, 172–90.

"The TFX Verdict." *Aviation Week and Space Technology,* January 4, 1971, 7.
Trimble, William F. "The Naval Aircraft Factory, the American Aviation Industry, and Government Competition, 1919–1928." *Business History Review* 60 (summer 1986): 175–98.
U.S. Civil Aviation Manufacturing Industry Panel. *The Competitive Status of the U.S. Civil Aviation Manufacturing Industry.* Washington, D.C.: National Academy Press, 1985.
Velocci, Anthony L., Jr. "Declining Business Base Spells Industry Consolidation." *Aviation Week and Space Technology,* March 18, 1991, 63–65.
———. "Douglas Eyes MD-12X Alliance as Financial Health Improves. *Aviation Week and Space Technology,* September 23, 1991, 42–45.
———. "Ill-Defined U.S. Defense Priorities Making Industry a 'Gambler's Paradise.'" *Aviation Week and Space Technology,* June 16, 1991, 141–45.
———. "Megamerger Points to Industry's Future." *Aviation Week and Space Technology,* September 5, 1994, 36–38.
———. "Northrop Braces for Martin Response." *Aviation Week and Space Technology,* March 28, 1994, 24–25.
Wartzman, Rick. "Defense Firms Gird for End of Cold War." *Wall Street Journal,* November 29, 1989, A8.
———. "Lockheed May Abandon Major Facility." *Wall Street Journal,* May 9, 1990, A4.
———. "Lockheed's Chief Faces His Biggest Test." *Wall Street Journal,* January 16, 1990, A6.
Wartzman, Rick, and Roy J. Harris Jr. "Northrop Draws Panel's Rebuke; Chairman Quits." *Wall Street Journal,* September 20, 1990, A3.
Wartzman, Rick, and David J. Jefferson. "Boeing Tries to Vector Its Defense Operations Out of Recent Clouds." *Wall Street Journal,* July 30, 1991, A1, A10.
Whalen, Richard J. "Banshee, Demon, Voodoo, Phantom—and Bingo!" *Fortune,* November 1964, 136–39, 256–82.
Williams, Nicholas M. "The Greening of the Helicopter." *Aviation Heritage,* March 1991, 18–25.
———. "The X-Rays." *American Aviation Historical Society Journal* 22 (winter 1977): 242–60.
Wilson, J. R. "McDonnell Douglas: Recovering from Critical Surgery." *Interavia,* November 1991, 26–31.
Wise, T. A. "How McDonnell Won Douglas." *Fortune,* March 1967, 155–56, 214–34.
Woods, George Bryant. *The Aircraft Manufacturing Industry: Present and Future Prospects.* New York: White, Weld, 1946.
Wright, Milton. "Builders of the Aviation Industry." *Scientific American,* March 1929, 228–33.
Wright, T. P. "50,000 Planes a Year." *Aviation,* July 1940, 100.
Zeigler, S. J., and John W. Meader. *An Investigation of the Financial Condi-*

tion and Recent Earnings of Aircraft Contractors. Prepared for the Assistant Secretary of the Navy for Air, May 14, 1942.

Books

Company Histories

Allen, Hugh. *Goodyear Aircraft.* Cleveland: Corday and Gross, 1947.

Allen, Richard Sanders. *The Northrop Story, 1929–1939.* New York: Orion, 1990.

Anderson, Fred. *Northrop: An Aeronautical History.* Los Angeles: Northrop, 1976.

AVCO Corporation: The First Fifty Years. Greenwich, CT: AVCO, 1979.

Bauer, Eugene E. *Boeing in Peace and War.* Enumclaw, WA: TABA Publishing, 1990.

Coleman, Ted. *Jack Northrop and the Flying Wing.* New York: Paragon House, 1988.

Corporate and Legal History of United Air Lines and Its Predecessors and Subsidiaries. 2 vols. Chicago: United Air Lines, 1953, 1965.

Cray, Ed. *Chrome Colossus: General Motors and Its Times.* New York: McGraw-Hill, 1980.

Cunningham, Frank. *Sky Master: The Story of Donald Douglas.* Philadelphia: Dorrance, 1943.

Davenport, William Wyatt. *Gyro! The Life and Times of Lawrence Sperry.* New York: Charles Scribner's Sons, 1978.

Fausel, Robert W. *Whatever Happened to Curtiss-Wright?* Manhattan, KS: Sunflower University Press, 1990.

Francillon, René. *McDonnell Douglas Aircraft since 1920.* London: Putnam, 1979.

Franklin, Roger. *The Defender: The Story of General Dynamics.* New York: Harper and Row, 1986.

Gunston, Bill. *One of a Kind: The Story of Grumman.* Bethpage, NY: Grumman, 1988.

General Dynamics Corporation. *Dynamic America: A History of General Dynamics Corporation and Its Predecessor Companies.* New York: General Dynamics, 1960.

A History of Eastern Aircraft Division, General Motors Corporation. Linden, NJ: Eastern Aircraft Division, 1944.

Ingells, Douglas J. *L-1011 Tristar and the Lockheed Story.* Fallbrook, CA: Aero Publishers, 1973.

———. *The McDonnell Douglas Story.* Fallbrook, CA: Aero Publishers, 1979.

———. *747: Story of the Boeing Super Jet.* Fallbrook, CA: Aero Publishers, 1971.

Kaman, Charles H. *Kaman: Our Early Years.* Indianapolis: Curtis Publishing, 1985.

Langworth, Richard M. *Kaiser-Frazer: The Last Onslaught on Detroit.* Kutztown, PA: Automobile Quarterly Publications, 1975.
Lockheed Corporation. *A History of Lockheed.* In *Lockheed Horizons.* Burbank, CA: Lockheed Corporation, 1983.
Loening, Grover. *Our Wings Grow Faster.* New York: Doubleday, Doran, 1935.
Maloney, Edward T. *Sever the Sky: Evolution of Seversky Aircraft.* Corona del Mar, CA: Planes of Fame, 1979.
Mansfield, Harold. *Vision: The Story of Boeing.* New York: Duell, Sloane, and Pearce, 1986.
Maynard, Crosby, ed. *Flight Plan for Tomorrow. The Douglas Story: A Condensed History.* 2d ed. Santa Monica, CA: Douglas Aircraft, 1966.
Moran, Gerard P. *Aeroplanes Vought, 1917–1977.* Temple City, CA: Historical Aviation Album, 1978.
Murray, Russ. *Lee Atwood . . . Dean of Aerospace.* Los Angeles: Rockwell International, 1980.
Norton, Donald J. *Larry: A Biography of Lawrence D. Bell.* Chicago: Nelson-Hall, 1981.
Of Men and Stars: A History of Lockheed Aircraft Corporation, 1913–1957. Burbank, CA: Lockheed Aircraft, 1957.
Pedigree of Champions: Boeing since 1916. 6th ed. Seattle: Boeing, 1985.
The Pratt and Whitney Aircraft Story. East Hartford, CT: Pratt and Whitney Aircraft Division, United Aircraft, 1950.
Rentschler, Frederick B. *An Account of Pratt and Whitney Aircraft Company, 1925–1950.* East Hartford, CT: Pratt and Whitney, 1950.
Schoen, Arthur L. *Vought: Six Decades of Aviation History.* Plano, TX: Aviation Quarterly Publishers, 1978.
Sikorsky, Igor I. *The Story of the Winged-S.* New York: Dodd, Mead, 1938.
Smith, Frank Kingston. Legacy of Wings: The Story of Harold Pitcairn. New York: J. Aronson, 1981.
Stoff, Joshua. *The Thunder Factory: An Illustrated History of the Republic Aviation Corporation.* London: Arms and Armour Press, 1990.
Thruelson, Richard. *The Grumman Story.* New York: Praeger, 1976.
Trimble, William F. *Wings for the Navy: A History of the Naval Aircraft Factory, 1917–1956.* Annapolis, MD: Naval Institute Press, 1990.
Wagner, William. *Reuben Fleet and the Story of Consolidated Aircraft.* Fallbrook, CA: Aero Publishers, 1976.
———. *Ryan, the Aviator.* New York: McGraw-Hill, 1971.
Wegg, John. *General Dynamics Aircraft and Their Predecessors.* Annapolis, MD: Naval Institute Press, 1990.
Wings for the Navy: A History of Chance Vought Aircraft. Stratford, CT: United Aircraft Corporation, 1943.
Yenne, Bill. *Rockwell: The Heritage of North American.* New York: Crescent Books, 1989.
Yesterday, Today, and Tomorrow: Fifty Years of Fairchild Aviation. Germantown, MD: Fairchild Hiller, 1970.

Other

Adams, Gordon. *The Iron Triangle: The Politics of Defense Contracting.* New York: Council on Economic Priorities, 1981.

Art, Robert J. *The TFX Decision: McNamara and the Military.* Boston: Little, Brown, 1968.

Barlett, Donald C., and James B. Steele. *Empire: The Life, Legend, and Madness of Howard Hughes.* New York: W. W. Norton, 1979.

Bilstein, Roger E. *Orders of Magnitude: A History of the NACA and NASA, 1915–1990.* Washington, D.C.: NASA, 1989.

Bluestone, Barry, Peter Jordan, and Mark Sullivan. *Aircraft Industry Dynamics: An Analysis of Competition, Capital, and Labor.* Boston: Auburn House, 1981.

Bowers, Peter M. *The DC-3: 50 Years of Legendary Flight.* Blue Ridge Summit, PA: Tab Books, 1986.

Bright, Charles D. *The Jet Makers: The Aerospace Industry from 1945 to 1972.* Lawrence, KS: Regent's Press, 1978.

Brown, Genevieve. *Development of Transport Airplanes and Air Transport Equipment.* Wright Field, OH: Air Technical Service Command, 1946.

Brown, Stanley H. *Ling: The Rise, Fall, and Return of a Texas Titan.* New York: Atheneum, 1972.

Cochrane, Dorothy, Von Hardesty, and Russell Lee. *The Aviation Careers of Igor Sikorsky.* Seattle: University of Washington Press, 1989.

Coffey, Thomas M. *Hap.* New York: Viking Press, 1982.

Coulam, Robert F. *Illusions of Choice: Robert McNamara, the F-111, and the Problem of Weapons Acquisition Reform.* Princeton: Princeton University Press, 1977.

Craven, W. F., and J. L. Cate, eds. *The U.S. Army Air Forces in World War II.* Vol. 6, *Men and Planes.* Chicago: University of Chicago Press, 1955.

Crouch, Tom. *The Bishop's Boys: A Life of Wilbur and Orville Wright.* New York: W. W. Norton, 1989.

Cunningham, William G. *The Aircraft Industry: A Study of Industrial Location.* Los Angeles: Lorrin L. Morrison, 1951.

Dade, George C., and George Vecsey, comps. *Getting Off the Ground: The Pioneers of Aviation Speak for Themselves.* New York: Dutton, 1979.

Dietrich, Noah. *Howard: The Amazing Mr. Hughes.* Greenwich, CT: Fawcett, 1972.

Durr, Clifford J. *The Early History of Defense Plant Corporation.* Washington, D.C.: Committee on Public Administration Cases, 1950.

Fitzgerald, Ernest. *The High Priests of Waste.* New York: W. W. Norton, 1972.

Foulois, Benjamin D., with C. V. Glines. *From the Wright Brothers to the Astronauts.* New York: McGraw-Hill, 1968.

Fox, J. Ronald. *Arming America: How the U.S. Buys Weapons.* Cambridge, MA: Harvard University Press, 1974.

Freudenthal, Elsbeth E. *The Aviation Business: From Kitty Hawk to Wall Street.* New York: Vanguard, 1940.

———. *Flight into History: The Wright Brothers and the Air Age.* Norman: University of Oklahoma Press, 1949.
Gansler, Jacques. *The Defense Industry.* Cambridge, MA: MIT Press, 1982.
Goldberg, Alfred, ed. *History of the United States Air Force.* Princeton, NJ: Van Nostrand, 1974.
Hallion, Richard P. *Designers and Test Pilots.* Epic of Flight series. New York: Time, Inc., 1983.
Hanle, Paul A. *Bringing Aerodynamics to America.* Cambridge, MA: MIT Press, 1982.
Harding, William Barclay. *The Aviation Industry.* New York: Charles D. Barney, 1937.
Hayward, Keith. *Government and British Civil Aerospace.* Manchester, UK: University of Manchester Press, 1983.
———. *International Collaboration in Civil Aerospace.* New York: St. Martin's Press, 1986.
Heinemann, Edward H., and Rosario Rausa. *Ed Heinemann: Combat Aircraft Designer.* Annapolis, MD: U.S. Naval Institute, 1980.
Higham, Robin. *Air Power: A Concise History.* New York: St. Martin's Press, 1972.
History of the Aircraft War Production Council, East Coast, Inc. N.p., 1947.
History of the Bureau of Aircraft Production. Wright-Patterson Air Force Base, Ohio: Air Materiel Command, Historical Office, 1951.
Holley, I. B., Jr. *Buying Aircraft: Materiel Procurement for the AAF.* U.S. Army in World War II Special Studies. Washington, D.C.: Office of Chief of Military History, 1964.
———. *Ideas and Weapons.* New Haven: Yale University Press, 1953.
Horgan, James J. *City of Flight: The History of Aviation in St. Louis.* Gerald, MO: Patrice Press, 1984.
Howard, Fred. *Wilbur and Orville: A Biography of the Wright Brothers.* New York: Knopf, 1987.
Hubler, Richard G. *Big Eight: A Biography of an Airplane.* New York: Duell, Sloane, and Pearce, 1960.
Ingells, Douglas J. *The Plane That Changed the World.* Fallbrook, CA: Aero Publishers, 1966.
Kaiser, William K., ed. *The Development of the Aerospace Industry on Long Island, 1904–1964.* 3 vols. Hempstead, NY: Hofstra University, 1968.
Kelly, Fred. *The Wright Brothers.* New York: Harcourt, Brace, 1943.
Kelsey, Benjamin S. *The Dragon's Teeth? The Creation of United States Air Power for World War II.* Washington, D.C.: Smithsonian Institution Press, 1982.
Kucera, Randolph P. *The Aerospace Industry and the Military: Structural and Political Relationships.* Beverly Hills, CA: Sage Publications, 1974.
Kuter, Lawrence S. *The Great Gamble: The Boeing 747.* University: University of Alabama Press, 1973.
Lambermont, Paul, with Anthony Pirie. *Helicopters and Autogyros of the World.* Rev. ed. New York: A. S. Barnes, 1970.

Loening, Grover C. *Takeoff into Greatness: How American Aviation Grew So Big So Fast.* New York: Putnam, 1968.

Maurer, Maurer, ed. *The U.S. Air Service in World War I.* Vol. 2. Washington, D.C.: Office of Air Force History, 1978.

McFarland, Marian W., ed. *The Papers of Wilbur and Orville Wright.* New York: McGraw-Hill, 1953.

McMurtrie, Mary L. *History of the AAF Materiel Command, 1926 through 1941.* Wright Field, OH: Historical Office, Air Materiel Command, 1943.

Miller, Ronald, and David Sawers. *The Technical Development of Modern Aviation.* New York: Praeger, 1970.

Miller, Thomas G., Jr. *Strategies for Survival in the Aerospace Industry.* Cambridge, MA: Arthur D. Little, 1964.

Mingos, Howard. *The Birth of An Industry.* New York: W. B. Conkey, 1930.

Mowery, David C. *Alliance Politics and Economics: Multinational Joint Ventures in Commercial Aircraft.* Cambridge, MA: Ballinger Publishers, 1987.

Mowery, David C., and Nathan Rosenberg. *Technology and the Pursuit of Economic Growth.* New York: Cambridge University Press, 1989.

Naylor, J. L., and E. Ower. *Aviation: Its Technical Development.* Philadelphia: Dufour Editions, 1965.

Nelson, Donald M. *Arsenal of Democracy: The Story of American War Production.* New York: Harcourt, Brace, 1946.

Newhouse, John. *The Sporty Game: The High-Risk Competitive Business of Making and Selling Commercial Airliners.* New York: Knopf, 1983.

Peck, Merton J., and F. M. Scherer. *The Weapons Acquisition Process: An Economic Analysis.* Boston: Division of Research, Harvard Business School, 1962.

Phillips, Almarin. *Technology and Market Structure: A Study of the Aircraft Industry.* Lexington, MA: Heath Lexington, 1971.

Purtee, Edward O. *History of the Army Air Service, 1907–1926.* Wright-Patterson Air Force Base, OH: Air Materiel Command, 1948.

Rae, John B. *Climb to Greatness: The American Aircraft Industry, 1920–1960.* Cambridge, MA: MIT Press, 1968.

Reguero, Miguel Angel. *An Economic Study of the Military Airframe Industry.* Washington, D.C.: Department of the Air Force, 1957.

Renstrom, Arthur G. *Wilbur and Orville Wright: Chronology.* Washington, D.C.: Library of Congress, 1975.

Riddle, Donald H. *The Truman Committee: A Study in Congressional Responsibility.* New Brunswick, NJ: Rutgers University Press, 1964.

Roseberry, C. R. *Glenn Curtiss: Pioneer of Flight.* Garden City, NY: Doubleday, 1972.

Scherer, F. M. *Industrial Market Structure and Economic Performance.* Chicago: Rand-McNally, 1980.

———. *The Weapons Acquisition Process: Economic Incentives.* Boston: Division of Research, Harvard Business School, 1964.
Serling, Robert J. *The Jet Age.* Epic of Flight series. New York: Time, Inc., 1982.
Simonson, Gene R., ed. *The History of the American Aircraft Industry: An Anthology.* Cambridge, MA: MIT Press, 1968.
Smith, Peter C. *Vengeance! The Vultee Vengeance Dive Bomber.* Washington, D.C.: Smithsonian Institution Press, 1986.
Smith, R. Elberton. *U.S. Army in World War II. The Army and Economic Mobilization.* Washington, D.C.: Office of the Chief of Military History, Department of the Army, 1959.
Smith, Richard Austin. *Corporations in Crisis.* Garden City, NY: Doubleday, 1963.
Stekler, Herman O. *The Structure and Performance of the Aerospace Industry.* Berkeley and Los Angeles: University of California Press, 1965.
Stewart, James T. *A Man for the Time: An Oral History Interview with Lieutenant General James T. Stewart.* Wright-Patterson Air Force Base, OH: History Office, Aeronautical Systems Division, Air Force Systems Command, 1990.
Still, Henry. *To Ride the Wind: A Biography of Glenn L. Martin.* New York: Julian Messner, 1964.
Stoff, Joshua. *The Aerospace Heritage of Long Island.* Interlaken, NY: Heart of the Lakes Publishing, 1989.
Taylor, Frank J., and Lawton Wright. *Democracy's Air Arsenal.* New York: Duell, Sloane, and Pearce, 1947.
Trimble, William F. *High Frontier: A History of Aeronautics in Pennsylvania.* Pittsburgh: University of Pittsburgh Press, 1982.
Vander Meulen, Jacob. *The Politics of Aircraft: Building an American Military Industry.* Lawrence: University Press of Kansas, 1991.
Van Vleet, Clarke, and William J. Armstrong. *United States Naval Aviation, 1910–1980.* Washington, D.C.: Naval Air Systems Command, 1981.
Van Wyen, Adrian O. *Naval Aviation in World War I.* Washington, D.C.: Chief of Naval Operations, 1969.
Walton, Francis. *Miracle of World War II: How American Industry Made Victory Possible.* New York: Macmillan, 1956.
White, Gerald T. *Billions for Defense.* University: University of Alabama Press, 1980.
Wilson, Eugene E. *Slipstream: Memoirs of an Air Craftsman.* New York: McGraw-Hill, 1950.
Wood, Derek. *Project Cancelled: The Disaster of Britain's Abandoned Aircraft Projects.* Rev. ed. London: Jane's Publishing, 1986.
Zimmerman, S. A. *Procurement in the United States Air Force, 1938–1948.* 2 vols. Wright-Patterson Air Force Base, OH: Air Materiel Command, 1950.

Dissertations

Kucera, Randolph P. "Grumman Aircraft Engineering Corporation and Its 'Familial' Relationship with the U.S. Navy." Syracuse University, 1973.

Materna, Robert D. "An Examination of Internationalization in the U.S. Aircraft Industry in the 1980s: A Case Study of McDonnell Douglas Corporation." Georgia State University, 1988.

Shin, Tai Saeng. "A Financial Analysis of the Airframe-Turned-Aerospace Industry." University of Illinois, 1969.

Vander Meulen, Jacob. "The American Aircraft Industry to 1936: A Political History of a Military Industry." University of Toronto, 1989.

Wilburn, James Richard. "Social and Economic Aspects of the Aircraft Industry in Metropolitan Los Angeles during World War II." UCLA, 1971.

Annuals and Directories

Aerospace Industries Association. *Aircraft Facts and Figures* and *Aerospace Facts and Figures.* Annual editions. Washington, D.C.: Aerospace Industries Association.

Aircraft Year Book and *Aerospace Year Book.* Annual editions, 1919–present. Washington, D.C.: Aerospace Industries Association.

The Aviation Annual of 1944. Garden City, NY: Doubleday Doran, 1945.

Derdak, Thomas, ed. *International Directory of Company Histories.* Vol. 1. Chicago: St. James Press, 1988.

Fahey, James C., ed. *U.S. Army Aircraft, 1908–1946.* New York: Ships and Aircraft, 1946.

Fifty Years of Aviation Progress. National Committee to Observe the Fiftieth Anniversary of Powered Flight, J. H. Doolittle, Chairman. 1953.

Financial Handbook of the American Aviation Industry. New York: Commercial National Bank and Trust Company, July 1929.

Frey, Royal D. *USAF Postwar Chronology.* Wright-Patterson Air Force Base, OH: Historical Office, Air Materiel Command, December, 1950.

Greenwood, John T., ed. *Milestones of Aviation.* Washington, D.C.: Smithsonian Institution Press, 1989.

Hallion, Richard P. *The Literature of Aeronautics, Astronautics, and Air Power.* Washington, D.C.: Office of Air Force History, 1984.

Hoagland, Roland W., ed. *The Blue Book of Aviation: A Biographical History of American Aviation.* Los Angeles: Hoagland, 1932.

Jane's All the World's Aircraft. Annual editions, 1909–present. Current editors, Reginald N. Cleveland and and Frederick P. Graham. London: Sampson Low, Marston and Co.

Leary, William M., Jr., ed. *Encyclopedia of American Business History and Biography. The Airline Industry.* New York: Facts on File, 1992.

NASA. *NASA Aeronautics.* Washington, D.C.: Office of Aeronautics and Space Technology, NASA, 1991.

Shrader, Welman A. *Fifty Years of Flight: A Chronicle of the Aviation Industry in America, 1903–1953.* Cleveland: Eaton, 1953.

Todd, Daniel, and Ronald Humble. *World Aerospace: A Statistical Handbook.* New York: Croom Helm, 1987.

Manuscript Collections

Henry H. Arnold Papers. Library of Congress.

Sherman Fairchild Papers. Garber Facility, National Air and Space Museum, Smithsonian Institution.

Robert E. Gross Papers. Library of Congress.

Clement Keys Papers. Garber Facility, National Air and Space Museum, Smithsonian Institution.

Glenn L. Martin Papers. Library of Congress.

U.S. Army Air Corps and Army Air Force Records. Record Group 18, National Archives.

General Index

Acosta, Bert, 65
Advanced Research Projects Agency (ARPA), 353. *See also* DARPA
Aerial Experiment Association (AEA), 11–12, 14, 16
Aeritalia, 314
Aermacchi, 348
Aero Club of America, 15, 35
Aerojet Engineering Corp./Aerojet-General, 127, 205
Aeromarine Company, 26, 34, 42–43, 56
Aeromarine-Klemm Co., 56
Aeromarine West Indian Airways, 39
Aeronautical Chamber of Commerce (ACC), 36, 121–22, 149, 151, 159
Aeronautical Corporation of America (Aeronca), 124, 136, 192
Aeronautical Division, U.S. Army, 31
Aeronautical Manufacturers Association, 15
Aeronautical Society, U.S., 15
Aeronautical Society of New York, 17
Aeronautical Systems Division (ASD) of AFSC, 248, 256, 274, 326
Aerospatiale Helicopter Corp., 311
AGA Aviation Corp., 141–42. *See also* Pitcairn Companies
Agusta, 189, 208, 348
Airbus Industrie, 283, 314, 318, 320, 346, 363, 365
Air Commerce Act of 1926, 59, 76, 101, 207
Aircooled Motors Co., 187
Air Coordinating Committee, Standing Subcommittee on Demobilization of the Aircraft Industry, U.S. Senate, 207
Air Corps, U.S. Army, 49, 51, 59–60, 75, 80, 87–88, 90, 98–99, 103, 105–6, 109, 111, 117–18, 120
Air Corps Act of 1926, 64–65, 89, 96
Aircraft Board, 31–33, 37
Aircraft Industries Assn./Aerospace Industries Assn. (AIA),151, 154, 186, 200, 207, 249, 353–54, 357
Aircraft Manufacturers Association, 35
Aircraft Production Board, 30, 34
Aircraft Production, Bureau of, 31–33
Aircraft Production Council, 194
Aircraft War Production Council (AWPC), 121–22, 127
Air Force, U.S., 195, 242, 344
Air Force Logistics Command (AFLC), 248, 352
Air Force Materiel Command (AFMC), 352
Air Force Systems Command (AFSC), 248, 274, 352
Air Holdings, Ltd., 280–81
Airline industry development, 41–42, 55, 68, 76, 81, 90–93, 101–3, 161–64
Air Mail, 37, 48, 60, 76, 79, 87–88
Air Mail Act of 1934, 78, 82–83
Air Mail (Kelly) Act of 1925, 60, 76
Air Mail (Watres) Act of 1930, 87
Air Materiel Command, 153, 194
Air National Guard, 193
AiRover Company, 109. *See also* Lockheed Aircraft Corp.
Airplane Development Corp. of AVCO, 73–74, 80
Air Research and Development Command, 247
Air Service Act, 37
Air Service, U.S. Army, 31, 33, 37, 46, 49, 51, 60, 75, 82
Air Service Command, USAAF, 128
Air Technical Service Command (ATSC), USAAF, 128, 194
Air Traffic Control (ATC), 101, 266
Air Transport Association (ATA), 102
Air Transport Command, USAAF, 140
Airways Modernization Act of 1957, 207
Alcock, John W., and Brown, Arthur Whitten, 39
Alco Hydro-Aeroplane Co., 54. *See also* Lockheed Aircraft Corp.
ALCOR Co., 86
Alger, Russell, 7

438 • General Index

Alhambra Airport and Air Transportation Co., 86
Allen, William M., 76, 135, 177, 215–16, 224, 233, 262, 276–77, 305, 317, 334, 367, 369
Allis, James A., 183
Allis-Chalmers Corp., 144
Allison Engineering Corp. (division of General Motors), 83, 129, 144, 200, 362
All Nippon Airways (ANA), 298
American Airplane and Engine Corp., 79–80
American Airways/American Airlines, 79, 92, 116, 136, 163, 186, 219, 281–83
American Aviation Corp., 124, 165
American Aviation (new), 301
American Aviation Mission of 1919, 35
American Engineering Council, 59
American-Marietta Corp., 223, 262. *See also* Glenn L. Martin Co.
American Stock Exchange, 306
Ames, Joseph F., 118
Ames Laboratory of NACA/NASA, 118, 207
Anaconda Copper Corp., 32
Anders, William A., 335, 359
Anderson, Robert, 266, 309–10, 339
Anderson, Roy A., 306–7, 319
Antitrust policy, 221, 241. *See also* U.S. Department of Justice
Apt, Milburn G., 198
Arcier, A. Francis, 55
ARGOSystems Corp., 333–34. *See also* Boeing Airplane Company
Armed Services Procurement Act, 195
Army, U.S., 50, 208–9
Army Air Corps Mail Operation (AACMO), 87
Army Air Forces, U.S., 120, 123, 125, 132, 135–36, 140, 146, 159
Army Aviation, 196, 208
Army Corps of Engineers, 123
Army Flying Service, 16
Army Ground Forces, 120
Arnold, Bruce, 118
Arnold, Henry H., 94, 97, 99, 111, 113, 117–19, 129, 136, 143, 152, 159
Atlantic Aircraft Corp., 55
Atlas Corp., 114, 168
Atwood, J. Leland, 62, 83, 137, 171, 230–31, 234, 262, 265–66, 309, 371

Augustine, Norman, 362
Automobile industry, 23, 30, 33, 72–73, 124, 138–40, 205, 261
Automotive Council for War Production, 121
Aviation Corporation (AVCO), 73–74, 78–79, 81, 87, 106–8, 115, 138, 168, 170, 342
Aviation Development Corp., 74
Aviation Manufacturing Corp. (of AVCO), 80, 106
Avion Corp., 77

Baker, Newton D., 89
Baker Board, 89–90
Balbo, Italo, 98
Baldwin, Thomas F., 11–12
Baldwin Aircraft Co., 43
Banner Industries, 330
Barling, Walter, 41
Barnes, Alphaeus, 17–18
Beachey, Lincoln, 14, 16
Beall, Donald R., 310, 339
Beall, Wellwood E., 101, 109, 214, 233, 268
Bedford, Clay P., 165
Beech Aircraft Corp., 124, 160, 323, 337, 347
Beisel, Rex B., 44, 77, 110, 182, 268
Bell, Alexander Graham, 11
Bell, Grover, 20, 369
Bell, Lawrence D., 20–21, 44, 47, 62, 66, 113–14, 117, 136, 143, 147, 159, 180, 199, 222, 368–70
Bell Aerospace Corp., 222
Bell Aerosystems Corp., 222
Bell Aircraft Corp., 67, 93, 114, 123, 132, 135–36, 144, 152, 155, 159, 179, 189, 192, 194, 204, 208, 221
Bellanca, Giuseppe, 8
Bellanca Aircraft Corp., 83, 124, 131, 136, 209
Bell Helicopter Canada, 341
Bell Helicopter Corp./Bell Helicopter Textron, 208, 211, 222, 262, 267, 311, 328, 337, 341–42, 349, 363
Bell Intercontinental Corp., 222
Belmont, August, 7
Bendix Aircraft Corp., 83, 86, 137, 205
Bendix Trophy Race, 39
Berle, Adolf A., 162
Berlin, Donovan R., 86, 107, 139, 184, 209–10
Berliner, Henry A., 82, 104

Berliner-Joyce Co. (B/J), 82–83
Bern, Edward G., 146
Berner, T. Roland, 173
Bernhard, Prince, 297–98
Bertrand, Lloyd, 40
Bethlehem Steel Corp., 226
Betti, John A., 335
Bierwirth, Jack, 180, 302–3, 332
Bishop, Courtland Field, 11–12
Bishop, Mrs. Lillian Fleet, 50
Black, Eli, 296
Black, Hugo, 87
Bleriot, Louis, 14
Blue Ribbon Commission on Defense Management (Packard Commission), 328–29
Boeing, Bertha, 215
Boeing, William E., 10, 22, 43, 76–78, 91, 145, 200, 215, 368, 370
Boeing Aircraft Company, 78
Boeing Aircraft Company of Canada, Ltd., 78
Boeing Airplane and Transport Corp., 76
Boeing Airplane Company/Boeing Company, 22, 34–35, 41–43, 51, 60, 67, 77–78, 83, 90, 93–94, 98, 100, 102, 107, 109, 117–18, 122–23, 132, 143, 150, 152–53, 155, 165, 176–78, 256–57, 261–64, 268, 270–71, 275–77, 279, 285, 292, 298, 304–5, 314, 319–20, 323, 328, 330, 332–33, 336, 346–47, 349, 351, 354, 357–58, 360–61, 363, 365–67, 371
Boeing Air Transport, 42, 76–77
Boeing Vertol/Boeing Helicopters, 210, 223, 262, 304, 312, 334, 342, 363
Boileau, Oliver C., 333–35
Boland, Frank E., 26
Boland Aeroplane Company, 26
Bolkow-Entwicklungen KG, 224
Bolling, Raynall C., 30
Bollinger, Lynn L., 149
Bombardier, Inc., 334
Borg-Warner Co., 66
Bottom-Up Review of U.S. Department of Defense, 351–52
Boullioun, E. M. ("Tex"), 304, 333
Boutelle, Richard, 183, 239
Bradley, Samuel S., 36, 159
Brantly, N. O., 143
Brashears and Co., 85, 221
Breech, Ernest R., 82, 137
Breese, Vance, 80

Brewster Aeronautical Corp., 69–70, 113, 132–33, 125, 155
Brewster Corp., 70
British Aerospace Corp. (BAe), 221, 323, 346, 351–52, 363
British Airways (BA), 320–21
British Purchasing Commission, 1–3, 110, 119, 122, 128
Brizendine, John C., 282, 307, 318, 338
Brown, Dayton, 70, 145
Brown, Walter F., 87
Budd Company, 71, 141
Bunker, George M., 175–76, 223, 368
Bunker-Ramo Industries, 301
Bureau of Aeronautics (BuAer), U.S. Navy, 37, 44, 60, 77, 88, 97, 118, 145, 249
Bureau of Air Commerce, U.S. Department of Commerce, 59, 85, 87, 91, 106
Bureau of Aircraft Production, 31–33
Bureau of Naval Weapons, U.S. Navy, 248
Burgess, Starling, 12–13, 16, 23
Burgess Company, 14, 23
Burgess Company and Curtis, Inc., 13
Burnelli, Vincent, 41, 56
Burnelli Aircraft Corp., 56
Burnelli Aircraft, Ltd., 56
Bush, George, 357
Buy American Act of 1933, 250
Byrd, Richard E., 65

California Institute of Technology (Caltech), 61, 127–28
Caminez Engine Co., 53
Canadair Corp. (division of General Dynamics), 169, 305
Caporali, Renso, 332, 360
Carlyle Group, 330, 358, 361
Carmichael, James V., 136, 177
Carmichael, J. H., 239
Carnegie, Andrew, 10
Carrier Corp., 312
Carrill, Don, 138
Carter, Jimmy, 273, 291, 299, 326–27
Castle, B. F., 71
Central Aerodynamics and Hydrodynamics Institute (TsAGI), 294
Central Intelligence Agency (CIA), 195, 227–28, 326
Cessna, Clyde V., 15
Cessna Aircraft Corp., 124, 196, 334, 348, 359

Chamberlin, Clarence, 65
Chamber of Commerce, U.S., 151
Champline, George, 145
Chance Vought Aircraft Corp., 41, 44, 60, 82
Chance Vought Division, UATC/UAC, 67–68, 78, 110, 144, 165, 182, 184, 200–201, 206, 241, 247, 254, 267, 311
Chandler, Harry B., 48
Chanute, Octave, 6–7, 16
Chappellet, Cyril, 84, 227
Chase Aircraft Corp., 165–66
Chase National Bank, 183
Chrysler Corp., 205, 266, 305
Church, Frank, 296
Churchill, Winston, 121
Cierva, Juan de la, 104
Civil Aeronautics Act of 1938, 101, 207
Civil Aeronautics Administration (CAA), 101, 161
Civil Aeronautics Authority (CAA), 101
Civil Aeronautics Board (CAB), 101, 160–61, 284, 318
Civil Aircraft Agreement of 1979, 318
Civil Reserve Air Fleet (CRAF), 162
Clark, Ramsey, 235
Clinton, Bill, 351, 357
Coast Guard, U.S., 175, 181
Cochran, Jacqueline, 168
Coffin, Howard E., 24, 30–31, 33, 35, 121
Cohu, La Motte T., 79, 115, 168, 186
Cold War, 117, 155, 158, 169, 192, 199–200, 238, 244–45, 261, 343, 348, 350, 363, 366
Collier, Robert J., 7, 18
Collier Trophy, 39, 68, 92, 100, 197, 307
Collins, Whitley G., 55, 185, 236
Collins Radio Co., 310
Columbia Aircraft Corp., 124, 134, 165
Columbia University, 24
Conant, Frederick W., 233
Congress of Industrial Organizations (CIO), 95
Consolidated Aircraft Corp., 35, 41, 44, 51, 60, 67–68, 70–71, 91, 94, 101, 105–6, 113–15, 117, 126, 133, 137, 139. *See also* Fleet, Reuben H.
Consolidated Foods Corp., 305
Consolidated-Vultee Aircraft Corp.
(Convair/Convair Division GD Corp.), 122, 138–39, 146, 150, 152–54, 167, 169–70, 176, 200–201, 215, 217–18, 221, 225–26, 232, 266, 305, 334, 345, 358–59
Continental Motors, 221
Convertiplane development, 212–13, 242. *See also* Helicopter development; Vertical-Takeoff-and-Landing (VTOL) aircraft
Coolidge, Calvin, 57
Cord, E. L., 73, 79, 106
Cord Corp., 79, 106
Cornell University, 24
Council of National Defense, 30, 120
Counterinsurgency (COIN) aircraft, 244
Cover, Carl, 136
Crane, Henry M., 24
Crowell Report, 35
Crown, Henry, 225, 258, 305, 334, 359
Cummings, Nathan M., 305
Curtis, Greely S., 12–14, 23
Curtiss, Glenn H., 6, 8–9, 10–11, 13–14, 16, 21–23, 26, 43, 45–46, 62, 68, 81, 371
Curtiss Aeroplane Company, 13, 15–16
Curtiss Aeroplane and Motor Company, 13–14, 23, 28, 30, 33–34, 35, 42–44, 51, 60, 62–63, 69, 80, 82, 106
Curtiss Aeroplane and Motor Company, Ltd., 13
Curtiss-Caproni Co., 80–81
Curtiss Engineering Company, 23
Curtiss Flying Service/Curtiss-Wright Flying Service, 53, 81–82
Curtiss Manufacturing Co., 12
Curtiss Motor Company, 11, 13
Curtiss-Robertson Co., 80
Curtiss-Wright Corporation, 49, 63, 67–68, 73–74, 78, 80–81, 83, 91, 93–94, 106–7, 116–17, 121, 123, 129, 133, 138–39, 144, 147, 150, 152–55, 171–73, 184, 194, 201, 213

Daland, Elliot, 49, 143
Damon, Ralph, 106, 136, 211
Daniell, Robert F., 342
Dassault, General Aeronautique Marcel/Dassault Aviation, 252, 286, 292, 315
Davison, F. Trubee, 60, 120

Da Vinci, Leonardo, 104
Davis, David R., 47, 105–6
Davis-Douglas Company, 47. *See also* Douglas Aircraft Co., Inc.
Day, Charles H., 20, 26
Dayton Engineering Laboratories Company (DELCO), 24, 30, 49
Dayton Wright Company, 24, 30, 33, 35, 44, 51, 72
Deeds, Edward A., 24, 30, 33, 75–76
Defense Act of 1916, 30
Defense Act of 1920, 37
Defense Advanced Research Projects Agency (DARPA), 326, 343–44, 351, 353–54
Defense Contract Administration Service (DCAS), 246
Defense industry, 99, 117, 290–95, 329
Defense Mobilization Base concept, 237, 252
Defense Plant Corporation (DPC), 123–25, 133–36, 137, 139, 145, 147, 172, 177, 180
Defense Production Act, 128
Defense Production Administration, 194
Defense Security Assistance Agency (DSAA), 291, 314
De Havilland Canada, Ltd., 333–34
Delaney hearings (1934), 90
Delta Air Lines, 49, 321
Demisch, Wolfgang, 354
de Seversky, Alexander P., 9, 70, 110–11, 201, 268
Detroit Aircraft Corp., 50, 70, 73–75
Detweiler, Frederick O., 182, 241
Deutsche Aerospace (DASA), 346, 348
Dickinson, Arnold, 52–53
Dietrich, Noah, 167
Dole, Bob, 273, 334
Doman, Glidden S., 189
Doman-Frazier Helicopters Co., 189
Doman Helicopters, Inc., 211
Doolittle, Jimmy, 40, 65, 117
Dornberger, Walter, 222
Dorsey-Logan Grant, 142
Douglas, Barbara, 118
Douglas, Donald, Jr., 233, 235, 265, 369
Douglas, Donald W., 10, 20–21, 44, 46–48, 54, 62, 65, 77, 81, 91, 95, 105, 115, 117–18, 127, 133, 147, 152, 162, 172–73, 201, 231–35, 262, 308, 317, 368–70

Douglas, William E., 48, 66
Douglas Aircraft Co., Inc., 65, 67, 73, 83, 86, 90, 92–94, 102, 107, 111, 114, 117, 121, 123, 131, 133, 135–36, 138, 144, 150, 152, 155, 163, 173, 179, 194, 200, 204–6, 214, 216, 220, 231, 240, 245, 247, 254, 265, 268, 271, 281–82, 307, 338, 348
Douglas Company, 48–49, 60, 66
Douhet, Giulio, 14
Dow Jones Industrial Average (DJIA), 334
DuPont, Felix, Jr., 188
Durant, William C., 44
Durham, Henry, 290

Eaker, Ira C., Gen., 65, 117, 167
Eastern Aircraft Division (of General Motors), 134, 138–39
Eastern Air Transport/Eastern Air Lines, 92–93, 172, 175, 215, 224, 279–80, 321
Echols, Oliver P., 113, 154, 186
Edison, Thomas A., 10, 26
Edwards, Glen, 185
Edwards Air Force Base, 185, 207
Egtvedt, Claire T., 22, 78, 98, 109, 135, 177–78, 224, 277
Eisenhower, Dwight, 117, 200, 202, 230, 245–46, 255
Electric Autolite Co. (ELTRA), 211, 239
Electric Boat Corp., 169, 225. *See also* General Dynamics Corp.
Ely, Eugene, 14, 16
Emanuel, Victor, 106, 108, 138, 168
Emanuel and Co., 106
Embraer (Brazil), 348
Emergency Plant Facilities (EPF) program, 114, 123
Emmons, Harold S., 74
Emtor Holding Co., 221
Engel, A. J., 26
Engel Aircraft Company, 26, 43
Enterprise aircraft carrier, 96
Equity Corp., 180
E-Systems Corp., 311
Evans, Edward S., 74
Evans, Llewellen J. ("Lew"), 239, 270, 302
Everest, Frank F., 253
Excess Profits Tax, 126
Export-Import Bank, 320

Fairchild, Sherman, 10, 53, 78–79, 107, 117, 153, 183, 201, 235, 239, 268, 301, 368, 371
Fairchild Aerial Camera Corp., 53–54
Fairchild Aerial Survey Co., 53–54
Fairchild Aircraft Corp., 107
Fairchild Airplane Manufacturing Corp., 54, 78
Fairchild Aviation Corp., 53, 68–69, 74, 78–80, 107
Fairchild Engine Corp. (Ranger Division), 53, 79
Fairchild Engine and Airplane Corp., 67, 94, 107, 124, 131, 136, 150, 152, 156, 165, 182–83, 201–2, 206, 220, 247
Fairchild Flying Corp., 54
Fairchild of Canada, Ltd., 69, 107
Fairchild Stratos Corp./Fairchild Hiller Corp./Fairchild Industries/Fairchild Corp., 239, 262, 271, 300–301, 323, 329–31, 357–58
Fairchild Republic Co. (division of Fairchild Industries), 261, 301, 329–30
Faneuf, Leston, 136, 222
Fechet, James E., 61
Federal Air Regulations (FARs), 207, 301
Federal Aviation Act, 207
Federal Aviation Administration (FAA), 262, 275–76
Federal Aviation Authority, 207
Federal Aviation (Howell) Commission, 87, 90, 101
Federal Trade Commission (FTC), 296
Federation Aeronautique Internationale (FAI), 7, 37, 40, 242
Fiat, 323
Finletter, Thomas K., 196
Finletter Commission. *See* President's Air Policy Commission (Finletter Commission)
Firestone Corp., 141, 150
First Aero Squadron, 28, 31
First Marine Aeronautical Company, 31
First World War, 8, 15, 21, 23–24, 26, 29–30, 33, 35, 37, 39, 55, 58, 62, 67, 72, 88, 105, 113, 121, 124, 139, 148, 155, 176, 246, 261, 355, 368
Fisher Body Corp. (division of General Motors), 35, 72, 123, 139, 184. *See also* General Motors Corp.

Fitzgerald, A. Ernest, 246, 271–72, 288, 293, 356
Fleet, Reuben H., 20, 37, 44, 50, 52, 53, 62, 66, 105, 117, 126, 137, 153, 168–70, 201, 268, 308, 367–68, 370. *See also* Consolidated Aircraft Corp.
Fleet Aircraft, Inc., 66, 113
Fleet Aircraft of Canada, Ltd., 66, 68, 105, 136
Fleetwings, Inc., 71, 131
Flettner, Anton, 190
Flxible Corp., 331
Flying Tiger Line, Inc., 161
FMA (Argentina), 347
Fokker, Anthony, 8, 55, 72
Fokker Aircraft Corp., 49, 55, 72, 82–83, 91
Fokker N. V. (Netherlands), 220, 237, 286, 298, 318, 323
Fonck, Rene, 52
Ford, Edsel, 72
Ford, Henry, 33, 72, 115, 137, 150
Ford Motor Company, 33, 72–74, 115, 124, 138–39, 150, 217, 306, 337
Ford Foundation, 152
Ford Instrument Co., 82–83
Foreign Corrupt Practices Act (FCPA), 299–300, 318, 340
Foreign Military Sales (FMS) program, 291
Forrestal, James V., 195
Foss, Eugene, 24
Foster Wheeler Corp., 230
Foulois, Benjamin F., 6, 87–90, 94, 98, 120
Fowler, Robert G., 26
Frye, Jack, 91, 102
Fthenakis, Emanuel, 329–30
Fuji Corp., 208
Fuqua, Don, 357

G & A Aircraft, Inc., 141, 189. *See also* Pitcairn Companies
Gallaudet, Edson, 9, 16, 19–20
Gallaudet Aircraft Corp., 20, 51
Gallaudet Engineering Company, 20
Gallaudet University, 19
Gansler, Jacques, 290, 293–95
Garrett AiResearch Corp., 101, 234
Gates, Artemis L., 125
Gavin, Joseph G., Jr., 302
Gaylord, Harvey, 180, 222
General Accounting Office (GAO), 194, 246, 263, 296

General Aeronautic Company, 21
General Agreement on Tariffs and Trade (GATT), 365
General Aviation Corp. (GAC), 72, 82–83, 91. *See also* General Motors Corp.
General Aviation Manufacturing Corp. (GAMC), 72, 82–83. *See also* General Motors Corp.
General Dynamics Corp., 19, 154, 169, 200–201, 205, 218–21, 224–26, 233–34, 245, 252–54, 256–58, 261, 265–66, 270, 272, 285–87, 298, 301, 305–6, 319, 328, 332–34, 336–37, 339, 348–49, 355, 357–59, 361, 364, 371. *See also* Consolidated Vultee/Convair
General Electric Co. (GE), 23, 44, 200–201, 206, 219, 275, 306, 315, 321, 335, 361, 364
General Headquarters Air Force (GHQAF), 57, 89, 120
General Motors Corp. (GM), 44, 62, 72–74, 79, 82, 87, 110, 120, 124, 129, 138, 150, 170–71, 194, 349, 355, 358, 362
George, Harold L., 167
Gilpatric, Roswell, 258
Girdler, Tom, 106, 138, 141
Glenn, John H., Jr., 199
Glenn L. Martin Co., 20–21, 35, 41, 43, 47, 67, 71, 77, 81, 83, 91, 93–94, 101, 112, 116–17, 123, 134–35, 150, 152, 155, 174, 183, 201–2, 204, 206, 222–23, 247, 262, 368, 370
Globe Aircraft Corp., 124, 165, 180, 194
Goddard, Robert H., 159
Goethals, G. W., and Co., 21
Goodyear Aircraft/Aerospace Corp., 112, 124, 132, 142, 150, 205, 211
Gordon Bennett Trophy, 14
Göring, Hermann, 93
Gott, Edgar N., 22, 43, 49
Grace Commission, 328
Graham, Benjamin, 179
Gray, Harold E., 277
Gray, Harry J., 312
Great Lakes Aircraft Corp., 71, 116
Green, Larry, 231
Gross, Courtlandt, 84–85, 108–9, 227, 252, 262, 307, 317, 369
Gross, Robert E., 84–85, 95, 107, 109, 117, 147, 152–53, 167, 227, 297, 367, 369. *See also* Lockheed Aircraft Corp.
Grumman, Leroy R., 25, 62, 69, 112, 126, 180, 201, 238, 249, 317, 331, 370
Grumman Aircraft Engineering Corp./Grumman Corp., 67–70, 94, 112–13, 134, 139, 145, 150, 152, 165, 180, 182, 200–201, 232, 237, 240, 249, 253–54, 256–58, 262, 264, 270, 301–3, 317, 323, 328, 330–32, 337, 347–49, 351, 357, 360, 362
Grumman American Aircraft Corp. (GAAC), 301
Guggenheim, Daniel, 60, 68; Daniel Guggenheim Fund for the Promotion of Aeronautics, Inc., 60, 65
Guggenheim Aeronautical Laboratories (of GALCIT), 127
Guggenheim School of Aeronautics, New York University, 61
Gulfstream Corp., 339
Gunderson, Harvey J., 175

Haack, Robert W., 306–7
Hagan, Henry R., 137
Hake, Richard von, 75, 84
Halaby, Najeeb, 277
Hall, Archibald, 105
Hall, Charles Ward, 70, 105
Hall-Aluminum Aircraft Corp., 70, 105
Hall-Scott Motor Co., 34
Hamilton Aero Co., 76, 115
Hamilton Metalplane Corp., 76
Hamilton-Standard Division (of UATC), 77–78, 182, 189
Harding, Richard Barclay, 95
Harriman, Averill, 20, 78–79
Harvard Business School, 149
Hassel, Kai-Uwe von, 251
Haughton, Daniel J., 177, 227, 265, 279–80, 282, 293, 306, 371
Hayden, Stone, and Co., 69
Hayes Aircraft Corp., 175
Healy-Aeromarine Bus Co., 50
Hearst, George R., 77
Heath, Edward, 280, 298
Heinemann, Edward H., 48, 86, 111, 133, 174, 197, 233, 317, 369
Helicopter development, 104, 141–43, 188–91
Hello, Bastian J. ("Buzz"), 273, 310, 339
Henry, Bill, 47–48

Herring, Augustus M., 6, 11, 12–13
Herring-Curtiss Company, 11–13, 16
Hibbard, Hall, 84, 214, 227
Higgins Industries, 143
Hiller, Stanley, Jr., 143, 190, 210, 239, 301, 368
Hiller Helicopters/Hiller Aircraft Corp., 206, 210, 239
Hiller Industries, 143
Honeywell Corp., 205
Hood, Robert H., Jr., 338
Hoover, Herbert, 87
Hopkins, John Jay, 169–70, 219
Horner, H. W. ("Jack"), 277
Houston, George H., 21
Howard Aircraft Corp., 136
Howell, Clark, 87, 95
Hoyt, Richard F., 69, 79–81, 107
Hubbard, Elbert, 76–77
Hudson Motor Co., 24, 30
Huff, Thomas H., 49
Huff Daland Airplane Co., 49, 53, 60, 115, 143
Huff Daland Dusters, Inc., 49
Hughes, Charles Evans, 31, 33
Hughes, Howard R., 100, 102, 107–8, 113, 145, 166, 190, 210, 217–19, 268, 313, 338, 368; Howard Hughes Medical Institute (HHMI), 167
Hughes Aircraft Co., 133, 145, 166–67, 189, 205, 209, 227, 333, 349
Hughes-Kaiser Corp., 146
Hughes Tool Co. (Toolco)/Hughes Helicopters, 102, 113, 166–67, 190, 210–11, 218, 239, 313, 338, 341, 343
Hunsaker, Jerome C., 22, 47–48, 127, 317
Hurley, Roy T., 173
Hydraulic Research and Mfg. Co. (division of Bell), 222
Hyland, Pat, 167

Iacobellis, Sam F., 339
Illinois Aero Club, 25
"Ill Wind" investigations, 328–29, 333, 354–55, 371
Imbrie and Co., 22–23
Intercontinent Aircraft Corp., 138
Intercontinent Corp. (of Curtiss-Wright), 81
Internal Revenue Service (IRS), 323
International Aircraft Engine Corp. (IAE), 323
International Air Transport Assn. (IATA), 161
International Association of Machinists (IAM), 95, 322, 334
International Business Machines Corp. (IBM), 53, 207, 312
International Civil Aviation Organization (ICAO), 161
International Logistics Negotiations (ILN) Office, 251, 291, 314
International Monetary Fund (IMF), 151
Interstate Commerce Commission (ICC), 87
Israeli Aircraft Industries (IAI), 87

Japan Aircraft Development Corp. (JADC), 323
Japan Aircraft Engine Corp., 323
Japan Commercial Transport Development Corp. (JCTDC), 323
Jay, W. Kenneth, 55, 86
Jet Assisted Takeoff (JATO), 127
Jet Propulsion Laboratory (JPL), 127, 207
Johns Hopkins University, 19
Johns-Manville Corp., 18
Johnson, Clarence L. ("Kelly"), 84, 108, 144, 177, 280, 307, 317
Johnson, Louis, 153
Johnson, Lyndon, 211, 258
Johnson, Phillip G., 22, 43, 78, 109, 135
Johnson, Robert, 221
Joint Aircraft Committee, 119, 123
Joint Committee on Civil Aviation, 59
Joint Congressional Air Policy Board, 192–93
Joint Technical Advisory Board, Army and Navy, 19, 30
Jones, E. L., 15
Jones, Thomas V., 236, 289, 298, 308–9, 340, 371
Jones & Laughlin Steel, 311
Joyce, Temple N., 82–83, 113
J. S. McDonnell and Assoc., 116

Kaiser, Edgar F., 165
Kaiser, Henry J., 145–46
Kaiser Cargo Corp., 131
Kaiser Corp., 131, 154, 165–66, 190
Kaiser-Frazer Corp., 169, 184
Kaman, Charles H., 189, 211, 368
Kaman Aircraft Corp./Kaman Corp., 190, 211, 312, 343, 349

Kartveli, Alexander, 71, 110, 187
Kassebaum, Nancy, 334
Kawasaki Corp., 307
Keeler, Fred S., 55, 75
Kellett, W. Wallace, 104, 111, 189, 369
Kellett Autogiro Co./Kellett Aircraft Corp., 141–43, 189
Kelly, Oakley G., 40
Kennedy, John F., 252, 254, 258
Kettering, Charles F., 24, 44, 76, 82
Keys, Clement M., 9, 23, 40, 43, 45, 62–63, 74, 80–82, 173, 370
Keystone Aircraft Corp., 49, 69, 71, 80–81
Kindleberger, James H. ("Dutch"), 48, 62, 83, 91–92, 110, 117, 128–29, 137, 140, 147, 171, 197, 230–31, 247, 367–68, 370–71. *See also* North American Aviation, Inc.
King, Ernest J., 88
Kitchen, Lawrence O., 306
Klemin, Alexander, 9, 60
KLM Royal Dutch Airlines, 297
Korean War, 151, 155, 158, 169, 176, 177, 180–81, 188, 193, 200, 202, 204, 207, 238, 247
Korth, Fred J., 258
Kotchian, A. Carl, 298, 306
Knudsen, Semon ("Bunkie"), 305
Knudsen, William S., 120, 305
Kreider, Ammon H., 53
Kreider-Reisner Aircraft Co., 53–54, 79, 107. *See also* Fairchild Aviation Corp.
Kresa, Kent, 340
Kucera, Randolph P., 237
Kuss, Henry J., Jr., 251

Laddon, I. M., 51, 105, 138
Lahm, Frank P., 6
Laird, Melvin, 285, 288, 302, 308
Lampert-Perkins Committee, 57–59
Landgraf Helicopter Co., 189
Langley Memorial Aeronautical Laboratory/Langley Research Center of NACA/NASA, 28, 118, 207, 267
Lanphier, Thomas G., 214, 218
Larsen, Agnew E., 175
Lassiter, William, 57
Lassiter Board, 57
Lawrance, Charles L., 75, 80
Lawrance Aeronautical Corp., 75
Lawrence Sperry Aircraft Co., 46, 63
Lawson, Alfred, 41
Lawson Airplane Co., 41

Lazard Freres, 234
Lee, Higginson and Co., 84
Lehman Brothers, 78–79
LeMay, Curtis E., 98, 256
Lend-Lease Act, 119, 129
LePage, W. Lawrence, 142
Levine, Charles, 65, 137
Lewis, Birdseye B., 25
Lewis, David S., 231, 235, 265, 305, 307, 334–35, 371
Lewis, Roger, 167, 226, 305
Lewis, Samuel C., 27
Lewis & Vought Co., 25, 43. *See also* Chance Vought
Lewis Laboratory (of NACA/NASA), 118, 207
Lilienthal, Otto, 7
Lilley, Tom, 149
Lindbergh, Charles A., 50, 52, 65, 100, 261, 276
Ling, James J., 211, 241, 310
Ling-Altec Electronics, 241
Ling-Temco Corp., 241
Ling-Temco-Vought/LTV Corp., 211, 241, 261–62, 285, 287, 301, 310, 331, 341, 357–58. *See also* Chance Vought Aircraft Corp.; Vought Aeronautics/Vought Corp.
Lippisch, Alexander, 170, 174, 268
Litton Industries, 312
Lockheed (Loughead), Allan, 75, 85–86, 267, 368, 371
Lockheed (Loughead), Malcolm, 55, 85–86, 267
Lockheed Advanced Development Projects Office/Company (LADC/"Skunk Works"), 144, 177, 227–28, 286, 325, 335–36, 348, 357
Lockheed Aircraft Corp./Lockheed Corp., 62, 73–75, 84, 94–95, 102, 107–9, 123, 134–35, 143, 150, 152, 160, 163, 167, 176, 179, 194, 200, 205–6, 214, 217, 220–21, 226–28, 232, 245, 249–51, 254, 262, 268, 270–71, 274, 279–81, 285, 290, 297–98, 301, 307, 309, 318–20, 324, 328, 338, 341, 347–48, 351, 357–58, 361, 362, 367, 371
Lockheed Air Service, Inc., 227, 306
Lockheed Hydraulic Brake Co., 54
Lockheed Martin Corp., 349, 351, 362–64, 366
Loening, Albert, 69

Loening, Grover C., 17, 24–25, 33, 35, 45, 62, 69, 85, 96–97, 117, 121, 183, 268, 371; Grover Loening Aircraft Co., 85
Loening Aeronautical Co., 43, 60, 68–69, 80
Loening brothers, 112, 369
Loening Laboratories, 85
Loral Corp., 349, 358, 363
Loral Vought Systems, 358
Loughead, Victor, 54
Loughead Aircraft Manufacturing Co., 54
Loughead brothers, 15, 54, 369. *See also* Lockheed (Loughead), Allan; Lockheed (Loughead), Malcolm; Loughead, Victor
Loughead Brothers Aircraft Corp., 85
Lowe, Edward, Jr., 26
Lovett, Robert A., 120
LTV Corp. *See* Ling-Temco-Vought/LTV Corp.
L-W-F Corp., 26, 42–43
Luftwaffe, 93
Lycoming Manufacturing Corp./Lycoming Div./Avco-Lycoming, 73, 80, 106, 108, 342

M-1 tank, 334
MacArthur, Douglas, 88, 94, 117
MacArthur-Pratt Agreement, 89
MacCracken, William P., 59
Macready, J. A., 40
Madelung, Georg, 159
Mahon, George., 258
Mahoney, Benjamin F., 49–50; B. F. Mahoney Aircraft Corp., 50
Mahoney Aircraft Corp., 50
Mahoney-Ryan Aircraft Corp., 50
Mamlock, Max, 54
Manufacturers Aircraft Association (MAA), 34–35, 39, 152, 159
Manville, Frank, 18
Marchev, Alfred, 136, 187
Marine Corps, U.S., 31, 60, 97, 132, 195, 209, 230, 242, 255, 287, 350
Marshall, George C., 117
Marshall Space Flight Center (NASA), 207
Martin, Glenn L., 14, 16, 18, 20, 22, 44, 47, 62, 65, 101, 112, 117, 122, 126, 147, 174–76, 199, 201, 368; Glenn L. Martin Co. (*see* Glenn L. Martin Co.)

Martin Marietta Corp., 205, 223, 231, 234, 264, 329, 358, 360–62
Massachusetts Institute of Technology (MIT), 22, 47, 49
Material Services Corp., 226, 258
Materna, Robert, 365
Mayo, William B., 74
McClellan, John, 258, 270
McCook Field, 37, 59–60
McCornick, Willis, 24
McCulloch, Robert, 241
McCurdy, John A. D., 11
McDonnell, James S., 62, 115–16, 131, 184, 231, 262, 265, 282, 307, 315, 317, 369–70
McDonnell, James S., III, 318
McDonnell, John F., 338–39, 361, 369
McDonnell, Sanford N., 235, 265, 307, 338, 369
McDonnell, William, 369
McDonnell Aircraft Corp./McDonnell Corp., 93, 116, 136, 142, 144, 152–53, 165, 172, 184, 189, 191, 200–201, 206, 211, 231–32, 234–35, 254, 263
McDonnell Douglas Corp., 201, 235, 242, 261–62, 264, 271, 273, 282, 287, 298, 301, 305, 307, 315, 318, 320, 323, 328, 332, 335, 337–39, 341, 343, 345, 351, 357, 359, 361–63, 365, 369, 371
McGowen, Jackson, 234–35, 265, 307
McNamara, Robert S., 217, 230, 232, 240, 244, 252, 254–57, 259, 254, 270, 272, 285, 300
McNarney, Joseph, 153, 218–19
Mead, George L., 75, 234
Meador, John W., 125
Meigs, Merrill C., 121
Mellon Bank, 78, 175
Menasco Co., 108
Menoher, Charles T., 33, 41
Messerschmitt-Bolkow-Blohm (MBB), 300, 342, 344, 348. *See also* Deutsche Aerospace (DASA)
Military Air Transport Service (MATS), 195
Military-industrial complex, 225, 245–46, 296
Millar, Richard, 106, 138, 153, 186, 309
Mingle, Harry B., 26, 35
Ministry of Aircraft Production (UK), 121

Ministry of International Trade and Industry (MITI-Japan), 351
Mitchell, William, 33, 41–42, 46, 57–60, 97
Mitsubishi Corp., 209, 308
Mitsui Company Ltd., 8, 26
Moffett, W. A., 37, 58, 60, 75, 88
Moffett Field, 118
Moisant, John, 16
Monocoupe Corp., 131
Moore, John R., 266, 309
Moore, Paul, 71, 111, 187
Moreland Aircraft Co., 86
Morgan, Thomas A., 107
Morrow Board, 57–59, 131
Morse, Frank L., 22, 66
Morse Chain Co., 22
Moss, Sanford, 100
Moth Aircraft Co., 80
Motorem und Turbinen Union (MTU), 323
Mott, Charles S., 74
Murphy, George, 309
Mutual Defense Assistance Program (MDAP), 236

Naish, John V., 219, 226
Nash-Kelvinator Corp., 142
National Advisory Committee for Aeronautics (NACA), 28–30, 56, 66, 68–69, 99, 118, 127, 156, 159, 197–98, 207, 351
National Aeronautic Association, 35
National Aeronautics and Space Administration (NASA), 207, 221, 230, 242, 254, 262, 264, 326, 329, 336, 344, 351, 353–54
National Aircraft War Production Council (NAWPC), 122, 151
National Air Transport Co., 77
National Cash Register Corp. (NCR), 30, 207
National City Bank of New York, 76
National Defense Advisory Commission (NDAC), 120, 123
National Security Act (1947), 195
National Security Council, 195
Naval Academy, U.S., 47
Naval Aeronautics, Office of, 28
Naval Aircraft Factory (NAF), 25, 34, 43–44, 47, 58–59, 64, 70, 90, 96, 112–13, 125, 131, 145, 165
Naval Aircraft Modification Unit, 145
Naval Air Materiel Center, 131, 165
Naval Air Systems Command (NAVAIR), 248, 269
Naval Air Transport Service, 140
Naval Appropriations Act, 28
Naval Flying Corps, 28–29, 34, 37
Naval Weapons Center (NWC), 205
Navy, Department of, 31, 58, 60, 122, 150
Navy, U.S. (Naval Aviation), 28, 69, 97, 117–18, 180, 195, 242, 249, 303, 344
Nelson, Donald, 121, 146
Netherlands Aircraft Corp., 55. *See also* Fokker N.V. (Netherlands)
Neutrality Act of 1935, 97
New York Aero Club, 11
New York Curb Exchange, 73
New York Trust Co., 180
New York University, 61
Niagara Aircraft Corp., 114. *See also* Bell Aircraft Corp.
Niles-Bement-Pond Co., 75
Ninety-Fourth Pursuit Squadron, 31
Nixon, Richard, 271, 285, 291, 298, 303, 318
Noorduyn, Bob, 55
North American Aviation, Inc., 62, 72–74, 79, 81–83, 87, 91, 94, 107, 109, 121–22, 128, 133, 137–38, 144, 150, 152–53, 155, 168, 170–71, 173, 176, 182, 186, 194, 200, 202, 206, 224, 228–32, 234, 247, 252, 254, 261–62, 264–65, 270, 272, 339, 367, 371
North American Rockwell, Inc., 264, 266, 271, 288, 300–301, 309–10. *See also* Rockwell-Standard Corp.
North Atlantic Treaty Organization (NATO), 204, 228, 236, 251–52, 273, 286–87, 292, 297, 305
Northrop, John K., 10, 48, 54, 62, 77, 86, 114–15, 117–18, 129, 153–54, 185–86, 201, 317, 368, 370
Northrop Aircraft Corp., Division of UATC, 77
Northrop Aircraft, Inc./Northrop Corp., 93, 114, 121–22, 132, 143, 152–53, 160, 165, 168, 176, 185, 200–201, 206, 232, 235–36, 262, 285, 287, 289, 298, 308, 328, 335, 337, 339–41, 348, 351, 357–60, 371
Northrop Aeronautical Institute/Northrop University, 186

Northrop Division, Douglas Aircraft, 86, 108, 114
Northrop Grumman Corp., 361, 364
Northwest Airlines, 282
Nuclear Energy for the Propulsion of Aircraft (NEPA) program, 156
Nunn, Sam, 349

O'Brien, John, 322–33
Odlum, Floyd, 115, 153, 168–70, 186, 218
Office of Production Management (OPM), 120–21
Ohain, Hans von, 104
Ordnance Engineering Corporation (Orenco), 27
Ostfriesland (battleship), 41
Otis Elevator Corp., 312

Pace, Frank, 196, 219, 226
Pace, Stanley C., 334–35, 359
Pacific Aero Products Co., 22. *See also* Boeing Airplane Company
Pacific Air Transport, 76
Packard, David, 285, 288–89, 308, 328
Packard Motor Car Co., 33–34, 72, 129
Paisley, Melvin, 328
Palmer, Richard W., 100, 138
Pan American Airways (Pan Am), 101–2, 112, 186, 215–16, 226, 276–79
Patent suits, 12–13, 16–19, 41, 62, 191
Patrick, Mason M., 37, 41, 45, 61
Patterson, Robert, 136
Patterson, W. A. ("Pat"), 77, 102
Peale, Mundy L., 187, 240
Pearson, C. C., 175–76
Perelle, Charles, 146
Perry, William, 357
Pershing, John J., 25, 28
Philco-Ford Co., 205
Phillips, Almarin, 250, 299
Piasecki, Frank, 143, 188, 201, 209, 368
Piasecki Aircraft Corp., 209
Piasecki Helicopter Corp., 188, 209
Pilatus (Switzerland), 347
Piper Aircraft Corp., 124, 168
Pirie, Robert B., 257
Pitcairn, Harold, 82, 104, 142, 175, 191
Pitcairn Companies, 82, 142, 189, 191
Platt, Haviland, 142
Platt-LePage Corp., 142–43, 189, 191
Pope-Hartford Co., 76

Porter, C. Talbot, 49
Post Office, U.S., 87–88
Post, Wiley, 67
Power Jets Ltd., 104, 144
Pratt, W. W., 89
Pratt & Whitney Aircraft Corp., 75–76, 78, 83, 90, 99, 117, 144, 172, 182, 200–201, 208, 219, 270, 275–78, 315, 323, 342, 345, 349, 364
Pratt & Whitney Co., 75
Pratt & Whitney of Canada, 68
President's Air Coordinating Committee, 192
President's Air Policy Commission (Finletter Commission), 152, 169, 192–93
Price, Nathan C., 101, 143
Pulitzer Trophy Race, 39
P-V Engineering Forum, Inc., 143, 188. *See also* Piasecki Helicopter Corp.

Queen Aeroplane Co., 24
Quimby, Harriett, 16

Rachmaninoff, Sergei, 52
Radar, 69
Radioplane Corp. (division of Northrop), 185, 236
Ramo, Simon, 167
Ramo-Wooldridge Corp., 167, 205. *See also* TRW Corp. (Thompson Ramo-Wooldridge)
RAND Corp., 152
Raymond, Arthur, 92, 127, 214, 233
Raytheon Corp., 205, 364
Reagan, Ronald, 309, 317, 320, 324, 327, 334, 343
Reconstruction Finance Corp. (RFC), 66, 108, 112, 123–24, 175
Reims Air Exhibition (1909), 8, 12
Reisner, Lewis, 53
Renegotiation Act of 1942, 126
Rentschler, Frederick, 62, 75–76, 117, 199, 368–70
Rentschler, George, 76
Republic Aviation Corp., 93, 111, 117, 144, 150, 152, 155, 165, 176, 186, 200–201, 206, 232, 236, 240, 247, 254
Republic Steel Corp., 106, 138, 141
Rex (ocean liner), 98
Rice, Ray, 171
Rivers, L. Mendel, 293
Rockefeller, John D., 10

Rockefeller, Laurance, 116, 131, 188, 209
Rocketdyne Division, North American Aviation, 228–29, 264, 266, 309–10, 329, 339, 348
Rockne, Knute, 91
Rockwell, Willard F., 266, 309–10
Rockwell, Willard F., Jr., 266, 309–10
Rockwell International Corp., 74, 273, 310, 327, 329, 339, 348, 357
Rockwell Manufacturing Co., 310
Rockwell-Standard Corp., 262, 265–66. *See also* North American Aviation Inc.
Rodgers, Calbraith, 14
Rodman, Burton, 54
Rogers Subcommittee, U.S. Congress, 89
Rohr Aircraft Corp., 124, 331
Rohrbach, Adolf, 40
Rolls-Royce Ltd., 129, 275, 279
Rolls-Royce (1971) Ltd./Rolls-Royce PLC, 280–81, 362, 364
Roosevelt, Elliott, 146
Roosevelt, Franklin D., 34, 87–88, 93, 97, 117–20
Rotorwings (division of Glenn L. Martin Corp.), 175
Rowland, H. T., 175
Royal Aircraft Establishment (RAE), 37
Royal Air Force (RAF), 69
Russell, Richard, 249
Ryan, John D., 32–33, 121
Ryan, T. Claude, 10, 48–50, 85, 126, 201, 266, 317, 368, 371
Ryan, Thomas F., 7
Ryan Aeronautical Co., 74, 85, 94, 112, 124, 137, 152, 184, 202, 204, 206, 212, 221, 239, 266, 311
Ryan Airlines, 48–50
Ryan Flying Service, 49
Ryan School of Aeronautics, 85

SAAB (Svenska Aeroplan A. B.), 323, 329–30
Safe Aircraft Competition, 61, 116
Sanders Associates, 335. *See also* Lockheed Aircraft Corp.
Schlesinger, James, 286
Schmued, Edgar, 9, 186
Schneider Cup Race, 39
Schriever, Bernard A., 248
Schumpeter, Joseph, 205
Schweizer Aircraft Corp., 343

Schwendler, William T. ("Bill"), 69–70, 239
Scott, J. F., 6
Second Revenue Act of 1940, 120
Second World War, 9, 18, 60, 67–69, 72, 89, 95, 97, 101, 106, 123, 131, 141, 149–50, 155, 158, 194–95, 220, 225, 245, 288, 296, 335, 349–50
Securities & Exchange Commission (SEC), 296, 306
Selfridge, Thomas E., 6, 11, 16
Service Life Extension Program (SLEP), 349, 352
Services of Supply, U.S. Army, 120
Seversky Aero Corp., 70
Seversky Aircraft Corp., 68, 70, 93, 110, 113
Shields, Paul, 173
Shin, Tai Sheng, 263
Short, Mac, 109
Short Brothers Co., 8–9, 348
Short-Takeoff-and-Landing (STOL) aircraft (also STOVL, ASTOVL), 61, 103, 112, 242, 253, 337, 344, 351
Shrontz, Frank A., 333
Siemens Co., 50, 85
Signal Corps, U.S. Army, 5–6, 16, 18, 20, 24, 26, 28–31, 47
Signal Oil & Gas Corp., 234–35
Sikorsky, Igor, 9, 51–53, 76, 104, 117, 188, 208, 268, 368, 371
Sikorsky Aero Engineering Corp., 52
Sikorsky Aviation Corp., 53, 67–68, 76
Sikorsky Division, UATC/UAC/UT, 77–78, 101, 110, 142, 182, 188–89, 208, 209, 274, 328, 341–42, 349, 363
Sikorsky Manufacturing Corp., 52
Simmons, Harold, 337
Simplex Auto Co., 21, 25
Skunk Works. *See* Lockheed Advanced Development Projects Office/Company
Slick Airways, Inc., 161
Sloan, John A., 26
Smith, Barney and Co., 107
Smith, C. R., 79, 92, 163, 219. *See also* American Airways/American Airlines
Smithson, J. S., 83
Smithsonian Institution, 19
SNECMA (France), 315
Society of British Aircraft Constructors (SBAC), 35
Sorenson, Charles, 139

Spaatz, Carl A., 65, 117
Sperry, Elmer, 46, 68
Sperry, Lawrence, 46, 65
Sperry Corp., 83, 107, 205
Sperry Gyroscope Co., 46, 82–83
Springfield Aircraft Corp., 26, 35, 43
Squier, Carl, 75, 84
Stack, John, 189, 197, 254
Stamper, Malcolm, 305
Standard Aero Corp., 8, 26, 30, 35, 41, 43, 49
Standard Aircraft Corp., 26, 37, 43
Standard Oil Co. of California, 86
Standard Steel Propellers Co., 77
Stavid Electronics, 227
Stealth technology, 307, 309, 327
Stearman, Lloyd, 77, 84–85, 227, 268
Stearman Aircraft Corp./Aircraft Division, Boeing, 77–78, 109
Steckler, Herman O., 246
Stewart, James T., 274, 286, 288, 326
Stinson, Eddie, 14, 40
Stinson Aircraft Division (of Aviation Manufacturing Corp./ AVCO), 73, 80, 106, 108, 124, 168
St. Louis Aircraft Co., 25, 35, 124, 136, 165
St. Louis Car Co., 25
Stonecipher, Harry, 362
Stout, William B., 72, 115, 138, 200
Stout Engineering Laboratories, 138
Stout Metal Airplane Co., 72. *See also* Ford Motor Company
Strategic Air Command (SAC), 199, 225, 254
Stringer, Eric, 47
Stroukoff, Michael, 9, 165–66
Stroukoff Aircraft Corp., 166
Sturtevant Aircraft Co., 24–25, 49, 62
Sturtevant Blower Works, 24
Summa Corp., 313
Sundstrand Corp., 362
Supplies, Priorities, and Allocations Board (SPAB), 120–21
Surplus Property Admininstration, 155
Sutro and Co., 85
Swearingen Aircraft Corp., 301, 329
Swirbul, Leon A. ("Jake"), 69–70, 180, 238–39
Symington, Stuart S., 153, 175, 186, 195

Tactical Air Command (TAC), 253
Taiwan Aerospace Corp., 361
Talbott, Harold E., Jr., 24, 44, 82, 91, 121, 167, 223, 246
Talbott, Harold E., Sr., 24, 44
Tanaka, Kakuei, 297
Taylorcraft Corp., 124
Teledyne Corp., 221, 266, 311, 349
Tellep, Daniel, 337, 362
Teterboro Airport, 27
Texas Engineering and Manufacturing Co. (TEMCO)/Temco Aircraft Corp., 182, 192, 241
Textron Corp., 222, 262, 306, 311, 341–42, 359
Thayer, W. Paul, 211, 310, 327, 341
Thieblot, Armand, 9, 137
Thiokol Corp., 205, 329
Thomas, B. Douglas, 13, 16, 22, 66
Thomas Brothers/Thomas Bros. Co., 8, 21–22
Thomas-Morse Corp., 22, 30, 35, 42–43, 66
Thompson, William B., 18
Thompson Trophy Race, 39
Tillinghast, Charles, 278
Total Package Procurement (TPP) system, 264, 270–72, 288, 302
Tower, John, 273, 340
Towers, John, 97
Towl, Clinton, 69, 239, 249, 302, 317
Trailmobile Co., 176
Trans-Canada Airlines, 78
Transcontinental Air Transport (TAT), 81–82
Transcontinental and Western Air (TWA), 81–91
Trans World Airlines (TWA), 83, 101–2, 115, 162, 168, 175, 186, 211, 217–18, 278–79
Travel Air Corp., 80
Trippe, Juan T., 101, 112, 216, 226, 276–77
Truman, Harry S, 166, 172, 175, 192, 195, 202
Truman Committee, 121, 147, 166, 172
TRW Corp. (Thompson Ramo-Wooldridge), 167, 334, 349
Twining, Nathan F., 115

Uhl, Edward G., 239, 301, 329
Union Carbide and Carbon Corp., 144
United Aircraft and Transport Corp. (UATC), 68, 73–74, 76–79, 82, 87–88, 90–91, 93, 369
United Aircraft Corp. (UAC), 67, 78,

83, 94, 104, 107, 110, 179, 142, 152, 182, 192
United Aircraft Exports Corp., 78
United Aircraft Manufacturing Co., 78
United Air Lines (UAL), 77–78, 102, 215, 219, 224, 282
United Auto Workers (UAW), 94, 179
United Brands, Inc., 296
United Helicopters, Inc., 190. *See also* Hiller Helicopters
United Technologies Corp. (UT), 74, 342. *See also* United Aircraft Corp.
Uppercu, Inglis M., 26, 39, 56
Uppercu-Burnelli Co., 56
U.S. Department of Commerce, 56, 59, 67, 101, 150, 207, 291
U.S. Department of Defense, 57, 162, 194, 246–47, 252, 270, 281, 291–93, 298, 328, 344, 350, 357, 361–62
U.S. Department of Justice, 221, 235, 265, 323, 361–62, 364
U.S. Department of State, 291
U.S. Department of Transportation, 262
U.S. Department of War, 29, 31, 57–58, 89, 99, 111, 122, 139, 150, 194
USS *Langley,* 44
U.S. Senate Subcommittee on Multinational Corporations, 296, 299

Vandenberg, Hoyt S., 115
Vanderbilt, Cornelius, 7
Varney, Walter, 84
Varney Speed Lines, 84
Vaughan, Guy A., 21, 80, 107, 147, 172–73, 201
Vega Aircraft Co. (division of Lockheed), 109, 135
Venzie, Harold, 143
Vertical-Takeoff-and-Landing (VTOL) aircraft, 212, 301
Vertol Corp./Vertol Div. Boeing, 208–9, 211, 223, 262. *See also* Boeing Airplane Company; Piasecki Helicopter Corp.
Verville, Alfred, 46
Viking Flying Boat Co., 84
Villa, Pancho, 28
Vincent, Jesse G., 34
Vinson-Trammell Treaty Naval Act, 96–97, 119, 193
V-Loans, 194, 234
von Kármán, Theodor, 9, 127, 132, 152, 156

Voisin, Gabriel, 9
Voisin Brothers, 9
Vought, Chance M., 16, 25, 43, 45, 68, 76–77, 369, 371. *See also* Chance Vought Aircraft Corp.
Vought Aeronautics/Vought Corp. (LTV), 202, 311, 337, 341, 347, 349, 358, 361
Vought-Sikorsky Division, UAC, 110, 142
Vultee, Gerald ("Jerry"), 48, 55, 73–75, 80, 106, 371
Vultee Aircraft, Inc., 94, 106, 121, 123, 137–38, 171
Vultee Aircraft Division, Aviation Mfg. Corp./AVCO, 93, 100, 106, 108

Waco Aircraft Co., 124
Walcott, Charles D., 56
War Assets Administration, 172
War Mobilization and Reconstruction Act, 149
War Production Board (WPB), 121–22, 146, 151
Ward, J. Carlton, 107, 182–83
Warner, Edward P., 60, 87, 161
Weapon system concept, 202, 225
Western Air Express, 82
Western Airlines, 83
Western Electric Corp., 205
Westervelt, Conrad V., 22, 145
Westinghouse Corp., 144, 200, 244
Westland Aircraft PLC, 188, 208–9, 342
Westover, Oscar B., 97
Wetzel, Harry, 234
Whitcomb, Richard T., 198, 267
White, Weld and Co., 111
Whitman, Ray, 114
Whittle, Frank, 69, 104, 144
Willard, Charles F., 16, 20, 26
Willow Run plant, Ford Co., 73, 137, 139–40, 154, 165, 184
Willys, John N., 23, 43
Willys-Overland Company, 23
Wilson, Charles E., 200
Wilson, Eugene E., 77
Wilson, Thornton A. ("T"), 277, 304–5, 327, 333, 371
Wilson, Woodrow, 31
Wilson, W. R., 71
Wilson Packing, 311
Wittemann Aircraft Corp., 27, 41, 43, 55
Wittemann brothers, 9, 27

Wittemann Brothers Co., 27
Wittemann-Lewis Co., 27
Woodhead, Harry, 138, 168
Woods, Robert J., 75, 114, 197
Wooldridge, Dean, 167
Work, James, 70, 145
World Bank (IBRD), 151
Wright, Katherine, 8
Wright, Jim, 258, 340
Wright, Orville, 6–8, 17–19, 21, 24, 28, 62, 81, 159
Wright, Theodore P., 43, 121
Wright, Wilbur, 5, 7–8, 14, 16–18
Wright Aeronautical Corp., 21, 43, 49, 69, 80, 83, 90, 99, 106–7, 117, 124, 144, 147, 200, 244
Wright brothers, 5–7, 9–12, 14, 16, 20, 45, 358–69
Wright Co., 7, 15–17, 20, 24, 62
Wright-Martin Corp., 21, 33, 35

Yale University, 19, 22
Yeager, Chuck, 159
Yorktown (aircraft carrier), 96
Young, Arthur, 143, 189

Ziegler, S. J., 125

Index of Aircraft, Engines, Missiles, and Space Vehicles

Note: Manufacturers are listed alphabetically, and specific types under each manufacturer are in roughly chronological order of appearance or development. Basic designs are listed under their original developer, even if licensed to or acquired by another. Some space and missile programs with multiple contractors appear under their project names. Also included are certain projects that may never have flown.

AEA: *White Wing:* 11; *Red Wing:* 11
AGA (Pitcairn) PA-36 autogiro, 141
Advanced STOVL (ASTOVL) project, 351
Aeronca L-16 Champion, 186
Airbus: A300: 277, 282, 283, 320–21; A320: 321, 324, 396; A330/A340: 324, 345–46
ALCOR C. 6. 1, 86
Allison: V-1710 engine, 100; V-3420 engine: 139; J35 engine: 157; T40 engine: 212; T56/501 engine: 177, 217; J71 engine: 184; TF41 engine: 243, 269, 362; AE2100 engine: 350
Apollo program, 231, 233, 261, 263–64, 273, 290, 309
Armstrong Siddeley Sapphire engine, 157, 172
Avion (Northrop) X216H. *See* Northrop (Avion) X216H

Barling (Wittemann) XNBL-1, 41, 43
Beech: Model 34 "Twin-Quad": 161; T-34 Mentor/T-34C Turbo-Mentor: 183, 347; small airliner project: 323
Beech-Pilatus Mark II, 348
Bell: FM-1 Airacuda: 114; P-39 Airacobra: 100, 114, 135, 180; P-59 Airacomet: 135–36, 144; P-63 Kingcobra: 136, 180; Model 30: 143, 180; (XS-1) X-1/X-1A: 159, 197; XP-83: 180; Model 47/H-13: 189, 196, 208, 210; Model 48/YH-12: 189; Model 61/HSL-1: 189; X-2: 198, 221; X-5: 182, 197, 254; 47J Ranger: 208; Model 204/204B/XH-40/UH-1: 208, 244, 341; Model 205: 208; 206 JetRanger/LOH/OH-58/206B JetRanger III/206L LongRanger: 208, 210, 311, 342; XV-3: 212, 267; XB-63/GAM-63 Rascal missile: 222; X-14: 212; X-16 project: 222; X-22A: 213; H-4: 208; Model 209/UH-1G Huey/AH-1 HueyCobra: 244, 311, 342; XV-15: 267, 311; YAH-63: 274; Model 212/UH-1N: 311; Model 222: 311; Model 412: 311
Bell Boeing XV-22 Osprey, 328, 341–42, 347
Berliner-Joyce P-16, 82
Boeing & Westervelt seaplane, 22
Boeing: B-1: 42; GAX/GA-1: 42; Model C: 22; XPW-9: 42; F4B/P-12: 43, 88; Model 40/40-A: 43, 90, 215; Model 80: 90, 215; Monomail: 68; B-9: 68, 90; Model 247: 68, 78, 85, 90–92, 102, 215; P-26: 88, 93; Model 299/B-17: 68, 89, 93, 98, 102–3, 105, 109, 123, 128, 133–35; XBLR-1/B-15: 98, 109, 133; 307 Stratoliner/C-75: 101–3, 215; 314 Clipper: 101, 103, 215; PBB-1: 123; Model 345/B-29: 109, 123–24, 128–29, 134–35, 157, 177–78, 186; Model 367/C-97/KC-97/377 Stratocruiser: 141, 160, 163, 178, 215; B-50/KB-50: 157–58, 179; F8B: 178; YL-15: 178; B-47: 158, 160, 173, 175, 177, 179, 185, 193, 215–16, 225, 268; Model 417 project: 178; Model 431 project: 178; B-52: 157–58, 160, 179, 199, 215–16, 223, 229, 239, 242, 254, 268, 272, 277, 304–5, 349; XF-99/IM-99/CIM-10 Bomarc missile: 179, 223, 255; Model 707/720/KC-135: 91, 164, 204, 215–20, 223–24, 232, 273, 276, 304–5, 332–33, 337, 345, 349;

453

Boeing *(continued)*
  Minuteman missile: 206, 223–24, 233, 277, 288, 304; 727: 204, 218, 224, 276, 304, 321–22, 333; X-20 Dyna-Soar project: 222, 224; 737/E-6/737-X: 224, 276, 305, 321–22, 333, 363; Model 2707 SST project: 268–69, 275, 304; 747/E-4/C-33: 220, 233, 273–74, 276–79, 282, 288, 304–5, 308, 318, 321–22, 324, 330, 345, 347, 349–50, 360, 362; AWACS/E-3 Sentry: 269, 273, 291, 304–5; AMST/YC-14: 274; SRAM missile: 288, 304; ALCM missile: 333; 757: 320–23, 358; 767: 320–22, 345, 349–50, 358; 7J7 project: 315, 323; 777: 345–46, 361; NLA project: 346; Sikorsky LHX/RAH-66 Comanche: 328, 334, 342–43, 347
Breguet 941S, 231
Breguet-Dorand *Gyroplane Laboratoire,* 104
Brewster: F2A Buffalo: 70, 110, 112–13, 145; SB2A Buccaneer/Bermuda: 70, 113, 145; XA-32 project: 145
Bristol Britannia, 216
Bristol Siddeley: BE. 53 engine: 242; Olympus engine: 236
British Aerospace: (Hawker Siddeley) Trident: 224; (Hawker) P. 1127 Kestrel: 242; Hawk: 337
British Aerospace/Aerospatiale Concorde, 243, 275, 283
Budd RB-1/C-93 Conestoga, 140–41
Burnelli: RB-1: 56; RB-2: 56

Cessna: L-19/OE-1 Bird Dog: 196, 210; T-37/A-37: 244, 330, 347, 349; CitationJet trainer variant: 348
Chance Vought. *See* Vought
Chase: CG-18 glider: 165; XC-122: 165, 212; CG-20 glider: 165; (Fairchild) C-123: 165–66; XC-123A: 165
Chrysler Jupiter missile, 255
Consolidated: PT-1: 51; NY: 51, 65; (Fleet) trainer: 66, 105; Fleetster: 66; XP3Y (PBY): 68, 105, 115, 131–32, 138, 181; (from Lockheed) P-30: 75, 105; XPB2Y Coronado, 105, 138; B-24: 89, 105–6, 109–33, 137–39; (Vultee) XP-54: 129
Consolidated Vultee: C-87, 138; PB4Y, 138; B-32: 138; B-36: 129, 153, 157, 160, 168, 175, 179, 184–85, 203
Convair: L-13: 168, 170; XP-81: 157; C-99/CV-37 project: 160, 163; Model 110 project: 168; Model 240/T-29: 160, 163, 168–69, 172; B-46: 168; XF-92A: 170; 340/440: 163, 169, 220, 225; XFY-1: 212, 244; XF2Y-1 Sea Dart: 170, 244; XP5Y-1/R3Y-2: 170; F-102: 157, 170, 198, 225, 236, 255, 268; F-106: 225, 243, 255, 268; Atlas missile: 204; Atlas-Centaur vehicle: 334; NX-2 project: 225; (WS-107A) B-58: 225, 256; 880 (Skylark 600): 218–19, 258, 319; 990: 218–19, 258, 319; (General Dynamics) Model 48 Charger: 248
Curtiss: *June Bug:* 11; A-1 Triad: 13; *America:* 13; HS-2L: 22, 54; Model J: 13; JN-4 Jenny: 13, 25, 29; Model D: 13; NC series: 23, 39; H-16: 34; P-1/P-6 Hawk series: 43; R-6 Racer: 40; CS-2: 44; Tanager: 61; (Robertson) Robin: 81; A-8: 81; A-12 Shrike: 81; B-2: 81; Condor: 81, 92; P-36: 103, 107, 111; CW-25/AT-9 Coupe: 106; CW-20/C-46: 106, 140, 172; P-40: 100, 128, 132–33, 139, 172; SB2C/A-25 Helldiver: 107, 133, 172; C-76: 140–41; SC-2: 172; XP-55: 129, 172; XP-62: 172; XBTC/XBT2C: 172; F15C: 157; CW-20E project: 172; XF-87 Blackhawk: 172; Wright X-19A: 213

Dassault: Mercure: 315; Mirage: 292
De Havilland: DH-4: 24, 27, 30–33, 37, 42, 57; Goblin engine: 157; Comet: 164, 214, 216–17, 313
De Havilland (Canada) Beaver, 197
Doman: LZ-5: 189, 211; D-10B: 211
Douglas: *Cloudster:* 47–48; DT: 48; O-2: 48; M-2: 37, 48; M-3 and M-4: 48; World Cruiser: 40, 48, 99; Dolphin: 91; XFD-1: 111; DC-1: 68, 85–86, 91–92, 216; DC-2(C-32): 78, 91–93, 215; DST/DC-3: 68, 92, 102–3, 111, 133, 161–62, 173, 215, 220; O-38 to O-46 series: 111; B-18: 92, 110, 112; TBD-1 Devastator: 71, 111; SBD/A-24 Dauntless: 111, 114–15, 133, 139; XBLR-2/XB-19: 98, 111, 129, 133; C-47 (from DC-3): 111, 131, 140, 160, 176; DC-4E: 102; DC-4/C-54: 102, 111, 133, 140, 162, 169; Model 7B/DB-7/A-20/Boston/

Havoc: 103, 111, 112, 128, 133; A-26 (later B-26): 128, 176, 244; XB-42: 173; XB-43: 173; AD/A-1 Skyraider: 133, 157, 173, 175, 232, 244, 247, 259, 269; D-558–1 Skystreak: 158, 173; C-74: 160, 163, 173; XC-112/DC-6/6A/6B: 160, 162–64, 173–74, 216–17; Super DC-3/R4D-8: 163; F3D Skyknight: 173; A2D Skyshark: 158; C-124 Globemaster II: 160, 173, 271; D-558–2 Skyrocket: 173, 197; F4D Skyray: 157, 174, 197, 232; A3D Skywarrior: 157, 174, 232; X-3: 197, 244; DC-7/7B/7C: 158, 163–64, 213, 215, 216–17; A4D/A-4 Skyhawk: 174, 199, 204, 232–33, 241, 259, 308; Nike missile: 204, 233, 255; B-66 Destroyer: 223, 232, 244; C-133: 232; DC-8: 164, 215–18, 232–33, 282, 346; C-132 project: 232, 271; F6D project: 232, 255; Thor missile: 233, 255; Skybolt missile project: 233, 245; DC-9/MD-80 series: 231–34, 282, 305, 321–22, 324, 348, 365; Manned Orbiting Laboratory (MOL) project, 233, 264, 268–69

EMBRAER (Brazil) Tucano, 348
Eastern (Grumman): TBM Avenger: 139; TBM Wildcat: 139

Fairchild: FC-1: 53; (Kreider-Reisner) XC-31: 79–80, 140; F-24: 80, 107, 136, 182; M-62/PT-19/PT-23/PT-26: 107, 137; AT-21 Gunner: 131, 136; Ranger engine: 183; C-82: 137, 141, 159, 182–83; C-119/C-119H: 183–84, 239; XC-120: 183; XNQ-1/T-31: 183; (from Chase) C-123 Provider: 183–84, 239; VZ-5FA: 213; F-27: 220, 239–40, 301, 323 (*see also* Fokker Friendship); Hiller FH-227: 220, 231; M185 project: 239; Hiller FH-1100: 239–40, 301; EWR VTOL project: 313; (Republic) AX/A-10: 269, 288, 301, 329, 341; (from SAAB) SF340: 329–30; NGT/T-46A: 330, 347, 358
Fairey FC. 1 project, 162
Fieseler Fi 156 Storch, 103
Fisher (GM) P-75/P-75A, 139
Fleetwings: Sea Bird: 71; BT-12: 131; XBTK-1: 131

Focke-Wulf: FW 61: 104, 142; FW 190: 99
Fokker: T-2: 40; F-10: 55, 90–91; F-32, 72; Friendship: 220 (*see also* Fairchild F-27); Republic Alliance project: 237; F-28: 301, 323; 100: 323
Ford (Stout) 4-AT trimotor, 72, 90

Gallaudet *Bullet,* 20
Gemini space program, 263
General Dynamics: TFX/F-111: 224, 228, 238, 247, 249, 252–55, 257, 264–65, 267–70, 278, 286, 288, 294, 304–5, 328, 332 (*see also* Grumman GD F-111B); FB-111: 259, 270; F-16: 243, 286–88, 305–6, 334–35, 340, 350–51, 356, 358, 362, 364; McDonnell Douglas A-12 project: 335, 337–39, 341, 347–48, 351, 359–60
General Electric: I-16: 144; J33: 157; J47: 157, 165; J79/CJ805: 177, 219, 225, 243; J85/CJ610: 243; J93: 230, 266, 275; GE4: 275; T64: 267; CF6: 275, 280, 315, 322; F400/F404: 267; F101: 315; F110: 332; F412: 334; SNECMA CFM56: 315, 322–23; GE90: 345
Goodyear: FG-1/F2G Corsair: 133; XR-9: 142
Great Lakes: 2T-1: 71; XBTG-1: 71
Grumman: FF-1: 70, 88; F2F: 70; F3F: 70, 112; G-21 Goose: 112, 134; J2F Duck: 134; F4F Wildcat: 112–13, 128, 134, 139; TBF Avenger: 112, 132, 134, 139; XF-5F/P-50 Bobcat: 134; F6F Hellcat: 128, 134, 180; F7F Tigercat: 134, 180; F8F Bearcat: 134, 157, 180–81; XTB3F-1/AF-1/AF-2: 181; JR2F/SA-16 Albatross: 181; Rigel missile: 181; F9F Panther: 181, 187; F9F-6/F-9 Cougar: 181, 187, 238, 268; S2F/S-2 Tracker: 181, 238; XF10F Jaguar: 182; F11F-1/F11F-1F/F-11 Tiger: 198, 228, 237–38, 249; WF-2/E-1B Tracer: 238; G-159 Gulfstream I: 238, 249, 323; Gulfstream II: 303; A2F/A-6 Intruder: 238, 259, 301, 332, 347, 349, 351, 360; OV-1 Mohawk: 238, 244; Lunar Module (LM) vehicle: 238, 264; Orbiting Astronomical Laboratory (OAL): 238; W2F/E-2 Hawkeye: 238, 301, 360; C-2A: 238; GD F-111B: 258–59, 269–71, 301;

Grumman *(continued)*
VFX/F-14 Tomcat: 182, 243, 269–71, 288, 291, 301, 332, 347, 350, 360; X-29: 344; E-8C/J-STARS: 322, 360

Hall-Aluminum PH-2, 71
Herring-Burgess design, 12
High Speed Civil Transport (HSCT) project, 344, 346
Hiller: X-235: 190; UH-5/Model 360/UH-12/HTE/H-23: 190, 210, 239–40; Hornet: 190; J-5: 190; X-18: 212–13; OH-5: 210
Huff Daland (Keystone) B-3/B-6, 49
Hughes: H-1 Racer: 100, 113; D-2 project: 133, 145; XF-11: 146–47, 166; Kaiser HK-1/Hughes H-4: 145, 166; (Kellett) XH-17: 189–90; Falcon guided missile: 167, 204; Model 269A/200/TH-55A: 210; Model 300/OH-6, 210: 343; Model 500/OH-6A: 313, 342–43; Phoenix missile: 303; (McDonnell Douglas) AAH/AH-64 Apache: 274, 313, 338, 343, 347, 360

IAE V2500 engine, 322–23

Jet Assisted Takeoff (JATO) rockets, 127
Joint Advanced Strike Technology/Joint Strike Fighter (JSF) program, 351, 362, 363
Joint Primary Aircraft Training System (JPATS) program, 347, 357–58, 360

Kaman: K-125: 190; HOK-1/HTK-1: 190; K-225: 190; H-43/HH-43B: 190, 211; HU2K-1/SH-2/UH-2 Sea Sprite: 211, 312–13, 343; Kmax: 349
Kellett: XR-10: 189; (Hughes) XH-17: 189–90

Langley Aerodrome, 19
Larsen (Junkers) J.L.6, 40
Lawson C-2 airliner, 41
Liberty engine, 30–31, 34, 40
Light Armed Reconnaissance Aircraft (LARA) program, 247
Lockheed: Vega: 54–55, 67, 75, 77, 84–85; Star series: 55, 75, 86; Altair: 55; Sirius: 55; Orion: 55; XP-900 (YP-24): 75; Model 10 Electra: 85, 92, 100, 108; XC-35: 100; Model 14/Hudson: 100, 108, 127; Model 18 Lodestar: 134; Excalibur project: 102; Model 322/P-38 Lightning: 99, 102, 108, 114, 134; (AiRover) Starliner: 109; 049/C-69/649/749 Constellation: 102, 134, 159, 162–63, 176; L-133 project, 143–44; PV-1/PV-2: 135; P2V/P-2 Neptune: 157–58, 176, 228; XP-80/F-80: 135, 144, 158, 171, 176–77, 180; R6O/R6V Constitution: 160; Saturn: 161, 176; AirTrooper: 176; T-33: 169, 177; L-1049 Super Constellation: 158, 160, 163, 215, 217, 297; XF-90: 176, 184; F-94: 177; X-7 ramjet missile:177; XFV-1: 212, 244; C-130 Hercules/C-130J Hercules II: 177, 184, 204, 217, 228, 251, 258, 274, 307, 335, 349–50, 362; F-104: 177, 197, 228, 230, 238, 242–43, 251–52, 255, 259, 285, 297; L-1649A Starliner: 213; U-2/TR-1: 222, 227–28, 307, 336; CL-400 project: 227; 188 Electra: 217, 220, 227–28, 237, 279; P3V/P-3 Orion: 217, 228, 298, 306–7, 336, 348; XV-4 Hummingbird: 213, 310; CL-475: 227; Polaris missile: 206, 227; XH-51: 227; C-140/JetStar/JetStar II: 228, 239, 307; A-11/A-12/YF-12A/SR-71: 227–28, 242–43, 267; C-141: 228, 271, 307, 349–50; AAFSS/AH-56 Cheyenne: 228, 274, 288; CX-HLS/C-5A Galaxy: 264–65, 268–69, 272, 274, 279–81,288–90, 293, 297, 304, 306, 335–36, 350; L-1011 TriStar: 265, 274–75, 278–83, 293, 298, 307, 318–19, 321, 324; S-3 Viking: 306–7; Trident missile: 327; F-117A: 307, 326, 336; LRAACA/P-7 project: 337; ATF/F-22: 334–36, 341, 347–48, 350, 358, 361–63
Loening: *Kitten:* 24; M-8/M8O: 25, 33, 43; amphibian: 65
Loughead: Model G: 54; seaplane: 54; S-1: 54. *See also* Lockheed
Loughead Brothers Duo-6, 85
LTV. *See* Vought
LTV-Hiller-Ryan XC-142A, 213
L-W-F Model V, 26

Macon airship, 88
Manned Orbiting Laboratory (MOL) space program. *See*

Douglas Manned Orbiting Laboratory project
Martin: TT: 20; MB-1/MB-2/NBS-1: 21, 33, 39, 41–42, 47; B-10: 68, 88–89, 98, 112; B-12: 98; M-130: 101; Maryland: 112; B-26 Marauder: 112, 128, 132–34; 187 Baltimore: 134; PBM Mariner: 134; XPB2M/JRM Mars: 134, 160; Model 202: 134, 163, 172, 175; Model 303 project: 175; AM-1 Mauler: 175; P4M Mercator: 157, 175; Model 404: 163, 175, 220; XB-48: 175; P5M Marlin: 157, 176, 223; Viking rocket: 176, 204; XB-51: 175; B-57 Canberra: 176, 223; P6M Seamaster: 222–23; XB-68 project: 223; Titan missile: 204, 223
McDonnell: *Doodlebug:* 116; XP-67: 131; XFD-1/FH-1 Phantom: 132, 144, 171, 184; XHJD-1: 191; XH-20: 191; F2H Banshee: 184; XF-85: 184; XF-88: 184; F3H Demon: 184, 231; F-101 Voodoo: 157, 184, 199, 231, 255, 268; F4H/F-110/F-4 Phantom II: 185, 204, 228, 231–32, 235, 237–38, 240, 242, 255, 257, 259, 268, 273, 286, 308; XV-1: 212; Model 119/220: 231; T85 project: 231
McDonnell Douglas: DC-10/KC-10A: 265, 274, 278, 282, 298, 305, 307, 315, 321, 324, 337–38, 346; FX/F-15: 230, 243, 269, 271–72, 285–86, 288, 291, 300, 302, 307–9, 329, 337, 350; AMST/YC-15: 274–75; AV-8B Harrier: 308, 337, 344, 351 (*see also* BAe/Hawker Siddeley Kestrel); F/A-18 Hornet and F/A-18E/F Super Hornet: 287, 309, 337, 339–40, 347, 350–51, 360, 363; ATMR/DCX-200 (DC-11) project: 315; MD-80 series (*see* Douglas DC-9/MD-80 series); MD-90: 321–22; MD-95: 346, 363; AH-64 (*see* Hughes (McDonnell Douglas) AAH/AH-64 Apache); MD-11: 338, 345–46, 348, 363; Fokker MDF 100 project: 324; Northrop YF-23 (*see* Northrop McDonnell Douglas YF-23); General Dynamics ATA/A-12 project (*see* General Dynamics McDonnell Douglas A-12 project); CX/C-17: 337–39, 347–48, 350, 358, 360, 363; VTXTS/T-45 Hawk, 337–38, 348, 360, 363 (*see also* British Aerospace Hawk); (from Hughes) MD 520/Explorer NOTAR: 343; MD-12 project: 346
Mercury space program, 263
Messerschmitt: Bf 109: 103; P. 1101 project: 197
Mikoyan: MiG-15: 199; MiG-23: 286; MiG-25: 230, 259, 271
Mitsubishi FS-X project, 364
Multi-Axis Thrust Vectoring (MATV) aircraft, 344
MX Peacekeeper missile, 327

NAF: N3N-1: 112–13, 131; PBN (PBY) Catalina: 131
National Aero-Space Plane (NASP/X-30) project, 344
North American: GA-15/O-47: 83–84, 109–10; GA-16: 83–84; BT-9: 109–110; NA-49/BC-1/AT-6/SNJ Harvard/Texan: 84, 110, 128, 137, 170–71; XB-21: 110; NA-40/B-25 Mitchell: 110, 122, 128, 137, 170; NA-73/P-51/A-36 Mustang: 128–29, 137, 170; P-82/F-82 Twin Mustang: 157, 170; (Ryan) Navion/L-17: 184, 196; FJ-1 Fury: 171; B-45: 158, 168, 170; AJ Savage: 157, 174; XP-86/F-86 Sabre: 156, 169, 171, 199, 243, 259; T-28 Trojan: 171; XA2J: 158; FJ-2/FJ-3/FJ-4 Fury: 171, 187, 230, 259; F-100: 157, 171–72, 197, 199, 229, 244, 255, 268; X-15: 222, 242, 267; T-39/Sabreliner: 230–31, 310; T2J/T-2 Buckeye: 230, 309, 347; A5J/A-5 Vigilante: 230, 309; F-107: 229; Navajo missile: 229; F-108 Rapier: 229, 243; (WS-110A) XB-70/RS-70: 229–30, 243, 254, 267; NA-300/OV-10 Bronco: 230, 244, 248, 309; NAC-100 Centuryliner project: 230; (Rocketdyne) F1 rocket engine: 264
Northrop: (Avion) X216H: 77; Alpha: 67, 86; Gamma: 86; Delta: 86; BT-1: 111; A-17: 103, 111; N-1M: 115, 129; N-3PB: 115, 152; XB-35: 115, 129, 132, 185; Turbodyne engine: 143; P-61: 122, 128, 132; XP-56: 129, 132; N-9M: 132; MX-324: 132, 185; XP-79B: 185; YB-49: 153, 185–86; X-4: 198; C-125 Raider/Pioneer: 160, 185; F-89 Scorpion: 172, 177, 185, 235; XB-62/SM-62 Snark

Northrop *(continued)*
  missile: 185, 235; T-38: 235–36, 309, 339; F-5/F-5E: 236, 291, 298, 309, 339; F-5G/F-20 Tigershark: 340; M2-F2/F3: 236; HL-10: 236; P-530 project: 285–86, 309; YA-9: 269; YF-17: 286–87; ATB/B-2: 317, 327–28, 334–35, 340–41, 347–48, 358, 360–61, 364; McDonnell Douglas YF-23: 336, 338, 341
NWC Sidewinder missile, 205

Orenco Model D, 43

Packard-LePere LUSAC, 40
Piasecki: PV-2: 143; PV-3/HRP-1: 143, 188; PV-22/H-21: 188–89; PV-14/HJP/HUP/H-25: 188; YH-16: 188–89; 16H Pathfinder: 210
Pilatus Porter, 240
Piper L-18 Super Cub, 196
Pitcairn. *See* AGA (Pitcairn) PA-36 autogiro
Platt-LePage XR-1, 142
Pratt & Whitney: Wasp engine, 76; R-2600: 208; R-4360 Wasp Major: 133, 160, 163, 186; J42: 157, 181; J48: 157, 181; J57: 156, 171, 215, 236, 243; J52: 243; J75: 225, 236, 243; TF33/JT3: 243; TF30/JT8: 243, 257–58, 270, 302–3; J60/JT12: 243; JTFD-12/T73: 209; J58: 266; JT9D: 275, 277, 279; F100: 267; PW2037: 321; PW4000 series: 322, 345

Reaction Motors XLR-99, 242
Republic: P-43: 136; P-47: 128, 136–39, 186; XF-12/Rainbow: 136, 186–87; Seabee: 136, 186–87; F-84 Thunderjet: 187; (F-96) R/F-84F: 187, 194, 236, 268; XF-91: 187–88; XF-103 project: 188, 236, 243; F-105: 188, 205, 228–29, 236–37, 240, 243, 253–55, 268; Rainbow project (new): 237
Rockwell: (North American) AMSA/B-1: 269, 272–73, 285–86, 293, 309, 315, 326–28, 333, 339, 348–49; (North American) XFV-12 project: 309–10; Orbiter: 329, 339 (*see also* Space Transportation System (STS-Space Shuttle) program); X-31A: 344, 348, 357
Rolls-Royce: Merlin engine: 129; jet engines: 157, 181, 220, 242–43, 275, 279–80, 320–22, 345
Ryan: M-1: 49; NYP *Spirit of St. Louis*: 67; S-T: 85, 112; S-C: 112; YO-51 Dragonfly: 112, 137; PT-22: 137; FR-1: 137, 157, 184; X-13: 212, 242; VZ-3RY: 212, 221; XV-5A: 213, 221; Firebee drone: 221; Firebird missile: 221

SAAB: 105: 330; 340/SF340: 323, 329–30. *See also* Fairchild (from SAAB) SF340
Saturn launch vehicle, 224, 231, 233, 261, 304
Seversky: SEV-3: 71; BT-8: 71; (Republic) P-35: 68, 71, 103, 110–11; NF-1: 110
Short S. 32 project, 162
Sikorsky: S-29-A: 52; S-35: 52–53; S-37: 53; S-38: 53; S-40: 78; S-42: 78, 101; S-44: 101; VS-300: 104, 142–43; XR-4: 142; R-5: 142; R-6: 142; S-51/H-5: 188, 208; S-55/H-19: 188, 197, 208; XH-35: 188; S-56/HR2S-1: 208; S-58, 208–9; S-61/HSS-2/SH-3: 209, 311, 342; S-60: 209; S-62: 209, 311; S-64/CH-54: 209, 311; S-65/CH-53A: 209, 311; CH-53E: 267, 342; S-67: 311; S-70 series/UTTAS/UH-60A/LAMPS/SH-60B: 312, 342, 347, 363; S-72: 312; S-76: 312, 342; S-92 project: 363; LHX/RAH-66. *See* Boeing
Skylab space program, 264
Space Station Freedom project, 344
Space Transportation System (STS-Space Shuttle) program, 236, 264, 329
SPAD fighter, 31
Sperry: Messenger: 46; Sparrow missile: 204
Standard: E-1: 46; SJ-1: 24
Stearman (Boeing) PT-13/PT-17, 109, 135
St. Louis PT-15, 124
Stroukoff YC-134, 166
Sud Aviation: Alouette: 237; Caravelle: 217–18
Supermarine: S-6B: 67; Spitfire: 103
Swearingen (Fairchild) Metro, 301, 330

Temco: T-35 Buckaroo: 183; TT-1 Pinto: 241

Thomas-Morse: MB-3A: 22, 33, 42; O-19: 66
Trans-Atmospheric Vehicles (TAVs), 344
Transporter Allianz Transall, 251

Unmanned Aerial Vehicles (UAVs), 344

V-2 missile, 127, 204
Vertol: Model 44: 210; Model 107/CH-46: 210; CH-47 Chinook: 210, 342; (Boeing) YAH-65: 274; (Boeing) UTTAS/YUH-61A: 312
Very Large Commercial Transport (VLCT) project, 346
Vickers: Vimy: 39; Viscount: 164, 217, 220
Vought: VE-7: 43, 45; VE-11: 44; trainer: 33; O2U-1 Corsair: 91; SBU: 110; SB2U/V-156F Vindicator: 110; OS2U Kingfisher: 110, 132; F4U Corsair: 110, 132, 145, 157, 182; XF5U: 182; F6U Pirate: 171, 182; F7U Cutlass: 182, 241, 244; F8U/F-8 Crusader: 157, 199, 231, 238, 241–42, 267, 285; Corvus missile: 241; VAX/A-7 Corsair II: 242, 255, 259, 269, 286, 294, 311, 341, 350
Vultee: V-1: 80; V-11: 80; V-72, A-31/A-35 Vengeance: 106, 115, 132; BT-13 Valiant: 131, 138

WACO C-62, 140
Westinghouse jet engines, 157, 181–82, 184
Whittle jet engine, 132, 143, 157
Wright: Flyer: 6, 47; Model A: 7; Model B: 18; Model C: 14; *Vin Fiz*: 14; Cyclone engine: 91; R-3350 Twin Cyclone engine: 133, 135; TC18 Turbo-Compound engine: 158, 163, 172; J65 (Sapphire) engine: 157, 172, 177; J67 engine: 236–37; XJR55 engine project, 237; Bellanca monoplane, 65